Carbon-Based Smart Materials

Also of Interest

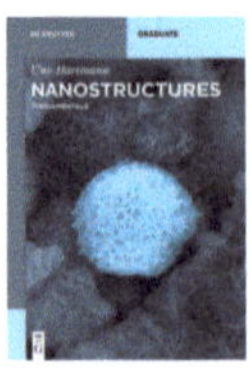

Nanostructures – Fundamentals
Uwe Hartmann, 2021
ISBN 978-3-11-036218-3, e-ISBN 978-3-11-045955-5

Nanomaterials in Joining
Constantinos A. Charitidis (Ed.), 2015
ISBN 978-3-11-033960-4, e-ISBN 978-3-11-033972-7

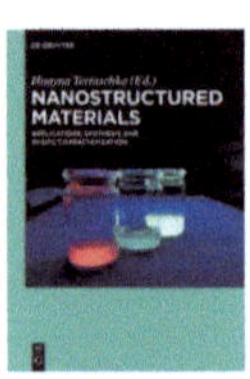

Nanostructured Materials – Applications, Synthesis and In-Situ Characterization
Huayna Terraschke (Ed.), 2020
ISBN 978-3-11-045829-9, e-ISBN 978-3-11-045909-8

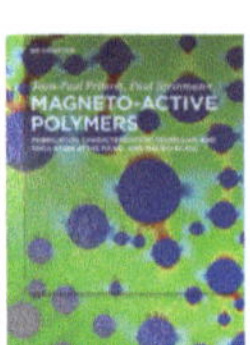

Magneto-Active Polymers – Fabrication, characterisation, modelling and simulation at the micro- and macro-scale
Jean-Paul Pelteret, Paul Steinmann, 2019
ISBN 978-3-11-041951-1, e-ISBN 978-3-11-041857-6

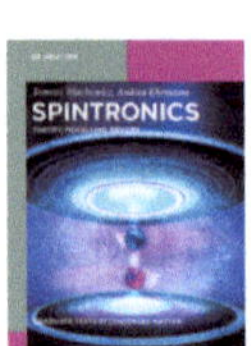

Spintronics – Theory, Modelling, Devices
Tomasz Blachowicz, Andrea Ehrmann, 2019
ISBN 978-3-11-049062-6, e-ISBN 978-3-11-049063-3

Carbon-Based Smart Materials

Edited by
Constantinos A. Charitidis, Elias P. Koumoulos
and Dimitrios A. Dragatogiannis

DE GRUYTER

Editors
Prof. Constantinos A. Charitidis
Department of Materials Science and Engineering
School of Chemical Engineering
National Technical University of Athens
Heroon Polytechniou St. 9
157 73 Athens
Greece
charitidis@chemeng.ntua.gr

Dr. Dimitrios A. Dragatogiannis
Department of Materials Science and Engineering
School of Chemical Engineering
National Technical University of Athens
Heroon Polytechniou St. 9
157 73 Athens
Greece
ddragato@chemeng.ntua.gr

Dr. Elias P. Koumoulos
Department of Materials Science and Engineering
School of Chemical Engineering
National Technical University of Athens
Heroon Polytechniou St. 9
157 73 Athens
Greece
elikoum@chemeng.ntua.gr

ISBN 978-3-11-047774-0
e-ISBN (PDF) 978-3-11-047913-3
e-ISBN (EPUB) 978-3-11-047775-7
DOI https://doi.org/10.1515/9783110479133

Library of Congress Control Number: 2019957724

Bibliographic information published by the Deutsche Nationalbibliothek
The Deutsche Nationalbibliothek lists this publication in the Deutsche Nationalbibliografie; detailed bibliographic data are available on the Internet at http://dnb.dnb.de.

Cover image: Dmitrty/iStock/Getty Images Plus
Typesetting: Integra Software Services Pvt. Ltd.
Printing and binding: CPI books GmbH, Leck

www.degruyter.com

Preface

The research in the field of advanced materials is significantly driven by the development of "smart" materials that are designed to react with environmental conditions and other stimuli. Smart materials that have attracted considerable attention in the last years have now become an intrinsic field in dealing with both societal challenges and significant engineering problems in various sectors such as energy and water consumption, recycling and engineering.

The field of advanced materials benefits from multidisciplinary basic and applied research in physics, chemistry, mathematical modeling and materials engineering. Technical challenges related to developing suitable smart materials and intelligent structures are applicable in a wide range of fields, such as sensors, spintronics, supercapacitors, biomedical applications (drug delivery and hyperthermia therapy), self-healing, flexible memories, construction technologies and other emerging technologies.

Carbon-based materials have gained major attention for replacing other conventional materials as a promising basis for developing novel smart materials due to their abundance and low cost, high-temperature stability and corrosion resistance, high electrical and thermal conductivity and enhanced mechanical properties, particularly in the microscopic and nanoscopic scale.

This book consists of chapters written by expert scientists and engineers from the international carbon-based materials community, who contribute important research work with emphasis to advanced applications. Contributions focus on novel applications of carbon-based smart materials, as well as on functionalization processes and multiscale modeling.

In the first chapter "New class of graphene-based devices for the next generation of nonvolatile memories," the main scientific works concerning the properties of memories based on graphene layers, graphene oxide, and reduced graphene oxide are reviewed by showing the promising results in terms of writing time and endurance/cyclability compared to existing technologies. Moreover, these materials can potentially allow developing a highly scalable complementary metal-oxide–semiconductor-compatible technology. This chapter suggests that graphene-based memory technology can be exploited for space applications, as it is being insensitive to radiation and easily reprogrammable.

Second chapter is devoted to the research undertaken at the University of Birmingham on the functionalization of carbon-based materials using the active screen plasma (ASP) technology. The chapter entitled "Plasma surface activation and functionalization of carbon-based Materials" exploits the typical physical and chemical phenomena associated with the plasma technology, further optimized by the additional control and degrees of freedom offered by the ASP technology. In this way, the highly reactive species in the plasma were used to reduce graphene oxide films, activate the surface of carbon paper membranes or functionalize the

https://doi.org/10.1515/9783110479133-202

surface of carbon fibers. All the above seem to be great opportunities to create patterned surfaces using the ASP technology, in combination with other processes, with potential application in multiple fields such as functional materials, sensors and biosensors, electrical, electronic and optoelectronic devices. Without question, the ASP technology remains a valuable enabling technology for the surface functionalization of engineering materials, showing broader prospects of practical applications than ever before.

Modeling of carbon-based smart materials is the subject of the third chapter. A better understanding of the thermophysical behavior of carbon-based polymer nanocomposites at different length- and time-scales could pave the way of discovering novel smart materials and could be facilitated by proper materials modeling. Advanced modeling approaches for the estimation of nanocomposite properties are based on linking and/or coupling various model types that are applicable at different length- and time-scales. This chapter review could be particularly useful for a broader audience of readers, whose research interest is focused on modeling issues in carbon-based materials. Atomistic, mesoscopic and macroscopic simulation tools that allow exploring the main properties of nanocomposites are extensively presented, and future perspectives of advanced modeling of carbon-based smart materials are discussed.

The fourth chapter entitled "Carbon nanotube-based materials for space applications" summarizes the activities performed in three sequential European Space Agency projects, highlighting the introduction of carbon nanotubes (CNTs) into composite materials for space applications. The first activity aimed at proving the feasibility of using CNT skeletons as an alternative to conventionally filled resins, overcoming high viscosity issues, which limits the processability of materials. For the second activity, the aim was to scale up the nanotube structures and integrate them in carbon fiber-reinforced composites. More recently, activities were focused on selecting two applications with highest potential for performance improvements, taking into account all knowledge gained, followed by the development and testing of two demonstrators. This chapter presents the development steps and demonstrator manufacturing and testing for one of these applications, a carbon fiber-reinforced polymer (CFRP) optical mirror. For this mirror demonstrator, a complete assembly consisting of two CFRP skins with outer CNT layers bonded to a carbon-based honeycomb center was manufactured. The principle of producing a highly accurate optical mirror in CFRP-CNT technology is rated as very promising, with important mass savings at required thermoelastic distortion values.

The next chapter deals with molecular dynamics (MD) simulations of graphene-based polymer nanocomposites. The unique, advanced properties of graphene justify why it is widely considered as the material of the future. Within this chapter, graphene's great potential for new applications is highlighted, and the recent progress in the design of graphene-based polymer nanocomposites from both an experimental and a computational point of view is discussed. The basic concepts of molecular simulations focusing particularly on the MD technique and simulation results for the structural, conformational and mechanical properties of a test nanocomposite system based on poly(methyl methacrylate) (PMMA) filled with two different types of graphene sheets are also presented: simple (i.e., unfunctionalized) graphene sheets and functionalized (i.e., graphene oxide). This study was driven by recent experimental reports according to which the addition of a small fraction of graphene sheets in a polar polymer matrix such as PMMA can lead to a remarkable enhancement of its elastic constants.

The last chapter "Carbon from waste source for Li-ion battery" suggests that the accomplishment of "12 principles of green chemistry" is a fundamental goal to be pursued by means of designing/optimizing environmentally "conscious" processes, materials and devices to reduce their environmental impact throughout all the phases of their life cycle, thus benefiting the economy, protecting people and achieving the real goal of a truly sustainable world. This goal can be achieved by exploiting carbonized nanocellulosic binders and water-based papermaking coupled with high-performing green composite polymer electrolytes.

The book is useful for everyone who is interested in applications of carbon-based smart materials and is especially oriented to the graduate students and young researchers to facilitate their introduction in this new area. The editors thank all the authors who contributed their work to this book and also acknowledge the superb assistance that the staff of De Gruyter has provided for the publication of this work.

Contents

Preface —— V

List of contributing authors —— XI

Paolo Bondavalli
1 New class of graphene-based devices for the next generation of nonvolatile memories —— 1

Hanshan Dong, Santiago Corujeira Gallo
2 Plasma surface activation and functionalization of carbon-based materials —— 17

Rajat Srivastava, Matteo Fasano, Shahin Mohammad Nejad, Hernán Chávez Thielemann, Eliodoro Chiavazzo, and Pietro Asinari
3 Modeling carbon-based smart materials —— 33

Marta Martins, Joana F. Guedes, Celeste Pereira, Stefan Forero, Volker Liedtke, Jiří Zelenka, Jan Václavík, Ernst Pfeiffer, Laurent Pambaguian, Advenit Makaya, Nuno Rocha
4 Carbon nanotube-based materials for space applications —— 81

Emmanuel N. Skountzos and Vlasis G. Mavrantzas
5 Molecular dynamics simulations of graphene-based polymer nanocomposites —— 115

Pravin Jagdale, Gemma Rius, Krishna Rajan, Jijeesh Ravi Nair, Massimo Rovere, Alberto Tagliaferro, Claudio Gerbaldi
6 Carbon from waste source for Li-ion battery —— 153

Index —— 181

List of contributing authors

Paolo Bondavalli
Thales Research and Technology
1 Av A. Fresnel
(site de Polytechnique)
Palaiseau Cedex F91767, France
Paolo.Bondavalli@thalesgroup.com

Hanshan Dong
School of Metallurgy and Materials
University of Birmingham
Edgbaston
Birmingham B15 2TT, UK
h.dong.20@bham.ac.uk

Santiago Corujeira Gallo
Deakin University Australia
santiago.corujeiragallo@deakin.edu.au

Pietro Asinari
Politecnico di Torino
Department of Energy
Corso Duca degli Abruzzi, 24
Turin 10129, Italy
pietro.asinari@polito.it

Rajat Srivastava
Politecnico di Torino
Department of Energy
Corso Duca degli Abruzzi, 24
Turin 10129, Italy

Shahin Mohammad Nejad
Politecnico di Torino
Department of Energy
Corso Duca degli Abruzzi, 24
Turin 10129, Italy

Matteo Fassano
Politecnico di Torino
Department of Energy
Corso Duca degli Abruzzi, 24
Turin 10129, Italy

Eliodoro Chiavazzo
Politecnico di Torino
Department of Energy
Corso Duca degli Abruzzi, 24
Turin 10129, Italy

Nuno Rocha
INEGI – Institute of Science and Innovation in
Mechanical and Industrial Engineering
Rua Dr. Roberto Frias 400
Porto 4200-465, Portugal
nuno.rocha@inegi.up.pt

Marta Martins
INEGI – Institute of Science and Innovation in
Mechanical and Industrial Engineering
Rua Dr. Roberto Frias 400
Porto 4200-465, Portugal

Joana F. Guedes
INEGI – Institute of Science and Innovation in
Mechanical and Industrial Engineering
Rua Dr. Roberto Frias 400
Porto 4200-465, Portugal

Celeste Pereira
INEGI – Institute of Science and Innovation in
Mechanical and Industrial Engineering
Rua Dr. Roberto Frias 400
Porto 4200-465, Portugal

Stefan Forero
FutureCarbon
Ritter-von-Eitzenberger-Straße 24
95448 Bayreuth, Germany

Volker Liedtke
AAC – Aerospace & Advanced Composites
Viktor-Kaplan-Straße 2, building F
2700 Wiener Neustadt, Austria

https://doi.org/10.1515/9783110479133-204

Jiří Zelenka
TOSEDA s.r.o.
U Panasonicu 376
530 06 Staré Čívice (areál TechnoPark Pardubice), Czech Republic

Jan Václavík
Institute of Plasma Physics of CAS
Za Slovankou 1782/3
Prague 182 00, Czech Republic

Ernst Pfeiffer
HPS GmbH
Hofmannstr. 25-27
81379 München, Germany

Laurent Pambaguian
ESA ESTEC
Keplerlaan 1
2201 AZ Noordwijk, The Netherlands

Advenit Makaya
ESA ESTEC
Keplerlaan 1
2201 AZ Noordwijk, The Netherlands

Emmanuel N. Skountzos
Department of Chemical Engineering
University of Patras
GR 26504 Patras, Greece

Vlasis G. Mavrantzas
Department of Chemical Engineering
University of Patras
GR 26504 Patras, Greece
vlasis@chemeng.upatras.gr

Pravin Jagdale
Department of Applied Science and technology (DISAT)
Politecnico di Torino
C.so Duca Degli Abruzzi, 24
Torino 10129, Italy

Gemma Rius
Institute of Microelectronics of Barcelona
IMB-CNM-CSIC, Campus UAB
Carrer dels Til.lers s/n
08193 Bellaterra, Spain

Pravin Jagdale
Department of Applied Science and technology (DISAT)
Politecnico di Torino
C.so Duca Degli Abruzzi, 24
Torino 10129, Italy

Krishna Rajan
Department of Applied Science and technology (DISAT)
Politecnico di Torino
C.so Duca Degli Abruzzi, 24
Torino 10129, Italy

Jijeesh Ravi Nair
Department of Applied Science and technology (DISAT)
Politecnico di Torino
C.so Duca Degli Abruzzi, 24
Torino 10129, Italy

Massimo Rovere
Department of Applied Science and technology (DISAT)
Politecnico di Torino
C.so Duca Degli Abruzzi, 24
Torino 10129, Italy

Alberto Tagliaferro
Department of Applied Science and technology (DISAT)
Politecnico di Torino
C.so Duca Degli Abruzzi, 24
Torino 10129, Italy

Claudio Gerbaldi
Department of Applied Science and technology (DISAT)
Politecnico di Torino
C.so Duca Degli Abruzzi, 24
Torino 10129, Italy
alberto.tagliaferro@polito.it

Paolo Bondavalli

1 New class of graphene-based devices for the next generation of nonvolatile memories

1.1 Introduction

Graphene is a one-atom thick layer of carbon atoms arranged in a hexagonal lattice. Graphene potentialities are attracting a lot of researchers to probe opportunities in a number of directions in the "more Moore" or "beyond CMOS" optics in order to identify the new future technologies [1–3]. Another promising utilization of graphene and related nanomaterials is to fabricate nonvolatile memories (NVM) exploiting their "memresistive" behavior storing a value of electrical resistance in a permanent way. This happens when a current passing through the materials changes the level of resistance. Therefore, resistive memory exploits the change in the resistance of a material under the effect of an electric field as an information write/erase principle for nonvolatile data storage. The reading of resistance states is nondestructive, and the memory devices can be operated without transistors in every cell [4, 5], as for flash-type memories [6–10] (see Section 1.2.2), thus achieving a classic cross-bar structure. This kind of memories is called resistive random-access memory (RRAM or ReRAM) and is only one of the possible types of nonvolatile ways to store information in a permanent way. One of the most important advantages of these new classes of 2D materials is that these materials can be implemented in flexible electronics [11–15], in the form of one-thick atom layers as for graphene or in the form of layers of flakes of graphene oxide (GO) or reduced GO (R-GO), thereby reducing the final cost of the final device exploiting roll-to-roll fabrication [16]. Another great advantage of ReRAMs is their potential to implement them by exploiting only two terminals to work (two contacts and not three as a common transistor, which has drain, source and gate, e.g., flash-type memories), which could dramatically reduce the circuitry and allow to implement easily in 3D architectures by using the roll-to-roll fabrication technique. Potentially, this is applied in various fields such as health monitoring [17–21], intelligent packaging [22, 23], cards, labels, badges, value paper and medical disposables. Another potential field is the radiofrequency identification (RFID). This is a technology to electronically record the presence of an object using radio signals. Indeed, an innovative alternative pathway to reduce RFID costs and integrate a memory chip to store data is to eliminate the silicon substrate completely, and produce RFID and memory on the same flexible plastic substrate as the antenna [24–26].

Paolo Bondavalli, Thales Research and Technology, Palaiseau Cedex, France

https://doi.org/10.1515/9783110479133-001

Thanks to graphitic layers, the antenna and chip can be built on the same low-cost substrate, and attachment costs can be removed.

1.2 Graphene-based NVM

As outlined in the International Technology Roadmap for Semiconductors 2011 section concerning Emerging Research Devices, "ultrathin graphite layers are interesting materials for macromolecular memories thanks to the potential fabrication costs that are considered as the primary driver for this type of memory, while extreme scaling is de-emphasized." The main drawback is related to the fact that memory operation mechanisms and the physics are still unclear and that a deeper research in this field is necessary to improve the comprehension of the phenomenon and the efficiency of the devices. These are not the same physical mechanisms exploited in memories based on graphene-related materials, which will be discussed in the next paragraphs.

1.2.1 Graphene and graphitic layers

The first paper highlighting the utilization of graphene for resistive NVM exploiting two terminal structures was issued in 2008 by Stadley et al. at Caltech [27]. They reported the development of an NVM element based on graphene break junctions. These junctions were obtained by achieving two-terminal devices, transferring graphene sheets on SiO_2/Si substrate and depositing metal electrodes on them using e-beam lithography process. After applying a specific voltage under ultravacuum condition (10^{-7} torr), they were able to create a break in the graphene layer (see Figure 1.1). In these papers, all the measurements were performed under ultravacuum to avoid environmental gas interferences, which strongly reduces the potential utilization of the device.

After cycling the memories, they observed that the resistance values clearly changed as a function of the voltage (see Figure 1.1b): they moved from a high resistance state (off state) to a lower one as a function of the voltage (on state). This effect was nonvolatile and the devices were cycled thousands of writing cycles (10^5), showing long retention times (more than 24 h, only limited by measurements).

The reproducibility of the switching behavior was very promising. One of the more coherent explanations was related to the fact that the conductance occurred along a small graphene ribbon that bridges the contacts. To better understand the underlying physical mechanism, researchers studied the time-resolved behavior of the switch from the OFF to ON states, shown in Figure 1.2a. The conductance I/V showed well-defined steps, with magnitude ~G_Q. Here G_Q is $\mathbf{2e^2/h}$, the conductance quantum, where $\boldsymbol{e}$ is the

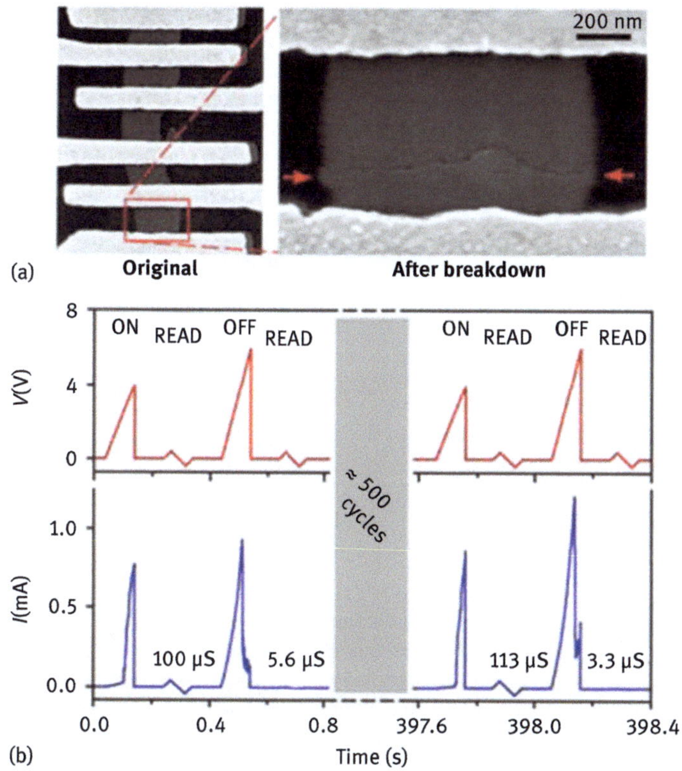

Figure 1.1: (a) SEM image of the device before (left panel) and after breakdown (right panel). The arrows indicate the edges of the break in the graphene layer. (b) Repeatable programming over hundreds of cycles. Upper left panel: Voltage applied to the junction versus time. A ramp with a peak value of ~4 V corresponds to an ON pulse, while a ramp with a peak value of ~6 V corresponds to an OFF pulse. A small sawtooth-shaped read-out pulse is applied after each write to determine the junction conductance. Lower left panel: Current flow through the junction, with the low-bias conductance labeled above each read-out pulse. Right panels: Similar data taken after approximately 500 intervening cycles, demonstrating the reproducibility of the switching behavior [27].

electron charge and h is Planck's constant. Since G_Q is the conductance of a spin-degenerate one-dimensional conductor, for example, a linear chain of gold atoms [28], observation of steps in the conductance suggests that the conductance states of a device are likely multiples of highly transmitting quantum channels. Authors proposed a model for device operation based on the formation and breaking of carbon atomic chains that bridge the junctions (see Figure 1.2a, right figure). This mode and the results seem to put in evidence the potential for multiresistive states.

The same physical explanation can be evoked for the results obtained in 2009 by Sinitski et al. [29] at Rice when they tested analogue carbon-based structures (see Figure 1.2b). The sensitive layer was consists of 10 nm thick graphitic disks

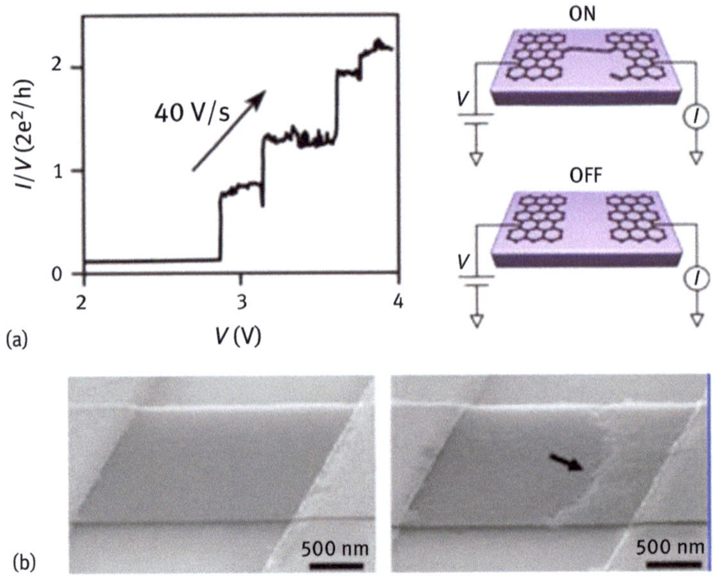

Figure 1.2: (a) Left side: Change in *I*/*V* as a function of *V* pointing out the quantic behavior linked to the formation of atomic chains. Right side: Proposed physical mechanism at the base of the nonvolatile resistive behavior (formation of atomic carbon chains). (b) Tilted-view SEM images of two different devices before and after applying the voltage difference that is necessary to break the layers defined as V_{break}. The arrow shows the fracture across the stripe due to V_{break} [29].

grown by chemical vapor deposition (CVD) process. Researchers evaluated the voltage that is necessary to break the layer. Calculating the Joule heating, researcher obtained that the breakdown voltage, the applied voltage difference necessary to break the layer, was

$$V_{break}\alpha[(C\rho\gamma T_{break})/\tau]^{1/2}\, l = A \tag{1.1}$$

where C, ρ, γ are, respectively, the specific heat, the density of CVD-grown graphitic sheets and resistivity, T_{break} is the breakdown temperature of the graphitic layer, A is independent of the devices and l the length of the device. Therefore, the voltage necessary to break the layer is directly proportional to the length. This is very positive because reducing the dimensions will allow to reduce the energy consumption during the writing step.

In 2009, the same team from the University of Rice [30] exploited another approach and grew graphitic layers on a freestanding silicon oxide nanowire (see Figure 1.3a). Indeed they used the nanowire to preserve the mechanical integrity of the device after applying a specific voltage breaking the surficial graphitic layers. The team from Rice fabricated two terminal devices consisting of discontinuous

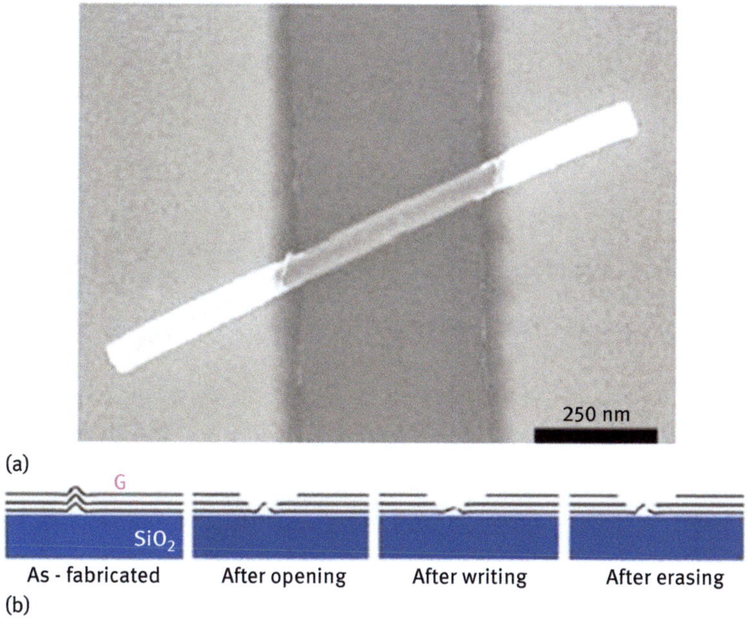

Figure 1.3: (a) Silicon oxide nanowire with graphitic layers grown on its surface and (b) physical principle at the base of the fabrication of nonvolatile graphitic memories exploited by Rice University [30].

5–10 nm thin films of graphitic sheets grown by CVD on nanowires having stable, rewritable, nonvolatile and nondestructive read memories with on/off ratios of up to 10^7 (see Figure 1.4a) and switching times of up to 1 μs (tested limit).

All the samples were not tested, as in the previous examples, at ambient conditions but in ultra-vacuum condition (5×10^{-5} torr) and showed an extremely good stability of the performances after several days and also after x-ray exposure (Figure 1.4b).

An interesting work by Wu et al. [31] in 2012 dealt with the electrical properties of graphene sheet suspended on a patterned indium tin oxide (ITO) electrode pair. They measured very interesting on/off ratio of 6 orders and retention time of at less 10^4 s in ambient conditions. In this case, data storage was achieved by applying voltage bias and rewritten after a simple heat treatment. Authors stated that the switch happens only in ambient atmosphere when oxygen changed the injection barrier at the contact. The degassing using a thermal process allowed coming back to the high conduction state.

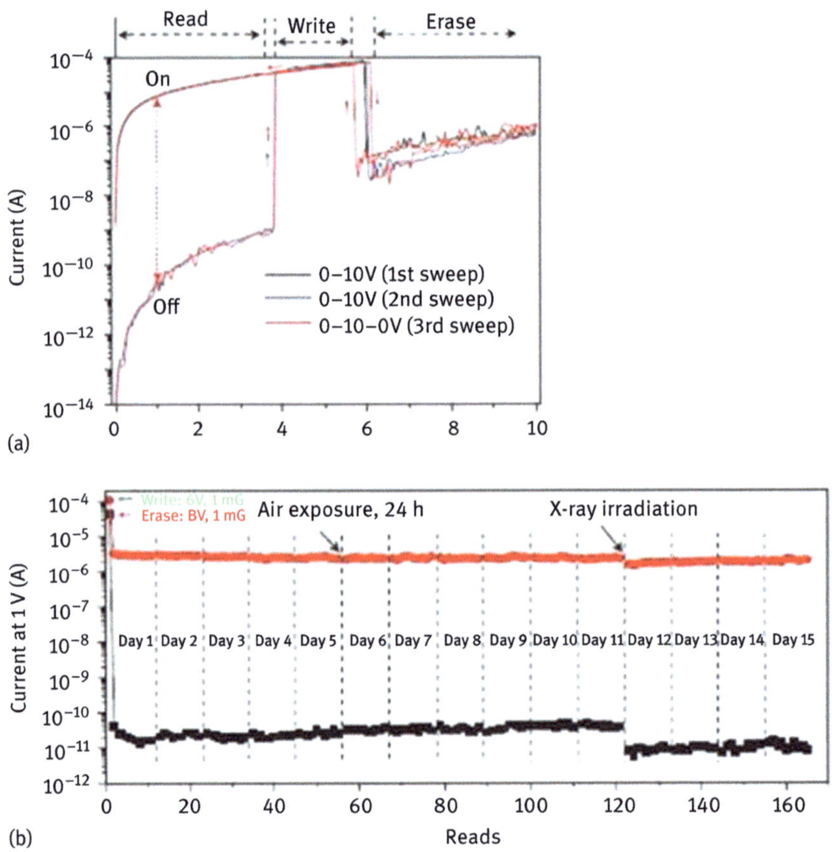

Figure 1.4: (a) The logarithmic *I–V* behavior of a G–SiO_2 nanoribbon; the "write" or "erase" regions can be achieved through a voltage pulse in the noted regions. After writing or erasing, the high- or low-conductivity state of the system is recorded in the "read." (b) Data retention by two other G–SiO_2 nanocable devices [30].

1.2.2 Nonvolatile resistive memories based on GO and R-GO oxide layers

Other teams focused their study on GO. GO is a wide band gap material (6 eV) with potential for modulation, thanks to oxidation/reduction process providing tunability of the electronics, mechanical and optical properties. GO [32] is commonly obtained by the oxidation of graphite using the modified Hummer's method, where the long oxidation time is combined with a highly effective method for the purification of reaction products [33, 34]. A great advantage of this material is its cost, which is at least one-tenth the price of graphene flakes and it is stable in water suspensions. In GO, the memory configuration exploited is similar to cross-bar memories

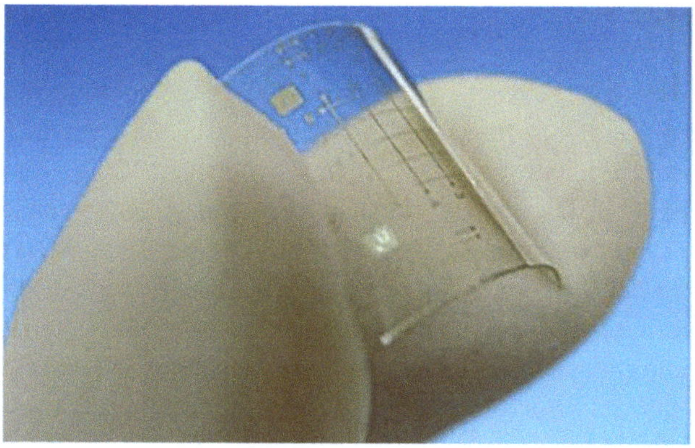

Figure 1.5: Cross-bar memory devices based on GO [39, 40].

[35, 36] (vertical junctions composed by piling up the different layers, sandwiching the active one as metal insulator metal-type structures). We can mention the pioneering work of He et al. in 2009 [37], where reliable and reproducible resistive switching behaviors were observed in GO thin films prepared by vacuum filtration method [38]. They fabricated Cu/GO/Pt structures showing an on/off ratio of about 20, a retention time of more than 10^4 s and switching threshold voltages of less than 1 V. The suggested physical principle at the origin of the switching effect was the desorption/absorption of oxygen-related groups on the GO sheets as well as the diffusion of the top electrodes. One of the major works on GO-based memories was issued in 2010 by Jeong et al. [39, 40]. This team demonstrated the nonvolatile effect on the resistance of a 70 nm layer thick GO in a layered structure composed of Al/GO/Al (Figure 1.5) in a cross-bar configuration.

Jeong et al. were one of the first teams that pointed out the performances of these kinds of layers on flexible substrates. The researchers from KAIST demonstrated that using GO layers they were able to achieve flexible nonvolatile resistive memories with impressive performances under traction and compression [they kept three orders between on (LRS) and off (HRS) levels]. Concerning the physics explanation of the phenomenon, Hong et al. [41, 42] performed a deep analysis of the switching mechanism for this kind of devices and demonstrated that these structures had performances dependent on the origin of the top contacts. For example, in case of Au-based top electrodes, there was no oxygen migration in opposition to Al electrodes. The effect of bottom contacts was also quite important. Indeed if the roughness was too high, the GO layer had such cracks and rough surface that the top electrode material could easily penetrate and then build filaments that hindered the switching operations. This drastically reduced the lifetime of the device (only 100 cycles) because of the formation of a permanent conductive path between the two electrodes. These results were highlighted by X-ray photoelectron spectroscopy measurements that pointed out the permanent presence of Al near the bottom electrode as the main reason for failure. Evaluating the effect of the migration of oxygen versus the formation of Al-based filaments, Hong and coworkers stated that the formation of conducting filaments was a local phenomenon and that the oxygen migration was the dominant mechanism. Indeed, they discovered that the leakage current between the bottom and lower electrodes was dependent on the cell dimension. In fact when the cells were larger the effect of the oxygen migration or reduction on the conduction was enhanced proportionally to the surface. In order to avoid the failure mechanism linked to the formation of permanent conductive filaments through the material cracks, one suggestion is to exploit deposition techniques that allow a more uniform distribution of the deposited material. One technique is the deposition by spray coating or heating the substrate to avoid the so-called coffee-ring effect [43, 44], which is responsible for the nonuniformity of the deposition [45]. This deposition could potentially prevent the formation of cracks linked to the roughness in the GO layers.

Another interesting study was performed by Vasu et al. at the University of Bangalore (India) in 2011 [46]. They reported very simple unipolar resistive switching device using ultrathin (~20 nm) films of R-GO with on/off ratios reaching five orders of magnitude. The thin films were formed at room temperature simply by drop-casting R-GO from suspension on ITO glass electrode, followed by aluminum or gold deposition (no difference was pointed out after the measurements). A very interesting result was also the switching time that could attain 10 μs with an on/off ratio of 100. In this case, the formation of nanofilaments of carbon atoms was evoked as the main physical working principle.

1.3 Other approaches to achieve nonvolatile memories using graphitic layers

In the previous sections, we have discussed about the resistive NVM. Other teams have decided to use different approaches to exploit the properties of graphene-based layers to fabricate NVM and have exploited the hysteretic electrical behavior of graphene, or multilayered graphene (MLG), when used as channel in a field effect transistor (FET)-type configuration. This is substantially based on the chemical modification of graphene under an electric field that will enhance environmental interactions. However, the more promising results, in terms of performances but also in terms of potential industrial exploitation, have been obtained in case of the fabrication of flash-type memory [47–48] using graphene or MLG as the floating gate (FG). Basically, a flash memory is an electronic nonvolatile computer storage medium that can be electrically erased and reprogrammed. In flash memory, each cell resembles a standard MOSFET [85], where the transistor has two gates instead of one. Indeed there is the traditional top gate, also called control gate and an FG embedded in an oxide layer; hence, it is electrically insulated. In FG, which is located between the top gate and the MOSFET channel, any electrons placed on it, activating the top gate, are trapped there and will not discharge for many years. If FG holds a charge, it screens (partially cancels) the electric field from the top gate; therefore, it modifies the threshold voltage (V_T) of the MOSFET. This operation is nonvolatile and is reversible if another adequate voltage is applied on the top gate. The present technology used for flash-type memories employs doped polysilicon as material for the FG. One of the main advantages of graphene/MLG is the reduction of voltage to achieve a correct memory window, which is the shift of the threshold voltage of the transistor when the memory is switched from 0 to 1 state. In the present flash memories, a voltage difference of ±20 V [49] is necessary to program/erase to achieve a memory window of 1.5 V, which is the industrial standard. Some works have demonstrated, as shown in the next paragraphs that thanks to the higher density of states (DOS) of graphene compared to degenerately doped polysilicon this difference can reduced up to 6 Volts. Another advantage is the higher work function of graphene [50–52] that is directly linked to the larger barrier height between the FG and the oxide [53–56], which embeds the FG. This potentially allows achieving longer retention time, potentially more than ten years. Finally, the possibility to achieve device using 2D devices with a lower dimensionality that will potentially allow an easier 3D implementation and so a higher memory density.

1.3.1 Graphitic-based nonvolatile memory using a transistor configuration

One of the first paper highlighting the interest of using graphene-based FET for NVM was issued by Wang et al. at the Nanyang Technological University in Singapore in 2010 [57]. Indeed they observed exactly the same phenomenon observed in case of FET with one carbon nanotube as a channel: unhysteretic phenomenon where they observed the formation of a loop of the current value between drain and source, I_{ds}, as a function of the gate voltage, V_{gs}, during the voltage sweep cycle from negative to positive values of V_{gs} and back. This phenomenon is caused by the presence of moisture interacting with the transistor dielectrics. This effect, it has been demonstrated, can be removed in vacuum [58, 59]. Moreover the switching time in case of activation of the hysteretic phenomenon in FET using graphene is of the order of some seconds, which is too high for memory applications.

In 2008, a pioneering work of Echtermeyer et al. of the Advanced Research Center in Aachen (Germany) and of A. K. Geim at the University of Manchester (UK) was performed adopting another point of view compared to the previous studies [60, 61]. Considering the difficulty to achieve nanoribbons with dimensions smaller than 10 nm to open a gap, they exploited another phenomenon. Indeed they fabricated two-gate transistor structures as shown in Figure 1.6, where the graphene has a width of 10 μm and was covered by an SiO_2 layer.

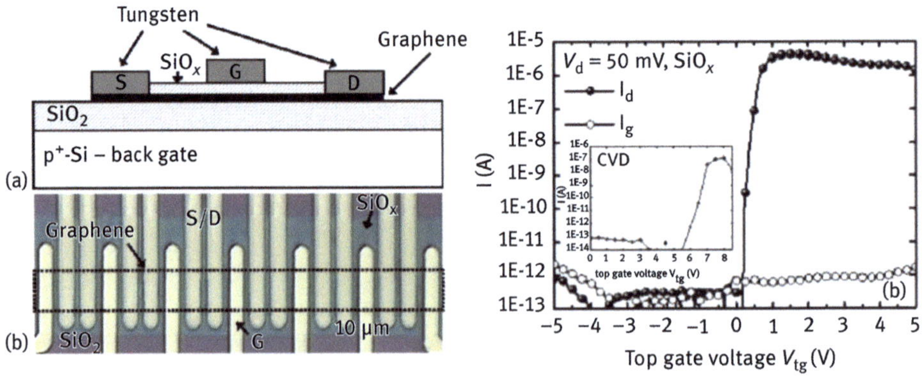

Figure 1.6: (a) Schematic of a double-gated graphene field effect device (FED) and (b) current between drain and source as a function of the top gate voltage value [60, 61].

They observed that applying a voltage spanning from −5 to +5 V, for positive voltage there was a strong enhancement in the current of 5/6 order, hence, a reduction in the resistance that was nonvolatile. Echtermeyer and coworkers suggested that considering that the measurements had been performed in ambient conditions, the water molecules were split in H^+ and OH^- (see Figure 1.6) by the electric field and tended to attach to the graphene, creating graphene or GO (that are insulating materials), passing through the highly porous silicon dioxide layer surface, and so drastically changing the conductivity of the channel in a nonvolatile way. This phenomenon allowed obtaining a strong on/off ratio, quite unusual for graphene (and not graphene nanoribbons) based FET. However, even if the results seem to be very promising, more studies have to be done on the switching time, which intuitively appears to be quite long considering the effects involved. We can add that the graphene layer has been obtained by mechanical cleavage, which is a technique that allows fabricating high-quality samples but that it is not industrially suitable. It is necessary to test this approach, for example, using CVD graphene in order to achieve a potential parallel fabrication of this kind of devices. It has also to be verified if it is fundamental to obtain a single graphene layer or if more stacked layers of graphene can reach the same performances.

1.3.2 Nonvolatile flash-type memories based on graphene/multilayered graphene

Another potential utilization of graphene is FG in flash-type memories. An extremely innovative approach has been proposed by Bertolazzi et al. of the Ecole Polytechnique de Lausanne (EPFL) in Switzerland. They were able to fabricate a flash memory using exactly the same architecture of the existing ones, exploiting only 2D materials [62].

Bertolazzi and coworkers were able to build a flash-type memory using MoS_2 as the channel and few graphene layers as the FG (see Figure 1.7), embedded in HfO_2 oxide, where the charge is trapped. It is clear that this kind of structure is extremely original but some performances especially considering the switching time (the lowest value is 100 ms) are not adequate. Some doubts can be raised on the fabrication technique, which includes three quite complex transfer steps and that are not suitable at the moment for an industrial exploitation. Hong et al. at IBM in Watson

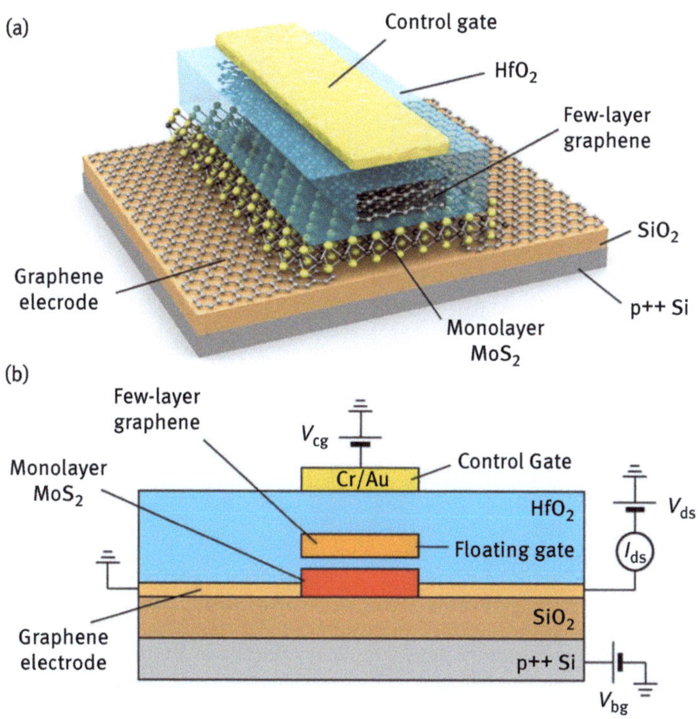

Figure 1.7: MoS_2/graphene heterostructure memory layout. (a) Three-dimensional schematic view of the memory device based on single-layer MoS_2. (b) Schematic view of the flash-type memory cell with a single-layer MoS_2 acting as a semiconducting channel and graphene contacts and multilayer graphene (MLG) as the FG [62].

(USA) fabricated a flash memory [63] using some graphitic layers, grown by CVD on Si as FG. In this case, the technological approach appears potentially scalable, CMOS compatible and so industrially suitable (see Figure 1.8) compared to the previous one. IBM researchers observed a potential retention time of 10 years, with a loss of only 8% of the charges in the FLG but also simulating the potential crosstalk between neighbor cells, who stated that this kind of memories showed negligible interference down to 10 nm (in case of common polysilicon FG, the interference raises dramatically under 25 nm). Briefly, considering the power reduction (as told previously, thanks to the higher DOS compared to common FG fabricated using polysilicon) and the increase in the storage density, Hong and coworkers estimated a potential reduction of 75% in the operating energy of this kind of memory. However, in this work there are no data on the switching time and therefore we cannot compare their performances with existing memories.

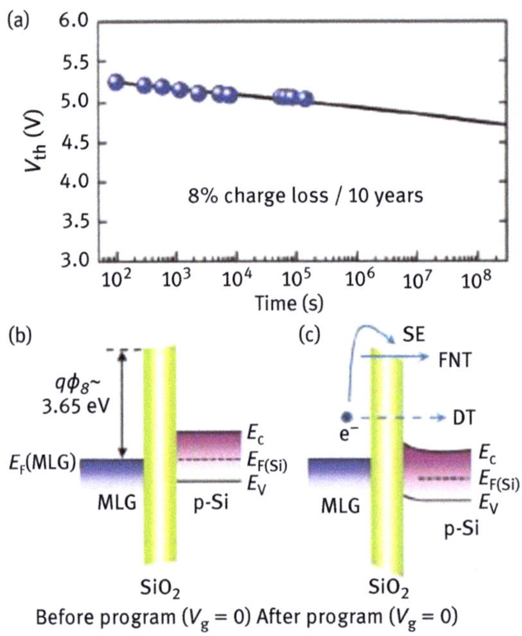

Figure 1.8: Retention characteristics of GFM. (a) Retention measurement of MLG-FM showing only 8% of charge loss in 10 years at room temperature. (b) Energy band diagram of MLG/SiO_2/Si junctions before programming. (a) Energy band diagram after programming, which shows three possible mechanisms for charge loss during retention state: SE, FNT and DT. All three mechanisms depend exponentially on the barrier height (φB) between the work function of graphene and the SiO_2 tunnel oxide [63].

1.4 Conclusions and potential applications

Graphene-based memories is a completely new technology that needs some time to demonstrate its utility especially in the difficult field of NVM for space application. These memories could constitute a real breakthrough compared to existing technologies also considering the dramatic potential reduction of energy consumption. These memories can be developed to bear harsh environments with a focus on radiation-resistant components. The stability (also in temperature), resilience, nonvolatility and on/off ratio that, in certain cases, can attain 10^7 or the switching time of 1 μs (competitive with flash-type memories), potential switching voltages of 3–4 V, combined with predictable fabrication-controlled I–V behavior, simple two-terminal geometry (no need for a gate and so dramatically reducing circuitry) and access to mass fabrication for the two approaches make them extremely attractive structures for NVM.

Indeed, thanks to thin-film technologies, electronic functionality can be foreseen in very large quantity and at very low cost on substrates such as plastic and paper. Additional functionality is also an attractive feature of carbon-based memories. For example, by using these kinds of materials, we open up the route for memories on flexible substrates, a key building block to enable the success of flexible electronics. Carbon-based resistive memories should also offer the capability for multilevel storage and "memristive-like" behavior, as seen in other resistive memory materials. Multilevel storage allows storing more than one bit per cell, so increasing data storage densities, while memristive-like behavior can be exploited to provide a remarkable range of signal processing/computing-type operations, including implementing logic, providing synaptic and neuron-like "mimics," and performing, in a very efficient way, analogue signal processing functions (such as multiply accumulate operations).

In case of graphitic-based memories based on other approaches, the transistor-type configuration exploiting the hysteretic effect, in our opinion, has a limited potential compared to new flash-type memories using graphene or MLG as the FG, because their architecture can be unlikely miniaturized and their performances especially considering the switching time are difficult to evaluate. In case of flash-type memories, these last exploit intrinsical characteristics of the materials such as the higher DOS (which allow reducing the voltage for the memory window) or the higher work function (which allows reaching more than 10 years of charge storage) that really allows improving the performances of this kind of memories. Moreover, it has been demonstrated to achieve high-density architecture with limited cross-talk. It is necessary to make an effort on the miniaturization of these devices but the recent works, for example, [51], make us think that this technology can be quite rapidly in competition with existing flash memories, also because it is CMOS compatible and easy to implement in existing production lines.

References

[1] Ohishi, M. et al. Spin injection into a graphene thin film at room temperature. Jpn. J. Appl. Phys 2007, 46, L605–L607

[2] Cho, S., Chen, Y. F., Fuhrer, M. S. Gate-tunable graphene spin valve. Appl. Phys. Lett 2007, 91, 123–105.

[3] Lee, M. J., Suh, D. S., Seo, S., Jung, R., Ahn, S. E., Lee, C. B., Kang, B. S., Ahn, S. E., Lee, C. B., Seo, D. H., Cha, Y. K., Yoo, I. K., Kim, J. S., Park, B. H. Two series oxide resistors applicable to high speed and high density nonvolatile memory. Adv. Mater 2007, 19, (22) 3919–3923, 0935–9648.

[4] Waser, R., Aono, M. Nanoionics-based resistive switching memories. Nat. Mater. 2007, 6 (11) 833, 840, 1476–1122.

[5] Bez, R., Camerlenghi, E., Modelli, A., Visconti, A., Introduction to Flash memory. Proceeding of the IEEE. 2003, 91 (4), 489.

[6] Aritome, S., Advanced flash memory technology and trends for file storage application. Tech. Dig. IEDM 2000, 763.

[7] Muller, G., Happ, T., Kund, M., Yong Lee, G., Nagel, N., Sezi, R., Status and outlook of emerging nonvolatile memory technologies, In proceeding of: Electron Devices Meeting, 2004. IEDM Technical Digest. IEEE International.

[8] Lai, S., Flash memories: Where we were and where we are going. IEDM Tech. Dig. 1998, 971–973.

[9] Pavan, P., Bez, R., Olivo, P., and Zanoni, E., Flash memory cells – An overview. Proc. IEEE 1997, 85, 1248–1271, Aug.

[10] Pavan, P., Bez, R., The industry standard Flash memory cell, in *Flash Memories*, P. Cappelletti et al., Ed. Norwell, MA: Kluwer, 1999.

[11] Chao Yan, Jeong Ho Cho, Jong-Hyun Ahn, Graphene-based flexible and stretchable thin film transistors. Nanoscale 2012, 4, 48–70.

[12] Bae, S. , Kim, H., Lee, Y., Xu, X., Park, J.-S., Zheng, Y., Balakrishnan, J., Lei, T., Ri Kim, H., Song, Y. I., Kim, Y.-J., Kim, K. S., Ozyilmaz, B., Ahn, J.-H., Hong, B. H., Iijima, S., Roll-to-roll production of 30-inch graphene films for transparent electrodes. Nat. Nanotechnol 2010, 5, 574–578.

[13] Wang, Y., Zheng, Y., Xu, X., Dubuisson, E., Bao, Q., Lu, J., Loh, K. P., Electrochemical delamination of CVD-grown graphene film: toward the recyclable use of copper catalyst. ACS Nano 2011, 5, 9927–9933.

[14] Lock, E. H., Baraket, M., Laskoski, M., Mulvaney, S. P., Lee, W. K., Sheehan, P. E., Hines, D. R., Robinson, J. T., Tosado, J., Fuhrer, M. S., Hernandez, S. C., Walton, S. G., High-quality uniform dry transfer of graphene to polymers. Nano Lett, 2011, 12, 102–107

[15] Xin, Wei, Liu, Zhi-Bo, Sheng, Qi-Wen, Feng, Ming, Huang, Li-Gang, Wang, Peng, Jiang, Wen-Shuai, Xing, Fei, Liu, Yan-Ge, Tian, Jian-Guo, Flexible graphene saturable absorber on two-layer structure for tunable mode-locked soliton fiber laser. Optics Express, 2014, 22 (9) DOI:10.1364/OE.22.010239

[16] Secor, E. B., Prabhumirashi, P. L., Puntambekar, K, Geier, M. L., Hersam, M. C., Inkjet printing of high conductivity, flexible graphene patterns. J Phys Chem Lett, 2013, 4 (8), 1347–1351.

[17] Schwartz, G., Tee, B., Mei, J. et al. Flexible polymer transistors with high pressure sensitivity for application in electronic skin and health monitoring. Nat Commun 4, 1859 (2013). https://doi.org/10.1038/ncomms2832

[18] Sekitani, T., Zschieschang, U., Klauk, H., Someya, T. Flexible organic transistors and circuits with extreme bending stability. Nature Mater 2010, 9, 1015–1022.

[19] Ramuz, M., Tee, B. C.-K., Tok, J. B.-H., Bao, Z. Transparent, optical, pressure-sensitive artificial skin for large-area sretchable electronics. Adv Mater, 2012, 24, 3223–3227.

[20] Wagner, S. et al. Electronic skin: Architecture and components. Phys E, 2004, 25, 326–334.

[21] Sokolov, A. N., Tee, B. C. K., Bettinger, C. J., Tok, J. B. H., Bao, Z. Chemical and engineering approaches to enable organic field-effect transistors for electronic skin applications. Acc. Chem. Res, 2012, 45, 361–371.
[22] Yam, L., Takhistov, Paul T., Miltz, Joseph, Intelligent packaging: Concepts and application kit. J Food Sci. 2005, 70, 1, pages R1–R10, January.
[23] Kololuoma, Terho K., Tuomikoski, Markus, Makela, Tapio, Heilmann, Jali, Haring, Tomi, Kallioinen, Jani, Hagberg, Juha, Kettunen, Ilkka Kopola, Harri K., Towards roll-to-roll fabrication of electronics, optics, and optoelectronics for smart and intelligent packaging. Proc. SPIE 2004, 77, 53–63, Emerging Optoelectronic Applications, (June 25).
[24] Kim, S., Cook, B., Le, T., Cooper, J., Lee, H., Lakafosis, V., Vyas, R., Moro, R., Bozzi, M., Georgiadis, A., Collado, A., Tentzeris, M M., Inkjet-printed antennas, sensors and circuits on paper substrate, IET microwaves, antennas & propagation,2013, 7 (10), 16 July, 858–868.
[25] Le, T., Ziyin Lin., Wong, C.P., Tentzeris, M.M., Novel enhancement techniques for ultra-high-performance conformal wireless sensors and "smart skins" utilizing inkjet-printed graphene. Electronic Components and Technology Conference (ECTC), 2013 IEEE 63rd, 2013, 1640, 1643, 28–31 May.
[26] Pham, H., Xiaochuan Xu, D.T., Chen, M. Y., Hosseini, A., Lu, Xuejun, Chen, R.T., Inkjet-printed two-dimensional phased-array antenna on a flexible substrate. Antennas and Wireless Propagation Letters, IEEE, 2013, 12, 170–173.
[27] Standley, B., Bao, W., Zhang, H. Bruck, J., Ning, Lau C., Bockrath, M., Graphene-based atomic-scale switches. Nanolett 2008, 8 (10), 3345–3349.
[28] Agrait, N., Levy-Yeyati, A., van Ruitenbeek, J. M. Quantum properties of atomic-sized conductor. Phys Reps 2003 377, 81–380.
[29] Sinitskii, A., Tour, J. M., Lithographic graphitic memories. ACS Nano 2009, 3 (9), 27–60.
[30] Li, Y., Sinitskii, A., Tour, J. M., Electronic two-terminal bistable graphitic memories. Nat Mat 2008, 7 966.
[31] Wu, C., Li, F., Zhang, Y., Guo, T. Recoverable electrical transition in a single graphene sheet for application in nonvolatile memories. Appl Phys Lett 2012, 100, 042–105.
[32] Son, D. I., Kim, T. W., Shim, J. H., Jung, J. H., Lee, D. U., Lee, J. M., Park, W. I., Choi, W. K., Flexible organic bistable devices based on graphene embedded in an insulating poly(methyl methacrylate) polymer layer. Nano Lett 2010, 10, 2441–2447.
[33] Dreyer, Daniel R., Park, S., Bielawski, C. W., Ruoff, R. S., The chemistry of graphene oxide. Chem Soc Rev 2010, 39, 228–240.
[34] Hummers, W. S., Offeman, R. E. Preparation of graphitic oxide. J. Am. Chem. Soc 1958, 80, 1339.
[35] Snider, Greg, Kuekes, Philip, Williams, R Stanley, CMOS-like logic in defective, nanoscale crossbars. Nano Tech 2004, 15, 881–891.
[36] Linn, Eike, Rosezin, Roland, Kügeler, Carsten, Waser. Rainer, Complementary resistive switches for passive nanocrossbar memories. Nat Mater 2010, 9, 403–406.
[37] He, C. L., Zhuge, F., Zhou, X. F., Li, M., Zhou, G. C., Nonvolatile resistive switching in graphene oxide thin films. Appl Phys Lett 2009, 95, 232101
[38] Wu, Z., Chen, Z., Du, X., Logan, J. M., Sippel, J., Nikolou, M., Kamaras, K., Reynolds, J. R., Tanner, D. B., Hebard, A. F., Transparent, conductive carbon nanotube films. Science 2004, 305, 1273–1276.
[39] Jeong, H. Y., Yun Kim, O., Won Kim, J., Hwang, J. O., Kim, J-E., Yong Lee J., Yoon, T. H., Cho, B. J., Kim, S. O., Ruoff, R. S., Choi, S-Y. Graphene oxide thin films for flexible nonvolatile memory. Nanolett 2010, 10, 4381–4386.
[40] http://spectrum.ieee.org/semiconductors/nanotechnology/flexible-graphene-memristors.
[41] Hong, S. K., Kim, J. Y., Kim, J. W., Choi, S. Y., Cho, B. J., Flexible Resistive Switching Memory. Device Based on Graphene Oxide. IEEE Electron Device Lett 2010, 31, 1005.

[42] Hong, S. K., Kim, J. E., Kim, S. O., Cho, B. J., Analysis on switching mechanism of graphene oxide resistive memory device. J Appl Phys 2010, 31 (1005), 2010

[43] Deegan, R. D., Bakajin, O., Dupont, T. F., Huber, G., Nagel, S. R. and Witten, T. A., Capillary flows as the cause of ring stains from dried liquid drops, Nature 1997, 1, 827.

[44] Deegan, R. D., Bakajin, O., Dupont, T. F., Huber, G., Nagel, S. R. and Witten, T. A., Contact line deposits in an evaporating drop. Phys Rev B 2000 62 (1), 756.

[45] Bondavalli, P, Delfaure, C., Legagneux, P., Pribat, D., Supercapacitor electrode based on mixtures of graphite and carbon nanotubes deposited using a new dynamic air-brush deposition technique. JECS 2013, 160 (4), A1–A6.

[46] Vasu, K. S., Sampath, S., Sood, A. K., Nonvolatile unipolar resistive switching in ultrathin films of graphene and carbon nanotubes. Solid State Communications 2011, 151 (16), 1084–1087

[47] Micheloni, Rino, Crippa, Luca, Marelli, Alessia, Inside NAND Flash Memories, Springer, e-book, ISBN 978-90-481-9431-5

[48] Masuoka, F., Iizuka H., US4531203 (A) Semiconductor memory device and method for manufacturing the same.

[49] Masuoka, F., Momodomi, M., Iwata, Y., Shirota, R., New ultra high density EPROM and flash EEPROM with NAND structure cell. *Electron Devices Meeting, 1987 International* 1987, IEEE.

[50] Van Houdt, J., Degraeve, R., Groeseneken, G., Maes, H. E., Nonvolatile memory technologies with emphasis on flash: A comprehensive guide to understanding and using NVM devices, Published Online: 23 MAR 2007 DOI: 10.1002/9780470181355.ch4

[51] Kingon, Angus, Device physics: Memories are made of . . ., Nature, 1999, 401, 658–659.

[52] Cheng Yuhua, Chenming Hu, MOSFET Classification and Operation. *MOSFET Modeling & BSIM3 User's Guide*. Springer. 1999, 13, ISBN 0-7923-8575-6.

[53] Song, Seung Min., Park, Jong Kyung., Sul, One Jae., Cho, Byung Jin. Determination of work function of graphene under a metal electrode and its role in contact resistance. Nano Letters 2012, 12 (8), 3887–3892.

[54] Yu, Y., Zhao, Y., Ryu, S., Brus, L. E., Kim, K. S., Kim, P., Tuning the graphene work function by electric field effect. Nano Lett. 2009, 9, 3430–3434.

[55] Yan, L., Punckt, C., Aksay, I. A., Mertin, W., Bacher, G., Local voltage drop in a single functionalized graphene sheet characterized by Kelvin probe force microscopy. Nano Lett 2011, 11, 3543–3549.

[56] Filleter, T., Emtsev, K. V., Seyller Th., Bennewitz R., Local work function measurements of epitaxial graphene. Appl. Phys. Lett. 2008, 93, 133–117.

[57] Wang, Haomin., Wu, Yihong., Chunxiao, Cong., Jingzhi, Shang., Yu, Ting., Hysteresis of electronic transport in graphene transistors. ACS Nano 2010, 4 (12), 7221–7228. DOI:10.1021/nn101950n

[58] Kim, W., Javey, A., Vermesh, O., Wang, Q., Li, Y., Dai, H., Hysteresis caused by water molecules in carbon nanotube field-effect transistors. Nano Lett 2003, 3, 193–198.

[59] Bondavalli, P., Legagneux, P., Pribat, D., CNTFET based gas sensors: State of the art and critical review. Sensors Actuat B 140, 1, 304–318.

[60] Echtermeyer, T., Lemme, M., Baus, M., Szafranek, B., Geim, A., Kurz, H. Nonvolatile switching in graphene field-effect devices. IEEE Electron Device Lett. 2008, 29, 8, 952–954.

[61] Echtermeyer, T. J., Lemme, M. C., M. Baus, B., Szafranek, N., Geim, A. K., and Kurz, H., A graphene-based electrochemical switch. arXiv:0712.2026v1.

[62] Bertolazzi, S., Krasnozhon, D., Kis, A. Nonvolatile memory cells based on MoS_2/Graphene heterostructures. ACS Nano 2013, 7 (4), 3246–3252.

[63] Hong, Augustin J., Song, Emil B., Yu, Hyung Sk., Matthew, J., Allen, J Kim., Fowler, J. D., Wassei, Jonathan K. , Park, Youngju., Wang, Yong., Zou, Jin., Kaner, R. B., Weiller, B. H., Wang, L., Graphene flash memory. ACS Nano 2011, 5 (10), 7812–7817.

Hanshan Dong, Santiago Corujeira Gallo

2 Plasma surface activation and functionalization of carbon-based materials

2.1 Introduction

Carbon-based materials have been extensively researched for an increasing number of applications because of their abundance and low cost, ease of processability, high-temperature stability and corrosion resistance, high electrical conductivity and good mechanical properties, particularly in the microscopic and nanoscopic scale [1]. The uses of carbon-based materials are very diverse, going from mechanical reinforcement in polymer matrix composites [2] to electrodes in supercapacitors and fuel cells [1]. However, their chemical inertness poses a common challenge in most applications, and several surface activation and surface functionalization methods have been developed to improve their performance, their compatibility with other materials or their interaction with the service environment [3].

The chemical inertness of carbon-based materials is attributed to the presence of graphene planes on the surface, having a low density of dangling bonds [5]. Graphene, carbon nanotubes (CNTs) and buckyballs are extreme examples of this microstructure, as they are composed of a single or a small number of graphene layers with a regular arrangement [6]. On the other hand, carbon fibers and amorphous carbon exhibit a turbostratic structure, with a higher degree of disorder and a larger number of dangling bonds on the surface (Figure 2.1). Several surface engineering techniques have been developed to activate and/or functionalize carbon-based materials [7]. These processes aim at increasing the number of dangling bonds and attaching functional groups to them to make the carbon-based material more compatible with the polymeric matrix in a composite, or to improve the wettability by an electrolyte in a supercapacitor and increase the charge exchange rate at the electrode.

The conventional surface engineering methods include wet chemical processes [8, 9], thermochemical [10] or electrochemical treatments [11]. Even though these methods are effective, they can be difficult to control and, in some cases, they raise environmental concerns. Therefore, recent studies have focused on the development of plasma processes for surface activation and functionalization of carbon-based materials [12, 13]. Plasma processes are conducted under vacuum or at atmospheric pressure and use highly reactive ionized gas molecules to create active sites and

Hanshan Dong, The University of Birmingham, UK
Santiago Corujeira Gallo, Deakin University, Australia

https://doi.org/10.1515/9783110479133-002

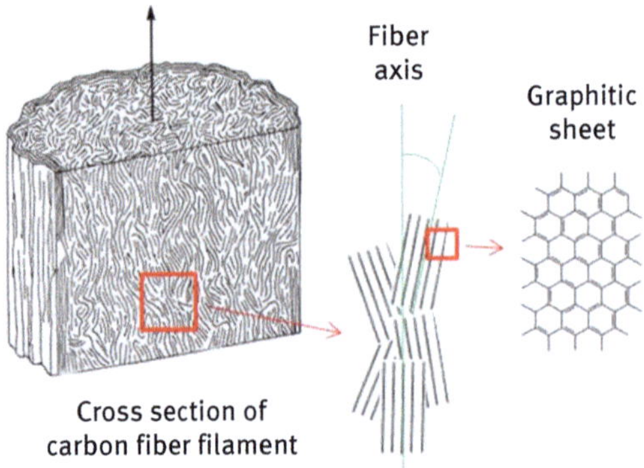

Figure 2.1: Turbostratic microstructure of carbon fibers, with inert graphene basal planes on their surface. Reprinted with permission from [4].

attach functional groups on the surface of carbon-based materials. In general, these processes are easier to control than conventional treatments, and they offer technical and environmental benefits.

2.2 Active screen plasma surface engineering

Active screen plasma (ASP) is a low-temperature and low-pressure plasma technology of surface engineering, which was developed at the University of Birmingham in the early 2000s [14–16]. The technology was initially applied to the thermochemical treatment of steels, but its potential for the functionalization of other materials was recognized immediately [17]. The schematic diagram in Figure 2.2 illustrates the main features of the ASP reactor in contrast to a conventional DC plasma reactor. The benefits of the ASP technology are mainly attributed to the creation of ions and active species on an auxiliary surface, the so-called *active screen*, as opposed to the surface of the treated material [18]. This allows more control over the type of plasma–surface interactions and the severity of the physical and chemical processes [3].

The active screen is a metallic mesh negatively biased to form ions and other active species from gas molecules, most typically N_2, H_2, Ar, CH_4. The surfaces to be treated can be simply left at floating potential[1] and exposed to the flow of active

1 Floating potential is the electric potential that an insulated object develops when exposed to the plasma. An object immersed in the plasma will be exposed to ions and electrons, but the electrons move much faster than the ions because of the difference in mass, so the number of electrons arriving at the surface of the object is higher than the number of ions. If the object is electrically insulated, the negative charge of the electrons accumulates, developing a self-induced bias that attracts positive ions and repels the less energetic electrons, until an equal number of ions and electrons arrive at the surface at any given time, that is, zero net current.

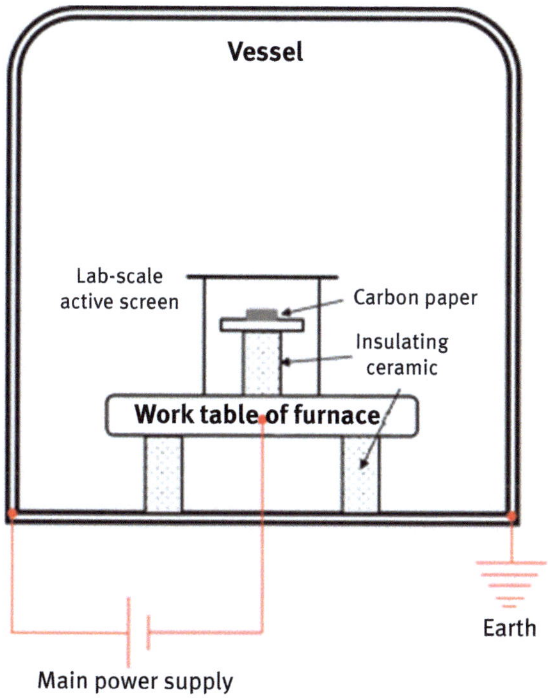

Figure 2.2: Schematic diagram of an active screen plasma reactor. Reprinted with permission from [19].

species, in which case the plasma–surface interactions are mainly of chemical nature, for example, attachment of functional groups. In other cases, a moderate bias is applied on the treated surface in order to attract the active species with increasing energy, making the physical action of the ions more pronounced. This ion bombardment can create new active sites and change the surface roughness. Alternatively, the material from the active screen can be sputtered[2] and deposited on the surfaces exposed to the plasma. This approach is particularly useful to functionalize surfaces with metallic materials.

In summary, the ASP technology has an additional degree of freedom compared to other plasma reactors, making it possible to create highly reactive ions and excited species on the *active screen*, and attracting them toward the work table with varying levels of energy (bias) to control the type of plasma–surface interaction, thus favoring physical or chemical phenomena. The following sections illustrate some of the recent

2 Sputtering refers to the ejection of material from a solid surface by the physical action of impacting ions.

uses of the ASP treatments to activate and functionalize carbon-based materials with a wide range of microstructures, from the highly ordered structure of graphene to the disordered microstructure of carbon paper.

2.3 Reduction of graphene oxide for optoelectronic devices

Graphene has attracted significant attention for optoelectronic devices, and graphene oxide (GO) is an interesting precursor for graphene coatings of high-quality and over large-surface areas [6, 20]. GO has a nanosheet-layered structure similar to graphene but contains different oxygen groups [21, 22]. The GO coatings can be produced by chemical methods [23], dissolving graphene powder and other salts in sulfuric acid, subsequently dispersing and exfoliating the GO nanosheets by sonication, and finally casting it onto the substrate. A reduction process is then required to remove the oxygen functionalities and obtain a graphene coating of high quality, and the ASP technology was used for this purpose [24]. The ASP treatments were conducted at 400 Pa in pure H_2 or N_2–H_2 mixtures, at temperatures between 100 and 200 °C for 1 h.

The appearance of coatings changed from a light brown color to silver-gray after the plasma reduction in pure hydrogen and nitrogen–hydrogen mixtures, and this change was accompanied by a decrease in optical transparency and electrical resistance of the coating (Figure 2.3). The reduction process with the plasma technology was more effective than the heat treatment of GO in a reducing atmosphere, and the results varied with the treatment temperature (100–200 °C) and the gas mixture (H_2 or H_2–N_2). The X-ray diffraction (XRD) traces revealed that the GO peak disappeared after the reduction process, and the Raman spectra indicated an increase in the ratio of disordered to graphitic bands (D/G), which was attributed to the introduction of defects in the graphene planes during the reduction process. The X-ray photoelectron spectroscopy (XPS) data showed a marked decrease in oxygen functional groups attached to the carbon network, with the carboxylic oxygen (**O**—C=O) being the most unstable functional group compared to other oxygen bonds. The XPS data also revealed the presence of Fe, Cr and Mo doping elements in a partially oxidized state. These metallic dopants are transferred from the mesh of the active screen equipment to the GO-coated substrates, particularly when using nitrogen containing gas mixtures. A small amount of nitrogen was also incorporated into the coatings.

The oxygen to carbon ratio (O/C), reported in Table 2.1, reflects the efficiency of each process to reduce the GO coating. The plasma treatment in pure hydrogen was the most effective, resulting in an extensive decomposition of carboxyl functionalities and the creation of carbonyl groups. On the other hand, the use of nitrogen in the gas mixture increased the doping of the GO coating with metal atoms from the active screen, with a marked drop in electric resistance. The level of doping can be

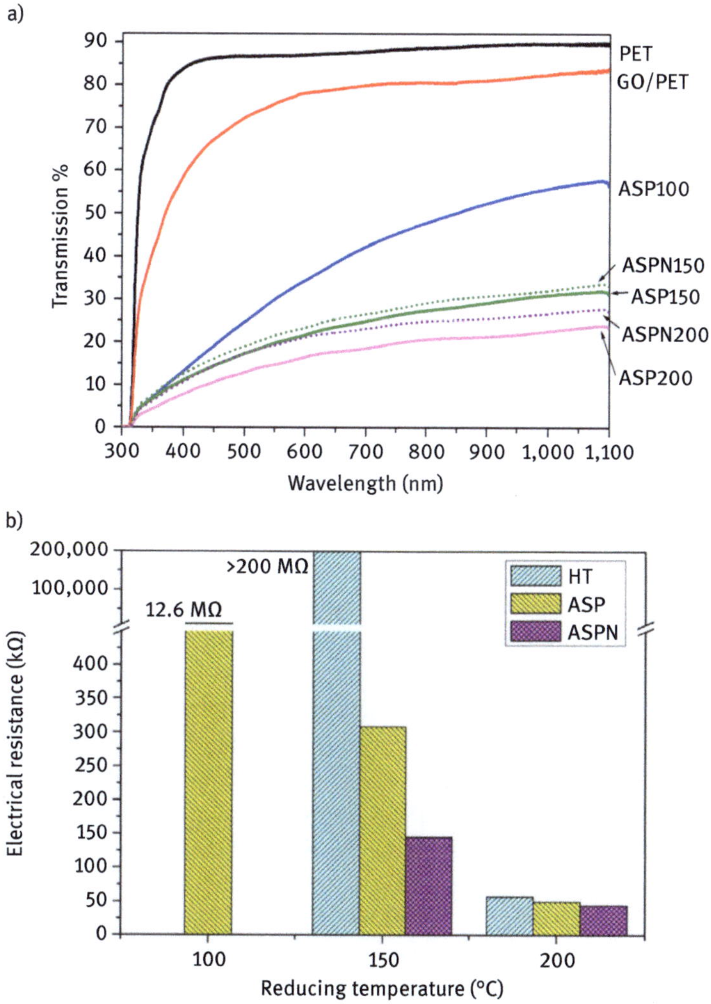

Figure 2.3: Properties of the untreated and reduced GO coatings: (a) optical transparency and (b) electrical resistance. Reprinted with permission from [24].
PET: poly(ethylene terephthalate) substrate; GO: graphene oxide coating; ASP: active screen plasma surface treatment.

controlled by varying the processing conditions (composition of the active screen, gas mixture, electric bias applied on the treated surface and duration of the process), opening new pathways to reduce and functionalize GO coatings.

Table 2.1: Chemical analysis from XPS: O components and O/C ratio. Reprinted with permission from [24].

Sample	Treatment	Gas	Temperature	O–Si	O–Me	O–C=O	O–C	O/C
GO	Untreated	N/A	N/A	2.6	0.0	9.2	9.0	0.235
HT200	Heat treated	H_2	200 °C	4.3	0.0	7.4	5.7	0.154
ASP150	ASP	H_2	150 °C	5.6	6.7	6.9	1.5	0.122
ASPN150	ASP	H_2–N_2	150 °C	2.1	7.4	7.4	1.3	0.125
ASP200	ASP	H_2	200 °C	1.7	6.2	6.2	1.1	0.102
ASPN200	ASP	H_2–N_2	200 °C	2.4	6.8	6.8	1.1	0.115

2.4 Functionalization of carbon nanofibers for supercapacitors

Carbon nanofibers (CNF) are good candidate materials for electrodes in energy storage devices because of the intrinsic properties of carbon (thermal and chemical stability, electric conductivity) and the ability to form controlled porous structures with the nanofibers [1]. This combination allows researchers to create stable electrodes with a very large surface area, and recent efforts have focused on increasing the surface conductivity and the charge exchange efficiency with the electrolyte. Silver and nickel are among the proposed electroactive materials, capable of increasing the pseudocapacitance and double-layer energy storage mechanisms,[3] and the ASP technology was used for this purpose [25]. CNFs were dispersed in isopropanol, the suspension was placed in Petri dishes and the solvent was left to evaporate before the plasma treatments.

The ASP treatments did not change the size or the structure of the CNFs, but their surface was covered with nickel and silver nanoparticles, respectively. The metal nanoparticles were approximately 10–50 nm in diameter (Figure 2.4), and their size and density increased with the ASP treatment time, which was reflected on the nickel and silver contents detected by energy-dispersive X-ray spectroscopy and XRD. The ASP functionalization with silver produced a rough surface, even after 120 min of plasma treatment, which is beneficial to increase the surface area. On the other hand, nickel tended to form a continuous coating on the CNFs after only 45 min of ASP treatment, thus limiting the amount of material and the morphology of the particles that could be deposited on CNFs.

3 Electrical double layer capacitors are based on non-Faradaic mechanisms, in which the electric charge is stored in the double layer formed at the interface between the inert electrode material and the electrolyte, without involving redox reactions. On the other hand, the pseudocapacitors rely on reversible redox reactions (Faradaic), which take place on the surface of the active electrode material in contact with the electrolyte [1].

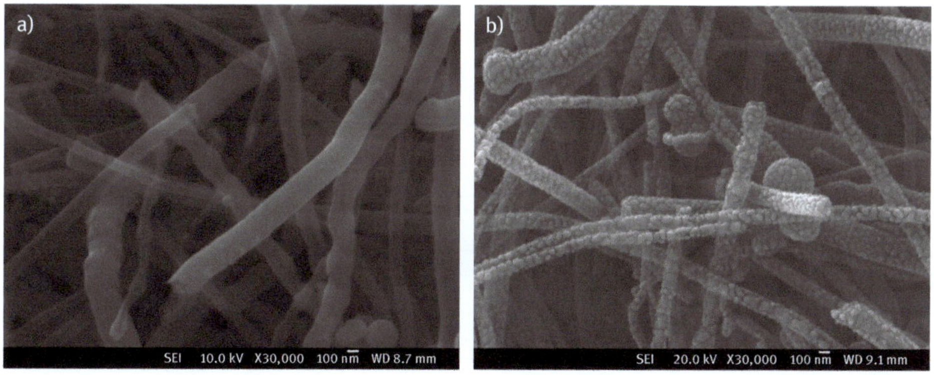

Figure 2.4: Surface morphology of carbon nanofibers: (a) untreated and (b) ASP functionalized with nickel. Reprinted with permission from [25] Copyright (2017) American Chemical Society.

The cyclic voltammetry (CV) tests conducted in 1 M sodium sulfate solution revealed a significant increase in the loop area after the ASP treatments, particularly with silver (Figure 2.5). More importantly, the CNFs functionalized with silver retained the rectangular shape of the CV curves, which was attributed to the high surface conductivity and ion transport efficiency. On the other hand, the nickel-functionalized CNFs showed severe distortions, particularly at higher scan rates, between 100 and 200 mV/s. The differences were also observed in the cyclic charge–discharge curves, with the nickel-functionalized CNF electrodes showing a twofold increase in capacitance and the silver-functionalized CNF electrodes exhibiting a fourfold increase in capacitance compared to the untreated CNFs. The electrochemical impedance spectra indicated that the increase in capacitance was mainly due to the double layer mechanism, rather than pseudocapacitance. However, recent reports have indicated that vanadium and other transition metal nitrides can improve both Faradaic and non-Faradaic charge storage mechanisms, which make them promising electrode materials for supercapacitors. This finding opens good prospects for the ASP technology, considering that it has been extensively used to form nitrides and carbides for surface functionalization and hardening.

2.5 Surface activation of carbon paper gas diffusion layers in fuel cells

Carbon paper is used as gas diffusion layer (GDL) in direct methanol fuel cells [26]. The electrocatalysts, which consist of reaction layers of precious metals with nanometer-size features, are frequently deposited directly on the GDL. These deposition processes are typically conducted in an aqueous environment and the poor

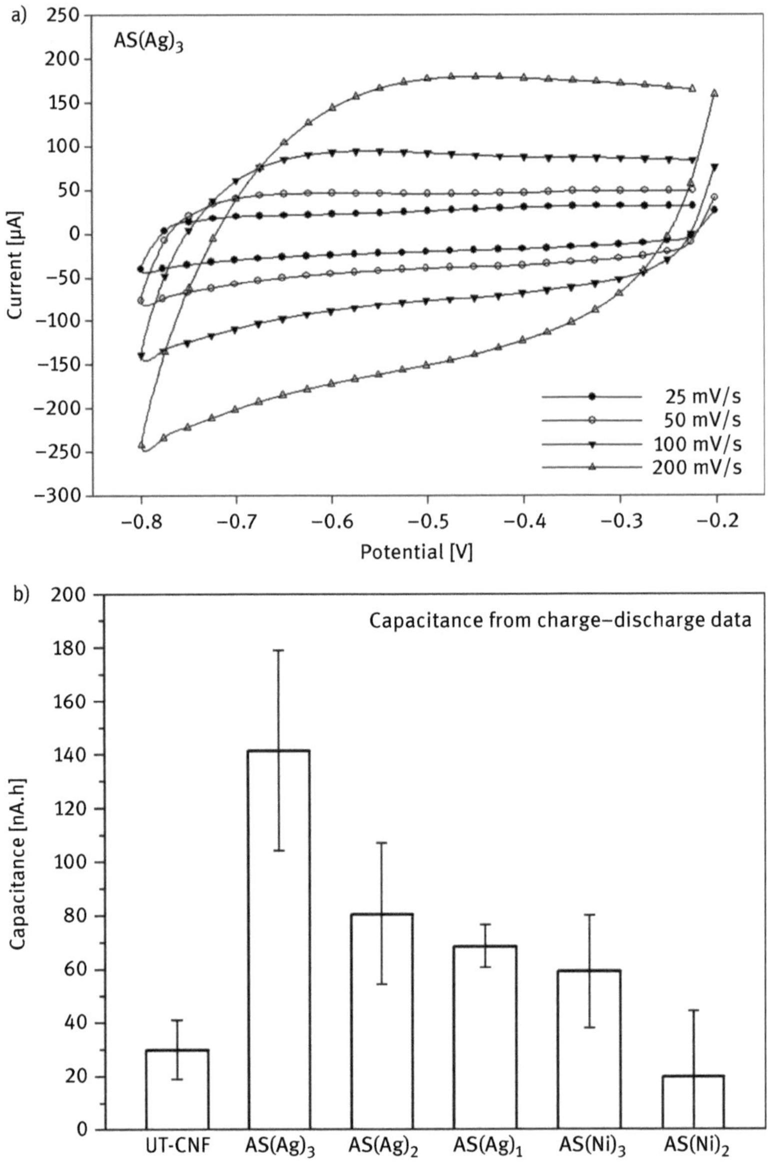

Figure 2.5: Cyclic voltammetry of carbon nanofibers functionalized with silver (a) and capacitance (b) calculated from charge–discharge experiments. Reprinted with permission from [25] Copyright (2017) American Chemical Society.

wettability of the hydrophobic carbon paper results in aggregation and uneven growth of the electrocatalyst, which reduces efficiency and increases cost [27, 28]. Therefore, the ASP technology was used to activate the surface of SIGRACET®

GDL 35BC carbon paper [19]. The activation treatment was conducted in a gas mixture of 25% N_2 and 75% H_2 at a pressure of 400 Pa. The temperature ranged from 100 to 210 °C and the duration of the treatment was between 10 and 30 min. After the activation treatment, the carbon papers were rinsed in ethanol and immersed in an aqueous solution of H_2PtCl_6 and formic acid to grow the platinum nanowires (2 mg/cm^2), and finally rinsed in water/ethanol/water and dried at 40 °C for 24 h. These carbon papers were then used as integrated cathodes in single cell tests, with a Nafion 117 membrane and the commercial JM anode (PtRu 4 mg/cm^2). The electrodes were then hot-pressed against the Nafion 117 membrane at 135 °C for 2 min under a constant pressure of 5 MPa. The reference cells were fabricated with Nafion 117 membrane, commercial JM cathode (Pt 4 mg/cm^2) and JM anode (PtRu 4 mg/cm^2). These electrode assemblies were sandwiched between two graphite flow field plates to form a single cell with an active electrode area of 5 cm^2.

The CV tests conducted on a single cell revealed that all the carbon papers activated with the ASP technology for 10 min had higher electrochemical surface area (ECSA) than the untreated carbon paper. What is more, the ECSA increased with the ASP treatment temperature up to 150 °C, and then decreased with further temperature increments (Figure 2.6). The maximum ECSA value of 25.8 m^2/g corresponded to the carbon paper activated at 150 °C with the ASP technology, which was 65% higher than the ECSA value of the untreated carbon paper. The polarization curves also revealed maximum power densities for moderate ASP treatment temperatures in the order of 120 and 150 °C. The highest power density measured in these experiments was 63.9 mW/cm^2, which was three times higher than the power density obtained with the untreated carbon paper.

This remarkable improvement in the performance of the single cells was attributed to the different surface morphology observed at high magnification. The surface of the untreated carbon paper was only partially covered by Pt nanowires, and they formed rather large cubic-shaped clusters. The uneven surface morphology was a result of the super hydrophobic nature of the untreated carbon paper and its poor wettability (Figure 2.7). In contrast, the surface of the ASP-activated carbon paper showed a more uniform distribution of the Pt nanowires, with a decreasing number of aggregates as the ASP treatment temperature increased up to 150 °C. At temperatures higher than 150 °C, the morphology of the Pt nanowires changed, becoming shorter and, consequently, reducing the ECSA and the cell performance. Longer ASP activation times, that is, 30 min, had a detrimental effect on the surface morphology of the catalyst layer and, consequently, on the cell performance. In summary, the ASP activation at moderate temperature (120–150 °C) and for a short time (10 min) can effectively promote the growth of the electroactive layer of Pt nanowires on the carbon paper, improving the utilization of this expensive catalyst and, ultimately, increasing the ECSA and the power density of the fuel cell.

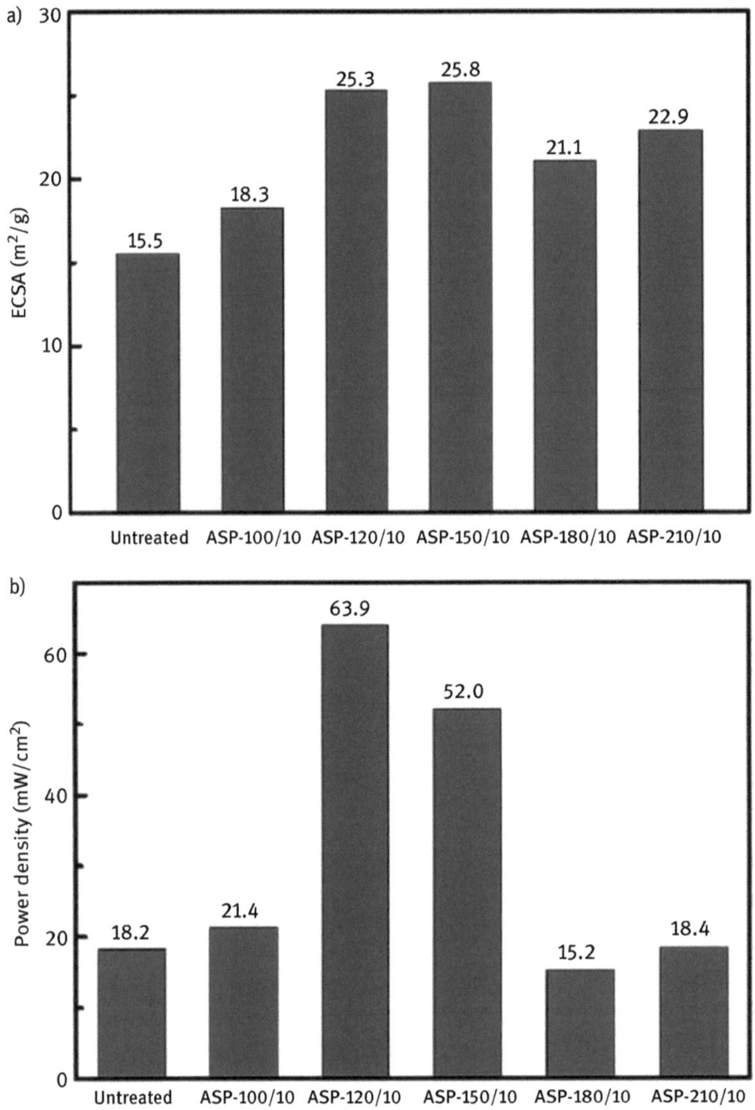

Figure 2.6: (a) Electrochemical surface area (ECSA) and (b) power density from polarization tests. Reprinted with permission from [19].

2.6 Improving the bond strength of carbon fibers in composite materials

Carbon fibers are exceptional reinforcements for polymer matrix composite materials, particularly because of their high strength, high stiffness and lightweight [29].

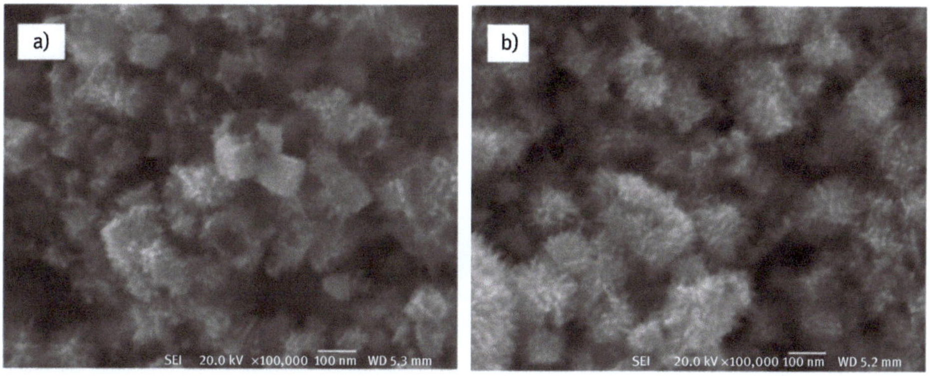

Figure 2.7: Surface morphology of the electrocatalyst layer deposited on the (a) untreated and (b) ASP functionalized carbon paper (ASP treatment conducted at 120 °C for 10 min) [19].

However, the graphitic basal planes present on the surface of the carbon fibers make them relatively inert and this reduces the wettability and the bond strength with the polymeric matrix, frequently an epoxy resin [3]. Plasma treatments have been used to introduce polar functional groups on the surface of carbon fibers. These functional groups help to modify the bond strength with the polymeric matrix and to tailor the mechanical properties of the composite material, particularly the strength and toughness [30, 31]. The ASP technology offers good control over the treatment conditions to avoid damaging the fibers, thus retaining their strength.

Carbon fibers were ASP treated in a gas mixture of 75% H_2 and 25% N_2 at a pressure of 50 Pa. This atmosphere was selected to introduce $-NH_x$ functional groups on the surface of the carbon fibers [32]. These functional groups are known to create strong chemical bonds with the epoxy resins used in these composites, and the physical etching creates a microroughness on the carbon fibers, which increases the mechanical interlocking with the matrix. Optical emission spectroscopy studies conducted during the ASP treatments revealed NH_x species in the plasma phase, and NH groups were also identified on the surface of the functionalized carbon fibers using XPS (Figure 2.8). The relative intensity of the NH species in the plasma decreased with increasing applied power, and this change correlated with a reduction in the nitrogen content on the surface of the functionalized fibers. A similar trend was observed on the contact angle of glassy carbon disks, functionalized alongside with the carbon fibers. All the ASP-treated carbon disks showed contact angles significantly lower (<60°) than the untreated disks (80°), but the lowest contact angle (40°) corresponded to the disk treated at low power and for short time.

Further wettability measurements were conducted on carbon disks immediately after the ASP treatment and over the subsequent 4 weeks. It was found that the contact angle dropped remarkably after the plasma treatments to values as low as 10°. However, the contact angle increased rapidly over the subsequent 7 days

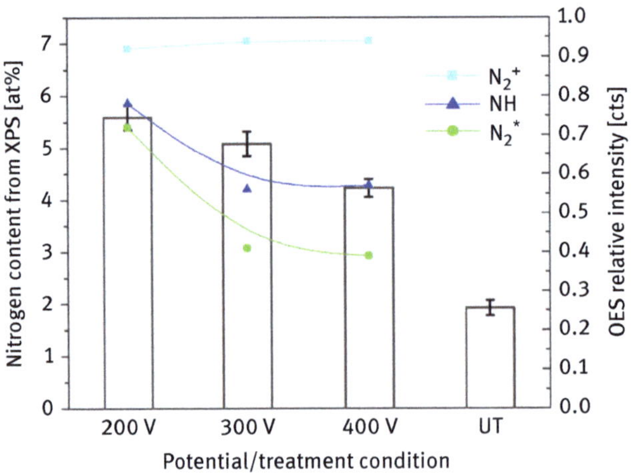

Figure 2.8: Correlation of active species in the plasma phase with functional groups on the surface of the carbon fibers as a function of the ASP treatment conditions (potential). Reprinted with permission from [30]. Copyright 2017, American Vacuum Society.

and remained relatively constant for the remaining 3 weeks (Figure 2.9). The chemical analysis by XPS showed a small but consistent reduction in the nitrogen content (from 5.8 to 5.1 at.%) over a period of 28 days. However, the initial XPS analysis was only conducted 3–4 weeks after the ASP treatment, so the wettability measurements could not be correlated with the XPS data. The change in the nitrogen content and in the wettability was consistent with an exponential decay, which could be attributed to a degradation of the nitrogen functionalities by oxidative scissions and evaporative loss upon their exposure to air. The elucidation of the degradation mechanism requires further study.

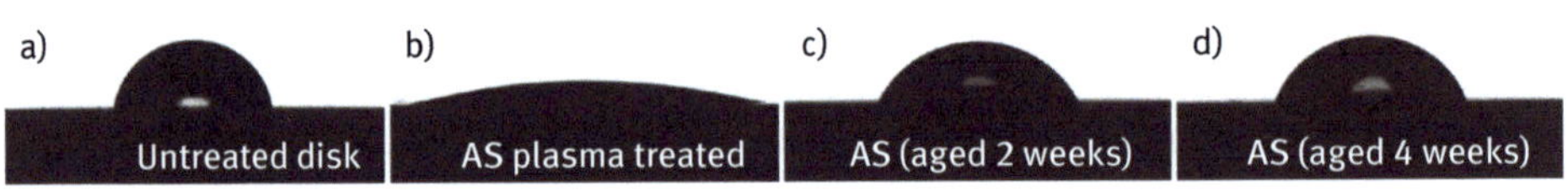

Figure 2.9: Evolution of the contact angle of an ASP-treated carbon surface with time. Reprinted with permission from [30]. Copyright 2017, American Vacuum Society.

2.7 Synthesis and deposition of carbon nanotube films

CNTs exhibit extraordinary electrical, thermal and mechanical properties, similar to CNFs. CNTs are frequently synthesized by arc discharges, laser ablation or chemical vapor deposition (CVD) [33–35]. The latter method is capable of producing vertically aligned CNT (VACNT), but the process requires prior deposition of metal nanoparticles on the substrate to act as catalysts and nucleation sites. In addition, the high temperature required for conventional CVD processes can be detrimental for the substrate. It is known that plasma-enhanced CVD (PE-CVD) can reduce the deposition temperature, but the energy of the ions needs to be carefully controlled to enhance rather than suppress the growth of CNT [36]. A multistep process to deposit metal nanoparticles and grow CNTs at low temperatures has been recently developed based on the ASP technology [33].

Silicon wafers (100) were used as substrates for the PE-CVD process with RF excitation (13.56 MHz, 30 W) and a modified active screen experimental setup (Figure 2.10). Metal nanoparticles with different compositions were deposited on the silicon wafers by changing material of the active screen, in this case nickel and stainless steel. The size, morphology and uniformity of the metal particles were critical for the success of the process, the density and vertical alignment of the CNTs. VACNTs were grown in the same active screen experimental arrangement using a gas mixture of acetylene, argon and hydrogen at 100 Pa (Figure 2.11). Transmission electron microscopic observations revealed a tip-growth mode, in which the metal nanoparticles were detached from the silicon wafer and remained at the tip (growing front) of the CNTs. For a given treatment time (100 min), the length of the CNTs increased with the processing temperature from 0.6 μm at 400 °C to 6 μm at 500 °C, and their diameter decreased from 28 to 16 nm, respectively (Figure 2.11). The structure of the CNTs did not change significantly with

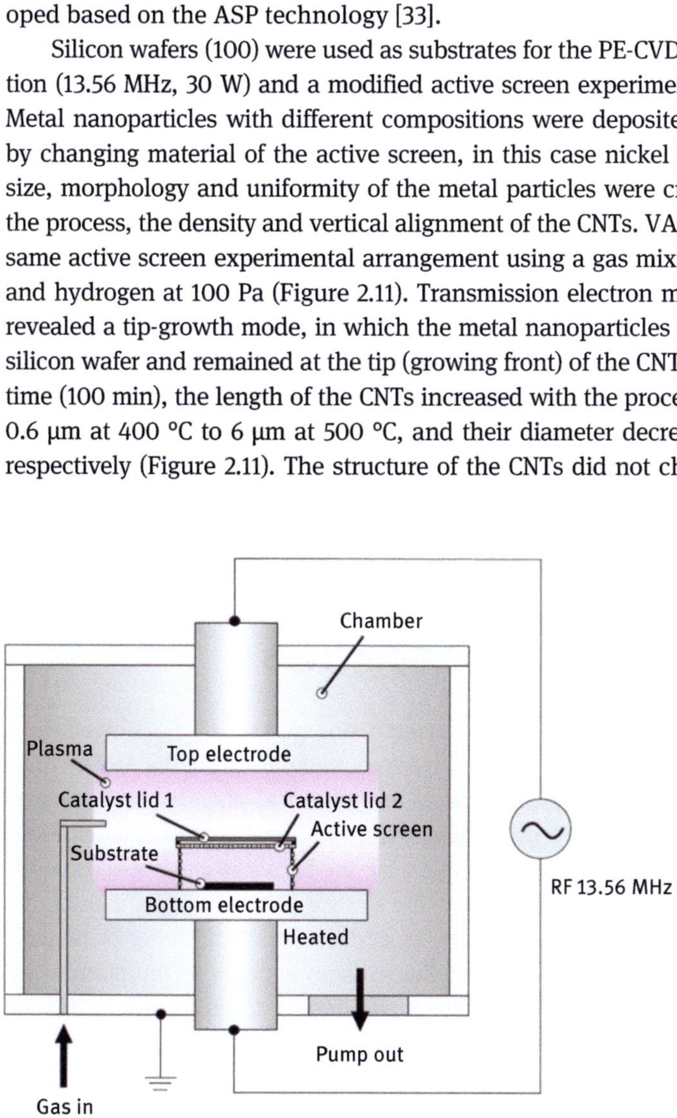

Figure 2.10: Schematic diagram of the active screen setup used for PE-CVD of vertically aligned carbon nanotubes. Reprinted with permission from [33].

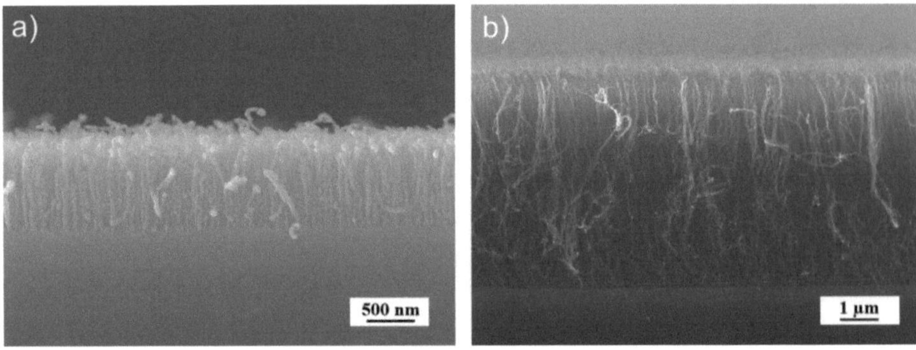

Figure 2.11: SEM images of CNTs grown at (a) 400 °C and (b) 500 °C. Reprinted with permission from [33].

the deposition temperature, but the ratio of D and G bands (I_D/I_G) in the Raman spectra changed from 0.7 to 0.97 with increasing content of acetylene in the gas mixture.

The multistep PE-CVD process, based on the ASP technology with RF excitation, could successfully deposit a layer of metal nanoparticles on a silicon wafer and subsequently grow vertically aligned CNTs. The size, morphology and microstructure of the VACNTs could be controlled by changing key processing parameters, such as size, distribution and uniformity of metal nanoparticles, gas mixture and processing temperature. This process represents a simple and cost-effective method to functionalize the surface of a wide range of substrates with CNTs.

2.8 Summary and future perspectives

This chapter presents a summary of the research undertaken at the University of Birmingham on the functionalization of carbon-based materials using the ASP technology. This research exploits the typical physical and chemical phenomena associated with the plasma technology, further optimized by the additional control and degrees of freedom offered by the ASP technology. In this way, the highly reactive species in the plasma were used to reduce GO films, activate the surface of carbon paper membranes or functionalize the surface of carbon fibers. In addition, the sputtering of material from the active screen and its deposition on the surfaces exposed to the plasma was used to increase the conductivity and surface area of CNFs and to create nucleation sites for CNTs. In all cases, the treatment conditions could be optimized, targeting specific plasma–surface interactions, to achieve the desired properties.

Future efforts will focus on expanding the range of gases and active screen materials used to functionalize temperature-sensitive substrates, as well as combining the ASP technology with pre- and post-treatments. There are also challenges related to the uniformity of the ASP treatment of porous and three-dimensional structures due to shadowing effects that need to be addressed. In addition, the interaction between active species in the plasma and different microstructures of carbon-based materials deserves further study, together with the evolution of the ASP functionalized surfaces over time. Lastly, there seem to be good opportunities to create patterned surfaces using the ASP technology, in combination with other processes with potential application in multiple fields such as functional materials, sensors and biosensors, electrical, electronic and optoelectronic devices. Without question, the ASP technology remains to be a valuable enabling technology for the surface functionalization of engineering materials, showing broader prospects of practical applications than ever before.

Acknowledgments: The research leading to these results has received funding from the European Union for Multi-Surf (Project ID 327750), FIBRALSPEC (GA604248) and MODCOMP (GA685844). The authors would like to thank their former and current colleagues Dr Xiaoying Li, Prof Jian Chen, Dr Kaijie Lin, Dr Shaojun Qi and Mr Xiaochao Ji for the reproduction of their original work.

References

[1] Pandolfo, A. G., A. F. Hollenkamp, Carbon properties and their role in supercapacitors. J Power Sources 2006, 157 (1), 11–27.
[2] Karger-Kocsis, J., Mahmood, H., Pegoretti, A., Recent advances in fiber/matrix interphase engineering for polymer composites. Prog. Mater. Sci 2015, 73, 1–43.
[3] Corujeira-Gallo, S., Dong, H., Effect of microstructure on the plasma surface treatment of carbon fibres. J Compos Mater 2017, 51 (23), 3239–3256.
[4] Behr, M. J., et al., Structure-property model for polyethylene-derived carbon fiber. Carbon 2016, 107, 525–535.
[5] Hoecker, F., Karger-Kocsis, J., Surface energetics of carbon fibers and its effects on the mechanical performance of CF/EP composites. J Appl Polym Sci 1996, 59 (1), 139–153.
[6] Novoselov, K. S., et al., A roadmap for graphene. Nature 2012, 490, 192.
[7] Tang, L. G., Karoos, J. L., A review of methods for improving the interfacial adhesion between carbon fiber and polymer matrix. Polym Compos 1997, 18 (1), 100–113.
[8] Zielke, U., Hüttinger, K. J., Hoffman, W. P., Surface oxidized carbon fibers: II. Chemical modification. Carbon 1996, 34 (8), 999–1005.
[9] Severini, F., et al., Chemical modification of carbon fiber surfaces. Carbon 2002, 40, (5), 735–741.
[10] Solís-Fernández, P., et al., A comparison between physically and chemically driven etching in the oxidation of graphite surfaces. J Colloid Interf Sci 2010, 344 (2), 451–459.
[11] Donnet, J. B., Guilpain, G., Surface treatments and properties of carbon fibers. Carbon 1989, 27 (5), 749–757.
[12] Käppler, I., Hund,R. D., Cherif, C., Surface modification of carbon fibres using plasma technique. AUTEX Res J 2014, 14 (1), 34–38.
[13] Tiwari, S., et al., Influence of cold remote nitrogen oxygen plasma treatment on carbon fabric and its composites with specialty polymers. J Mater Sci 2011, 46, (4), 964–974.

[14] Li, C. X., Bell, T., and Dong, H., A study of active screen plasma nitriding. Surf Eng, 2002, 18 (3), 174–181.
[15] Cheng, Z., et al., Study of the active screen plasma nitriding. Cailiao Rechuli Xuebao/ Transactions of Materials and Heat Treatment 2004, 25 (5), 330–333.
[16] Georges, J., Nitriding process and nitriding furnace therefor, U.P. Office, Editor. 1999: United States of America.
[17] Li, C. X., Bell, T., Potential of plasma nitriding of polymer for improved hardness and wear resistance. J. Mater. Process. Technol 2005, 168 (2), 219–224.
[18] Corujeira Gallo, S., H. Dong, Study of active screen plasma processing conditions for carburising and nitriding austenitic stainless steel. Surf Coat Tech 2009, 203 (24), 3669–3675.
[19] Lin, K., et al., The effect of active screen plasma treatment conditions on the growth and performance of Pt nanowire catalyst layer in DMFCs. Int. J. Hydrog. Energy 2016, 41 (18), 7622–7630.
[20] Lee, Y., Ahn, J.-H., Graphene-based transparent conductive films. Nano 2013, 08 (03), 1330001.
[21] Gómez-Navarro, C., et al., Atomic Structure of Reduced Graphene Oxide. Nano Letters 2010, 10 (4), 1144–1148.
[22] Dreyer, D. R., et al., The chemistry of graphene oxide. Chem. Soc. Rev 2010, 39 (1), 228–240.
[23] Marcano, D. C., et al., Improved Synthesis of Graphene Oxide. ACS Nano 2010, 4, (8), 4806–4814.
[24] Chen, J., et al., Reducing and multiple-element doping of graphene oxide using active screen plasma treatments. Carbon 2015, 95, 338–346.
[25] Corujeira Gallo, S., et al., Carbon nanofibers functionalized with active screen plasma-deposited metal nanoparticles for electrical energy storage devices. ACS Appl Mater Inter 2017, 9, 27, 23195–23201.
[26] Schweiss, R. and Öttinger, O., The importance of carbon materials in micro-porous layer in gas diffusion layers for proton exchange membrane fuel cells. Carbon 2015, 93, 1079.
[27] Rajalakshmi, N., Dhathathreyan, K. S., Nanostructured platinum catalyst layer prepared by pulsed electrodeposition for use in PEM fuel cells. Int. J. Hydrog. Energy 2008, 33 (20), 5672–5677.
[28] Du, S., Pollet, B. G., Catalyst loading for Pt-nanowire thin film electrodes in PEFCs. Int. J. Hydrog. Energy, 2012, 37 (23), 17892–17898.
[29] Sharma, M., et al., Carbon fiber surfaces and composite interphases. Compos Sci Technol, 2014, 102,35–50.
[30] Corujeira Gallo, S., Charitidis, C., Dong, H., Surface functionalization of carbon fibers with active screen plasma. J Vac Sci Technol A: Vacuum, Surfaces and Films 2017, 35 (2), 021404-1-10.
[31] Corujeira Gallo, S., et al. Active screen plasma treatment and characterisation of carbon fibres – A feasibility study. Carbon 2015, Dresden.
[32] Pittman Jr, C. U., et al., Chemical modification of carbon fiber surfaces by nitric acid oxidation followed by reaction with tetraethylenepentamine. Carbon, 1997, 35 (3),317–331.
[33] Ji, X., et al., A novel hybrid method combining ASP with PECVD for in-situ low temperature synthesis of vertically aligned carbon nanotube films. Diam. Relat. Mater 2017, 77, 16–24.
[34] Thess, A., et al., Crystalline ropes of metallic carbon nanotubes. Science 1996, 273 (5274), 483–487.
[35] Iijima, S., Helical microtubules of graphitic carbon. Nature 1991, 354, 56.
[36] Jang, I., et al., Characteristics of carbon nanotubes grown by mesh-inserted plasma-enhanced chemical vapor deposition. Carbon 2007, 45 (15),3015–3021.

Rajat Srivastava, Matteo Fasano, Shahin Mohammad Nejad, Hernán Chávez Thielemann, Eliodoro Chiavazzo, and Pietro Asinari

3 Modeling carbon-based smart materials

3.1 Introduction

Smart materials show changes in their properties in a predictable manner upon application of external stimuli, such as chemical, electrical, thermal, mechanical, magnetic and light ones. Our knowledge about smart materials exists from the era of Roman Empire: there are several evidences that building materials used at that time, mainly mixtures of several compounds including limestone, had self-healing capabilities, since they were able to repair cracks by the simple action of rainwater [1]. Modern smart materials are opening up new opportunities in several fields, for instance, healthcare, defense, waste management and packaging. In fact, these materials are widely used in applications such as sensors, actuators, robots, artificial muscles and drug delivery. Smart materials can be classified based on their responses to the external stimuli, that is, shape memory, piezoelectric, photoresponsive, electroresponsive, magnetoresponsive and thermochromic materials.

Shape memory materials have the unique capability of remembering their original shape after deformation, thereby returning to it upon application of external stimuli such as thermal, mechanical or magnetic variations [2, 3]. These materials have found application in automobiles, aircraft, biomedical devices, robots, civil structures and textiles [3–8]. Shape memory materials can be made out of alloys, ceramics, polymers, hybrid materials or gels. The first shape memory material dates back to 1932, when Arne Ölander first discovered a solid cadmium–gold alloy that, when deformed in a cold state, returned to its original shape upon heating [9]. Nickel–titanium (NiTi) was then a popular shape memory alloy (SMA) discovered by Buehler et al. [10] in 1963, which spread the use of SMA to the first industrial applications. NiTi alloys are preferable in most applications due to their good ductility, corrosion resistance, mechanical stability and biocompatibility [11, 12]. In recent years, carbon-based shape memory polymers attracted attention due to their capability of handling high strain and temperature, their lower density and recovery temperature and their flexibility and recyclability [13–15]. All these superior properties make them interesting shape memory materials, particularly for aerospace and biomedical fields.

Rajat Srivastava, Matteo Fasano, Shahin Mohammad Nejad, Hernán Chávez Thielemann, Eliodoro Chiavazzo, Pietro Asinari, Department of Energy, Politecnico di Torino, Torino, Italy

https://doi.org/10.1515/9783110479133-003

For example, Zhang et al. [16] fabricated multistimuli-responsive shape memory materials based on polymer composites reinforced by graphene nanoplatelets (GNPs), which could be triggered by electrical and infrared stimuli. Figure 3.1 shows the shape memory mechanism actuated by photothermal effect: after a deformation at low temperature (*Shape 2*), the original shape of the GNP-based polymeric material (*Shape 1*) can be recovered upon exposure to infrared radiation [16]. Recently, carbon-based shape memory polymer nanocomposites seem promising candidates for electrical actuators. In fact, polymer nanocomposites filled with carbon nanotubes (CNTs) have shown low electrical resistivity while fast thermal response to applied voltage. These composites can be tailored to specific actuation requests, since a desired shape recovery can be attained by printing CNTs with a specified pattern throughout the polymeric matrix [17].

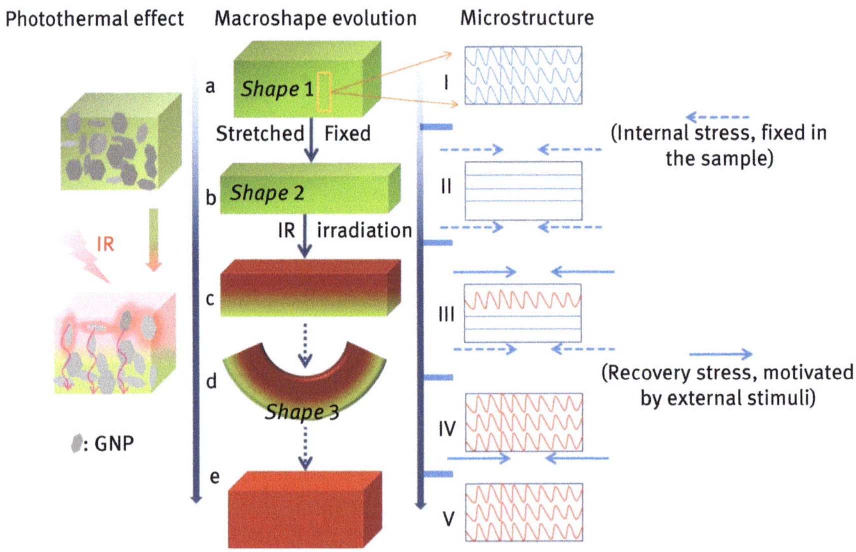

Figure 3.1: Schematic illustration of infrared radiation-actuated shape memory effect of GNP-based polymer blend. Reproduced from Ref. [16] with permission from the Royal Society of Chemistry.

Piezoelectric materials are capable of converting mechanical deformation into electricity, and vice versa [18, 19]. These materials are widely used as sensors, actuators, structural health detectors and energy generators [19–22]. Particularly, there is a growing demand of wearable sensors for monitoring physiological and biomechanical signals in sports and medical applications. However, bulky batteries are currently needed to power these sensors; therefore, smart garments generating their own power by human body motion would be highly beneficial (see Figure 3.2). This requirement can be, for instance, fulfilled by polymeric fiber-based piezoelectric generators, which are flexible and durable as well [23–25].

Figure 3.2: Sketch showing power generation from human movement using fiber-based piezoelectric generator. Reproduced from Refs. [23, 26] with permission from the Royal Society of Chemistry.

Zhong et al. [24] fabricated a piezoelectric fiber using cotton threads coated with CNTs and dipped into an aqueous suspension of polytetrafluoroethylene. This carbon-based polymeric piezoelectric fiber utilizes the biomechanical energy from human motion to generate electrical energy, with an average power output of $\sim 0.1\ \mu W/cm^2$, which would be enough to energize wearable sensors or smart shirts (see Figure 3.3).

Photoresponsive materials show changes in their physical and chemical behavior when exposed to light. Usages of these smart materials include anticounterfeiting applications [27], chemical sensors [28], optical switches [29], flexible rewritable optical memories [30] and toys (e.g., doll whose skin get tanned under the sun [31]). The most widely used photoactive agents are azobenzene, spiropyrans, spirooxazines, diarylethenes and fulgides, which are summarily depicted in Figure 3.4.

Recently, Renuka et al. [29] developed an optical switch using multiwalled CNTs (MWCNTs) and azo-based chromophores (Figure 3.5). The electrical conduction of the composite is controlled by azo-based system, which shrinks when exposed to UV irradiation (*cis* isomer), thereby increasing the electrical conduction, because of the reduced tunneling barriers through the MWCNTs. The conduction drops down again when UV irradiation stops, because azo molecules return to their *trans* isomer configuration.

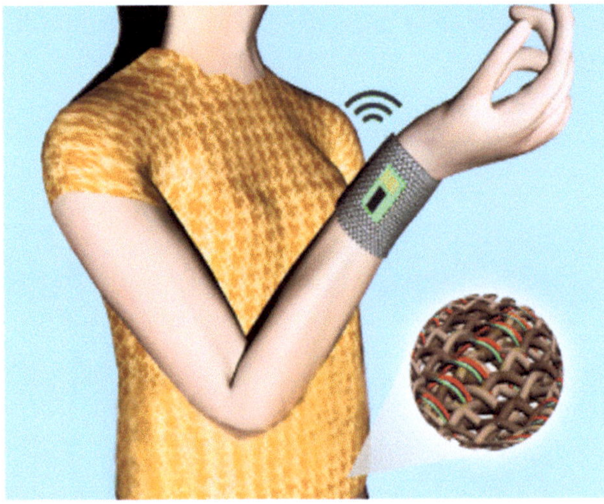

Figure 3.3: Schematic diagram of a smart shirt. Reprinted from Ref. [24] with permission from American Chemical Society.

Lately, *electroactive polymers* (EAPs) are gaining interest because of their unique property to change geometrical/structural properties in response to the application of an electric field. Due to this capability, EAPs can be employed in artificial muscles, sensors, robots, biomedical devices and energy storage systems [33, 34]. *Magnetoresponsive materials*, instead, can be shaped upon the application of a magnetic field. Magnetoresponsive fluids, for instance, can be used to construct dampers with adaptive response and thus enhanced suppression of vibrations. These can be fitted to buildings and bridges to mitigate the detrimental effect of strong winds or earthquakes. Among other applications, magnetoresponsive materials can be used also in nanomedicine [35, 36]. Finally, *thermochromic materials* change color in response to different temperatures. They have been used, for example, in bath plugs that change their color when water is too hot.

Carbon-based polymer nanocomposites seem to be among the most promising candidates for the development of smart materials. However, significant efforts are still needed to design smart materials with specific thermal, mechanical and electrical properties. Since these properties derive by mechanisms coupled from nano- to macroscale, multiphysics simulations could support a more rational research of smart materials with tailored properties. For example, the overall piezoelectric response of carbon-based polymeric smart materials is influenced by several mechanisms, including tunneling of electrons and formation of percolating networks of carbon fillers in the polymer composite [37–39], and should therefore be addressed by linking atomistic to mesoscopic models. In the following sections, the atomistic,

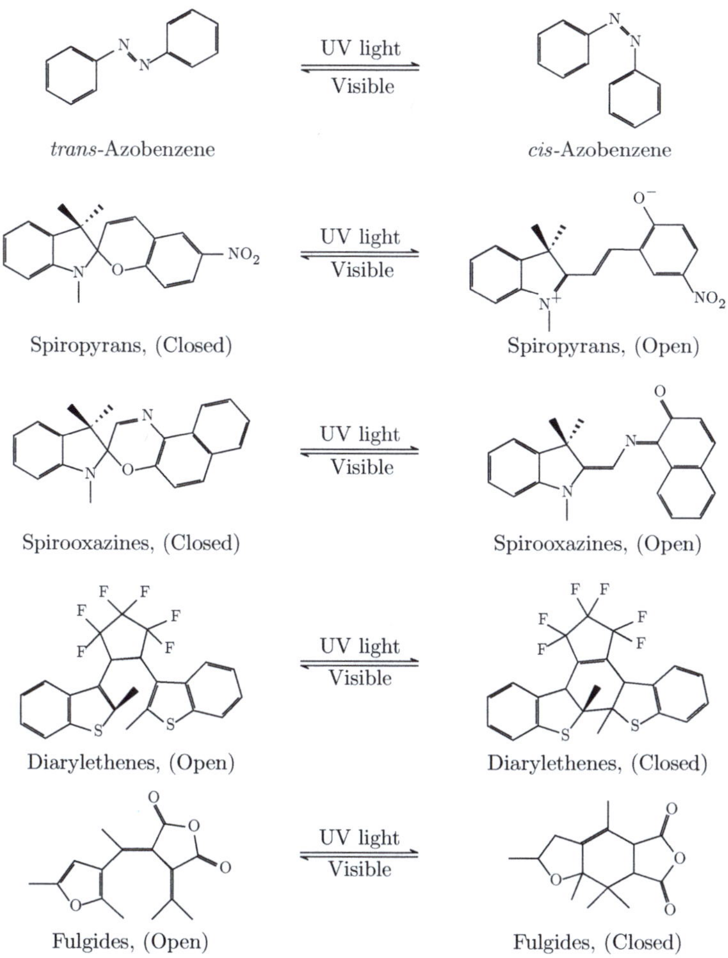

Figure 3.4: Chemical structure of photoresponsive molecules [32].

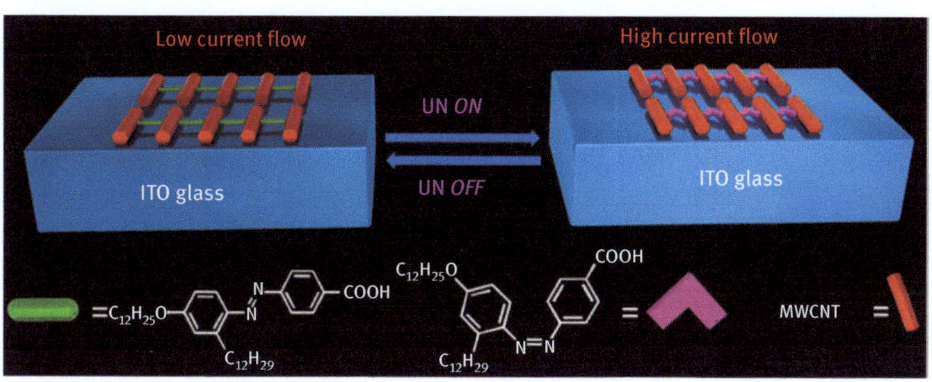

Figure 3.5: Schematic diagram of an optical switch based on MWCNT/azo composite. Reprinted from Ref. [29] with permission from American Chemical Society.

mesoscopic and macroscopic simulation tools that allow to explore the main properties of nanocomposites are extensively presented, and future perspectives of advanced modeling of carbon-based smart materials are discussed.

3.2 Multiphysics modeling

A better understanding of the thermophysical behavior of carbon-based polymer nanocomposites at different length- and timescales could ease the discovery of novel smart materials, and could be facilitated by proper material modeling approaches [40–47]. Advanced modeling approaches to the properties of nanocomposites are based on linking and/or coupling various model types that are applicable at different length- and timescales (see Figure 3.6).

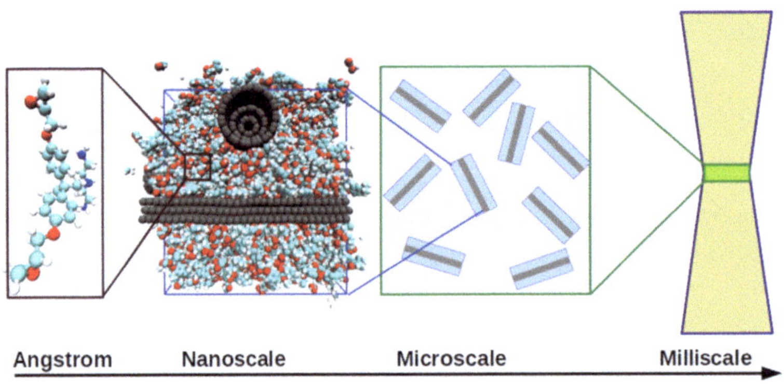

Figure 3.6: Schematic diagram of advanced modeling of polymer nanocomposite across multiple scales.

This situation is ubiquitous in many sectors, where modeling can support the effective development of advanced materials, but the heterogeneity of modeling techniques may prevent to highlight the computational synergies. For this reason, the European Commission published a Review of Materials Modelling (RoMM), now in its sixth edition [48], which provides a classification of materials modeling that enables a coherent description of model types and a standardized documentation of modeling data (called "MODA") of materials. Furthermore, the RoMM contains several examples of MODA documentation applied to a compendium of applications from EU H2020 LEIT NMBP Materials projects. Based on the above review, the European Materials Modelling Council (EMMC, http://www.emmc.info) proposed a CEN Workshop Agreement about "Materials modelling – terminology, classification and metadata" [49], endorsed by more than 15 European organizations with the objective of standardization of terminology, classification and documentation of materials modeling and simulation. Moreover, the EMMC already proposed a European Materials Modelling Ontology (EMMO) [50]. The EMMO aims at addressing the granularity levels of materials modeling (atomistic, electronic, mesoscopic and continuum), which do not really depend on the scale they are applied to. This is why multiscale modeling is somehow confusing and it would be better to talk about multiphysics modeling.

In particular, at atomistic/molecular level, molecular dynamics (MD) simulations allow understanding the behavior of the constitutive elements of nanocomposites (such as carbon fillers and polymer matrices) and, most importantly, their interaction at the interface [51]. The structural and interfacial characteristics of the constitutive elements of nanocomposites at molecular level are successively employed in coarser modeling descriptions (mesoscopic models). The mesoscopic structure of a smart composite is modeled by a representative volume element (RVE) of the material, and the properties computed at mesoscale can be finally homogenized to evaluate the effective thermophysical properties at macroscopic (continuum) level.

3.2.1 Atomistic models

Thermophysical properties of carbon-based nanocomposite materials are mainly governed by the interaction between nanofillers and surrounding matrix at molecular level. MD simulations are typically the most adequate tool to study such atomistic phenomena. In this section, the process of building the molecular geometry of nanocomposite is first discussed; then, the MD protocols for assessing thermal, mechanical and electrical properties are described, and main results from the literature are presented.

Building the atomistic setup

While the atomistic structure of carbon nanofillers (e.g., nanotubes, nanoplatelets and nanoribbons) can be conveniently generated by tools available in the literature [52–54], one of the major challenges in simulating polymer-based nanocomposites at the molecular level is to accurately describe the structure of polymeric matrix. Curing is a chemical reaction that happens between the components of thermosetting plastics (e.g., epoxy resins), namely the resin and hardener, which result in a final cross-linked (cured) structure. MD methods possess an inherent advantage over continuum approaches, since they are able to accurately simulate the mechanisms underlying curing reaction. Schulz and Frisch [55] first attempted to model curing procedure at the molecular level. They employed a lattice Monte Carlo simulation method to understand the reaction kinetics of curing process in terms of degree of polymerization and molecular weight distribution. Later works also included topological information of polymer networks in the description of curing process, therefore allowing the simulation of thermomechanical properties of epoxy resins by MD simulations [56–64].

Protocols for simulating the curing process and creating the atomistic setup of epoxy resins can be classified into two groups: (1) single-step methods, where all the potential chemical bonds between reactive sites within a specified cut-off distance are formed at once; (2) multistep approaches, where reactive pairs within a spatial distance are cross-linked together and the resulting structure relaxed by MD, iteratively, until the desired curing degree is met. Generally, single-step approaches lead to large artificial strains in the cured structure; whereas, multistep methods – even though they are more computationally expensive – provide structures closer to real curing reactions, since they operate through intermediate MD relaxations.

Jang et al. [64] compared the performance of these two approaches on a system consisting of bisphenol A diglycidyl ether (DGEBA) epoxy and poly(oxypropylene) diamine (POP) curing agent. In the single-step method, a Monte Carlo algorithm was adopted to allocate cross-linking bonds. For the multistep algorithm, an initial cut-off radius of 10 Å was assigned, and covalent bonds between all pairs of potentially reactive atoms within the cut-off radius generated. After that, the partially cured system was relaxed in the NPT and NVE ensembles, respectively. In addition, the partial charges of atoms were adjusted to neutralize the net charge in the system. Finally, in the case that no reactive pairs remained within the current cut-off radius, the reaction cut-off radius was increased iteratively. The thermal (thermal expansion, heat capacity and glass transition temperature), structural (dihedral angle distributions, radial distribution functions, minimum inter-nitrogen contour length distributions and fragment molecular weight distribution) and mechanical (Young's modulus) properties of the systems cured by either single- or multistep approach were then compared. Results showed that the fragment molecular weight distribution was different in the two cases, since fewer and larger fragments were obtained with the single-step approach; nevertheless, the other properties had no significant dependency on the choice of cross-linking method.

Thermal properties

Thermal conductivity

The thermal conductivity (λ) of nanostructured materials can be estimated by either equilibrium MD (EMD) or nonequilibrium MD (NEMD) simulations. It is noteworthy to mention that EMD simulations are best to study homogeneous systems, whereas NEMD techniques are more suitable for heterogeneous systems [65]. Green-Kubo formulation underlies EMD simulations, since it relates λ with the fluctuations of thermal current via the fluctuation-dissipation theorem [66, 67]:

$$\lambda = \frac{1}{Vk_BT^2}\int_0^\infty \langle \mathbf{J}(t)\mathbf{J}(0)\rangle dt, \tag{3.1}$$

where k_B is the Boltzmann constant, T and V are the temperature and volume of the system, respectively. In addition, $\mathbf{J}(t)$ is the heat current vector at time t and $\langle \mathbf{J}(t)\mathbf{J}(0)\rangle$ is the heat current autocorrelation function. In terms of MD entities, the heat current vector is commonly defined as

$$\mathbf{J}(t) = \frac{d}{dt}\sum_{i=1}^{N} E_i\mathbf{r}_i, \tag{3.2}$$

being $\mathbf{r}_i$ and E_i the position and total energy of atom i, respectively. E_i can be computed by summing up kinetic and potential energies of the atom, namely

$$E_i = \frac{1}{2}m_i|\mathrm{v}_i|^2 + \frac{1}{2}\sum_{j\neq i}^{N} U_p(r_{ij}), \tag{3.3}$$

where m_i and $\mathbf{v}_i$ indicate mass and velocity of atom i, respectively. $U_p(r_{ij})$ is the total potential energy of atom i, which depends on the type of bonded and non-bonded interaction potentials used in the simulations and r_{ij} is the distance between atoms i and j.

For example, Kumar et al. [68] used EMD to study the correlation between λ of cross-linked epoxies (DGEBA resin, 4,4′-diaminodiphenyl sulfone - DDS hardener) and temperature. Results demonstrate that the inclusion of long-range Coulomb interaction corrections leads to a better agreement between modeling and experimental results. Sirk et al. [69], instead, employed MD simulations to study the effect of epoxy composition on its thermal, structural and volumetric properties. Different cross-linked networks including DGEBA resin and various mixtures of flexible (POP) and stiff (MCA, 4,4′-Methylenebis(cyclohexylamine)) curing agents were built. They found out that the flexibility of cross-linker has considerable impact on the thermal and volumetric characteristics of the cross-linked network, especially at temperatures close to glass transition. Fasanella and Sundararaghavan [70] adopted the EMD approach to study the effect of temperature and functionalization on λ of DGEBA + DDS epoxy, possibly reinforced by single-walled CNTs (SWCNTs). In particular, the thermal conductivity was computed both perpendicularly and along the main axis of the nanotube. Results showed that, for what concerns the direction perpendicular to the CNT axis, the λ of pristine SWCNT/epoxy nanocomposites is lower than the neat epoxy one, whereas an improved thermal performance is shown for functionalized SWCNT/epoxy nanocomposites (see Figure 3.7). In addition, a significant enhancement in the λ along nanotube axis has been observed in the case of pristine SWCNT/epoxy in comparison to both neat epoxy and functionalized SWCNT/epoxy nanocomposite (see the inset in Figure 3.7).

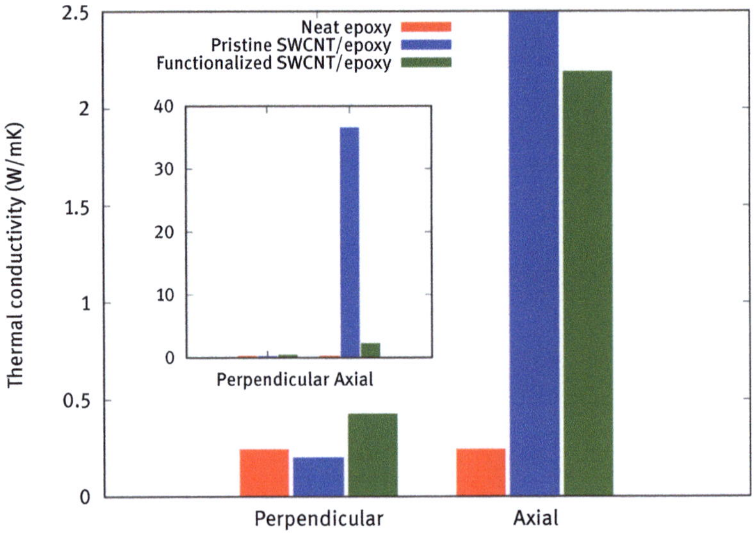

Figure 3.7: Average thermal conductivity in the temperature range of 220–420 K in the perpendicular and axial directions with respect to the nanotube axis for different simulated nanocomposites. Adapted with permission from Springer Nature [70].

NEMD methods for computing thermal conductivity of nanostructured materials are based on Fourier's law. For example, the thermal conductivity of an isotropic system with temperature gradient along z direction can be found as [71]

$$\lambda_z = -\frac{q_z}{dT/dz}, \tag{3.4}$$

where $q_z = Q_z/A$ is the specific heat flux, namely, the heat flux Q_z transferred through the surface A along the direction z. In addition, dT/dz is the steady-state temperature gradient along z-axis. The temperature profile throughout the simulation domain can be computed after dividing the box into N slabs (depending on the system size) along the desired direction. Equation (3.4) can be exploited in nonequilibrium simulations to compute λ according to two possible approaches.

The first approach is typically called NEMD (see an example in Figure 3.8), where a temperature gradient is imposed across the extremities of computational system. This is achieved by a proper design of regions with either high (red atoms in Figure 3.8) or low (blue atoms in Figure 3.8) controlled temperature. To sustain the temperature gradient during the simulation, energy is constantly supplied and extracted from the high- and the low-temperature regions by thermostats, respectively. Such addition or extraction of energy in the hot or cold regions is generally carried out by modifying the kinetic energy of atoms through a velocity rescaling procedure. The instantaneous local kinetic temperature (T_i) in each slab can be computed as

$$T_i = \frac{1}{3N_i k_B} \sum_{k=1}^{N_i} m_k |\mathbf{v}_k|^2, \tag{3.5}$$

where N_i is the number of atoms in the ith slab and m_k the mass of the kth atom. The energy exchange between hot and cold regions at steady state causes a specific thermal flux (q_z) through the system, which can be calculated as

$$q_z = \frac{Q_z}{2A} = \frac{1}{2A\Delta t} \left\langle \frac{1}{2} \sum_{k=1}^{N_B} m_k (|\mathbf{v}'_k|^2 - |\mathbf{v}_k|^2) \right\rangle, \tag{3.6}$$

where N_B is the number of atoms in the thermostated region. Besides, $\mathbf{v}_k$ and $\mathbf{v'}_k$ represent the atomic velocities in the thermosetted regions (hot or cold regions) before and after rescaling, respectively. Note that, in the considered geometry (Figure 3.8), the heat flux in eq. (3.6) is divided by 2 due to the symmetrical simulation box.

The second approach is named as reverse NEMD (RNEMD), and it was originally conceived by Müller-Plathe [72]. The main difference between RNEMD and NEMD methods is that the former generates nonequilibrium conditions through the computational domain by fixing heat flux (which gives rise to temperature gradient), while the latter by fixing the temperature gradient (which gives rise to heat flux). In case of RNEMD, the heat flux is induced by interchanging the atomic velocities between hot and cold regions. In the hot region, the coldest atom is selected and its velocity (v_{cold}, red region in Figure 3.8) is exchanged with the velocity of hottest

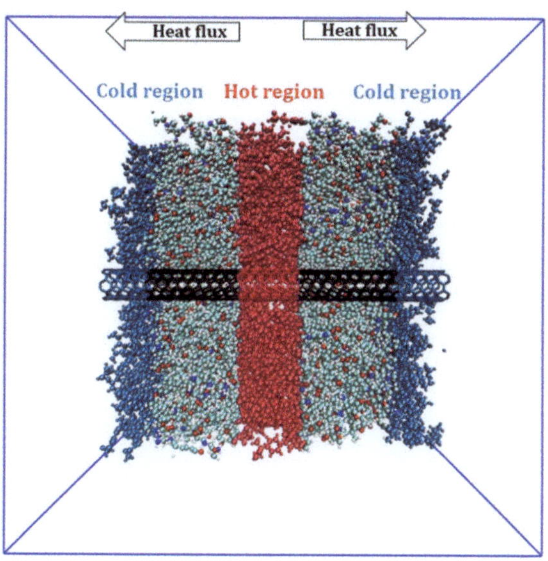

Figure 3.8: Specification of hot and cold baths for computing the thermal conductivity of a carbon-based nanocomposite by NEMD simulations.

atom in the cold region (v_{hot}, blue regions in Figure 3.8). This process eventually leads to an artificial specific heat flux (q_z) flowing from cold to hot regions. All velocity interchange procedure (i.e., *transfers*) follows the total energy conservation law; therefore, at steady state, equal amount of energy per area and time is induced from hot to cold regions via heat conduction, namely

$$q_z = \frac{1}{2A\Delta t} \sum_{transfers} \frac{m}{2} \left(|\mathbf{v}_{\text{hot}}|^2 - |\mathbf{v}_{\text{cold}}|^2 \right). \tag{3.7}$$

The most significant advantage of RNEMD over classical NEMD is the fact that heat flux, which is a slowly converging quantity with respect to temperature, is fixed from the beginning of simulation. Therefore, in general, RNEMD approaches provide faster convergence of λ computation with respect to classical NEMD ones.

Interestingly, Varshney et al. [73] evaluated the thermal conductivity of a cross-linked network of EPON-862 and curing agent W (diethyltoluenediamine, DETDA) using both EMD and NEMD approaches. In the case of EMD, a cubic simulation box ($53.6 \times 53.6 \times 53.6$ Å^3) was employed; in the case of NEMD, a thin slab with large aspect ratio along the heat flow direction ($21.4 \times 21.4 \times 373.3$ Å^3) was considered instead. The results obtained by both methods were in good agreement with experiments, even though EMD led to a slight overestimation of λ. Mortazavi et al. [74] employed NEMD to estimate the thermal conduction in graphene-reinforced nanocomposites. In their study, epoxy consisted of DGEBA resin and DETA (diethylenetriamine) or DDS curing agents. They found that the thermal conductivity of a single-layer graphene immersed in the epoxy matrix at atmospheric pressure and room temperature decreases by around 30% with respect to the value of pristine graphene, while the type of hardener has no considerable effect. Furthermore, they also investigated the impact of pressure and epoxy-graphene covalent bonds on λ of graphene: no significant change in the λ of graphene was observed by increasing the pressure up to 14 GPa, while the formation of 5% covalent bonds led to 65% reduction in the thermal conductivity of graphene. Clearly, such methodologies can be successfully adopted in a large variety of polymeric matrices: for example, Alaghemandi et al. [75] used RNEMD to investigate the λ of SWCNT–polyamide (PA) nanocomposites along both parallel and perpendicular directions with respect to the nanotube axis, finding a good agreement with experiments [76, 77].

Thermal boundary resistance

The thermal boundary resistance, also known as Kapitza resistance or interfacial thermal resistance (R_k) [78], at filler–filler and filler–matrix interfaces is one of the most significant bottlenecks in the overall thermal conduction through polymer nanocomposites [79, 80]. This thermal resistance arises mainly from the weak van der Waals interactions between adjacent fillers, and between fillers and the surrounding matrix.

Such weak interactions considerably hinder the specific heat flux (q) across the interface due to phonon scattering, which in turn results in a localized interfacial temperature jump (ΔT). The inverse of thermal boundary resistance, also known as thermal boundary conductance, $G_k = 1/R_k$, is defined as

$$G_k = \frac{q}{\Delta T}. \tag{3.8}$$

MD simulations are typically used to investigate thermal boundary resistance, since it arises from atomistic details of interfaces. Generally, two NEMD approaches have been used in the literature to study interfacial thermal properties of nanocomposites [81].

In the first approach, the simulated system is divided into multiple slabs according to the geometrical characteristics of the box. On the one side, a constant heat flux is added to the atoms of nanofiller (heat source); on the other side, a constant heat flux is removed from the outermost region of the computational domain occupied by the matrix (heat sink). This thermal power could be extracted or introduced by means of a velocity rescaling procedure, which must conserve the total linear momentum of the system. As an example, this method was adopted to compute the thermal boundary resistance between the CNT and octane liquid depicted in Figure 3.9. In those simulations [81], the continuous addition (CNT) and extraction (outermost region occupied by octane) of a constant heat flux leads to a steady-state temperature profile throughout the system, where the average temperature in each

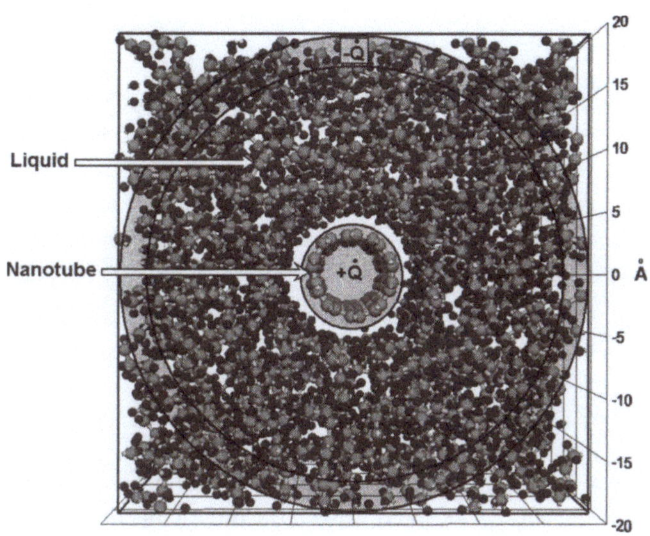

Figure 3.9: Schematic of division of carbon nanotube/octane system into concentric cylindrical slabs to investigate thermal interfacial effects. Reprinted from Ref. [81] with the permission of AIP Publishing.

slab can be computed by eq. (3.5). An example of temperature profile measured in this system is shown in Figure 3.10: the average temperature of the CNT is around 575 K, while the liquid temperature ranges from 320 K (close to the CNT surface) down to 300 K (heat sink). The significant temperature drop at the nanotube–octane interface proves the existence of Kapitza resistance, which is then computed by eq. (3.8) since both q and ΔT can be extracted from simulations.

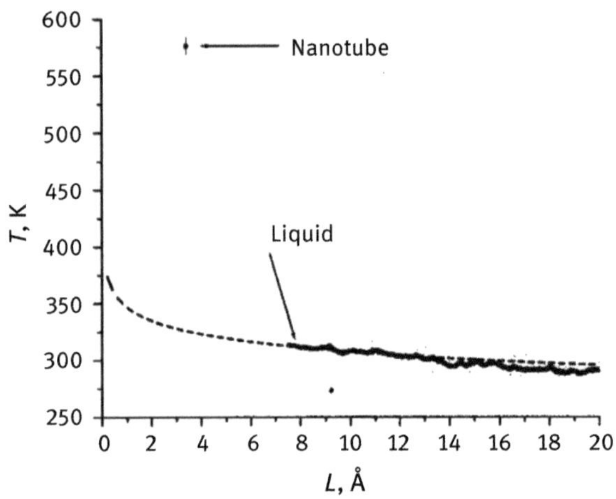

Figure 3.10: Relation between the temperature gradient and distance from the center of CNT at constant heat flow (see configuration in Figure 3.9). Reprinted from Ref. [81] with the permission of AIP Publishing.

In the second approach, the whole system is first equilibrated at the initial temperature $T(0)$; after that, the filler is heated up to a predetermined temperature $T_f(0) = T(0) + \Delta T(0)$, while the matrix temperature is kept constant at the $T_m(t) = T(0)$ value. Such a temperature gradient can be determined by a proper distribution of thermostats in the computational domain. The system is then allowed to relax without any heat source or sink. As depicted in Figure 3.11, the temperature difference between the filler and the surrounding matrix ($\Delta T(t) = T_f(t) - T_m(t)$) tends to decay exponentially with time according to the Newton's law of cooling, namely

$$\Delta T(t) = \Delta T(0) \exp\left(-\frac{t}{\tau}\right). \tag{3.9}$$

If the conduction resistance of the filler is considerably lower than the matrix one (which is typically the case of carbon nanofillers), the characteristic decay time (τ) can be related to the area of filler–matrix interface (A_{int}), the filler heat capacity (c) and the (R_k) at filler–matrix interface as [82]

$$R_k = \frac{A_{int}\tau}{c}. \tag{3.10}$$

Due to the paramount importance of heat transfer across nanoscale interfaces on the effective thermal properties of nanocomposites, several researches have used MD simulations to achieve a mechanistic understanding of the parameters that could affect thermal boundary resistance, such as curing degree of polymeric matrix, surface functionalization of fillers and presence of covalent or noncovalent bonds between fillers and the surrounding matrix [83–100].

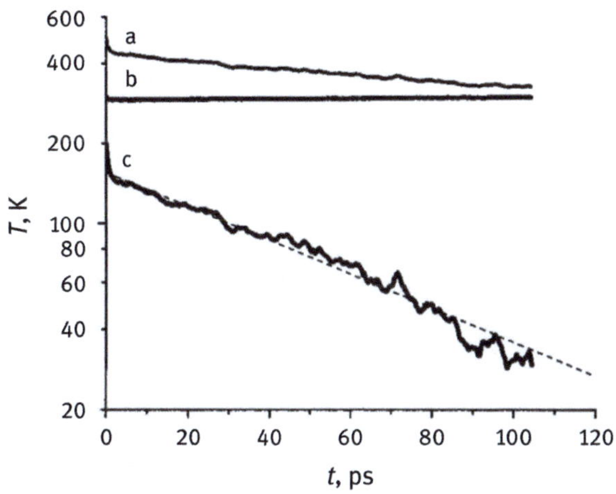

Figure 3.11: Instantaneously heated nanotube at a desired temperature is allowed to relax without heat source or sink, then (a) the temperature of the CNT, (b) liquid temperature and (c) difference between the temperature of CNT and liquid as a function of time are shown (see configuration in Figure 3.9). Reprinted from Ref. [81] with the permission of AIP Publishing.

As an example, Varshney et al. [83] analyzed by MD simulations a nanocomposite made of epoxy polymer (EPON-862/DETDA) reinforced with SWCNTs. They reported a 20% enhancement in the G_k at filler–matrix interface by increasing the curing degree of epoxy. Such behavior was explained by the increased structural rigidity and non-bonded interactions of the cured epoxies, since the curing process generates a denser matrix structure. Huang et al. [94], instead, investigated the impact of covalent bonds (cross-links) between SWCNT and PEK (poly ether ketone) on the thermal boundary conductance at their interface. Considering a simulation domain similar to the one in Figure 3.9, the number of cross-links between SWCNT and PEK was changed in the range 0–80 (0% – 6.25% functionalization degree), and the G_k was then computed in each case. Results in Figure 3.12 show that the temperature jump at CNT–PEK interface decreases with the number of cross-links; thus, G_k increases from 23.58 (no cross-links) to 236.44 MW/m^2 K (80 cross-links). Finally, Wang et al. [89] employed RNEMD to determine how different surface functionalizations of graphene could affect the

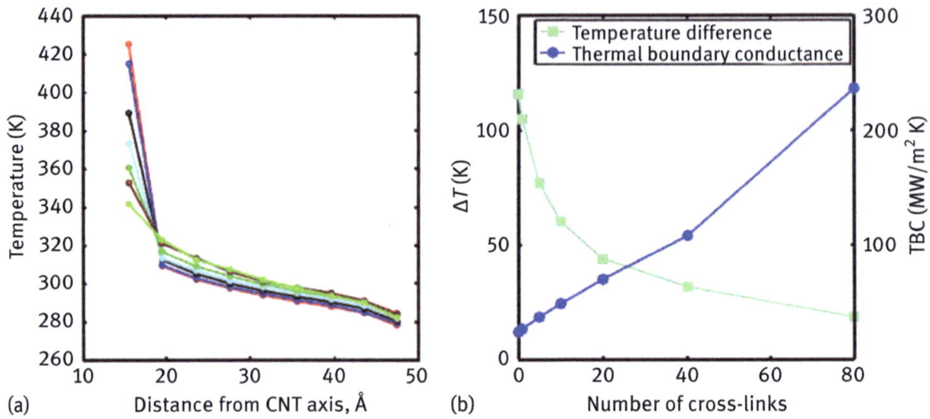

Figure 3.12: (a) Radial temperature distribution in the system under steady-state condition for seven simulated cases: no cross-links (red, 0%), 1 cross-link (blue, 0.08%), 5 cross-links (black, 0.39%), 10 cross-links (cyan, 0.78%), 20 cross-links (dark green, 1.56%), 40 cross-links (brown, 3.12%) and 80 cross-links (green, 6.25%). (b) Temperature difference and thermal boundary conductance at the CNT–PEK interface for different studied cases. Reprinted from Ref. [94] with the permission of AIP Publishing.

interfacial thermal resistance at graphene–polymer interface. To this aim, various types of covalent (butyl, carboxyl and hydroxyl) and noncovalent (1-pyrenebutyl, 1-pyrenebutyric acid and 1-pyrenebutylamine) functional groups were considered at different coverages, namely, the ratio between the number of functional groups to the total number of carbon atoms in the graphene sheet. Furthermore, the effect of isotope doping and acetylenic linkage in graphene was also assessed. Simulation results for covalent and noncovalent functionalizations are reported in Figures 3.13 and 3.14, respectively. It is noteworthy to mention that the thermal boundary resistance values in these figures are normalized using the value of R_k in the case of no functionalization on graphene, that is, $R_{k0} = (0.713 \pm 0.036) \times 10^{-8}$ m^2 K/W. While both increasing covalent and noncovalent functionalizations led to R_k decrease, no significant effects were instead observed with isotope doping or acetylenic linkages in the graphene filler [89].

Mechanical properties

MD simulations are also useful to compute mechanical properties of composite materials, such as the elastic moduli and the whole stress–strain response of a nanometric specimen [101, 102]. Discrepancies between MD and experimental results may be due to the faster strain rate simulated, the effect of dislocations at higher scales, the local high variability of temperature and the possible unwanted residual stresses from cross-link process [103].

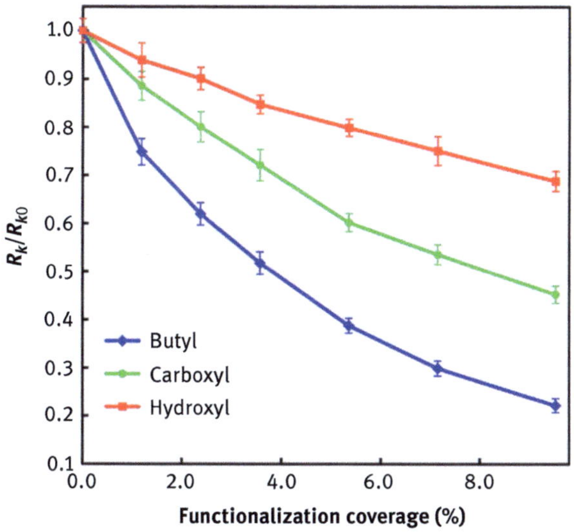

Figure 3.13: Dependence of the relative thermal boundary resistance on coverage percentage of different covalent functional molecules. Reprinted from Ref. [89] with the permission of American Chemical Society.

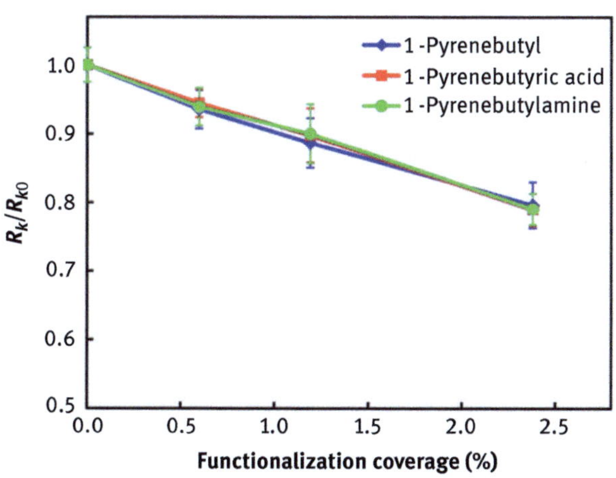

Figure 3.14: Dependence of the relative thermal boundary resistance on coverage percentage of different noncovalent functional molecules. Reprinted from Ref. [89] with the permission of American Chemical Society.

In general, the elastic properties of a material along one specific direction can be quantified by the Young's modulus (E_m) as

$$\sigma = E_\mathrm{m}\varepsilon, \tag{3.11}$$

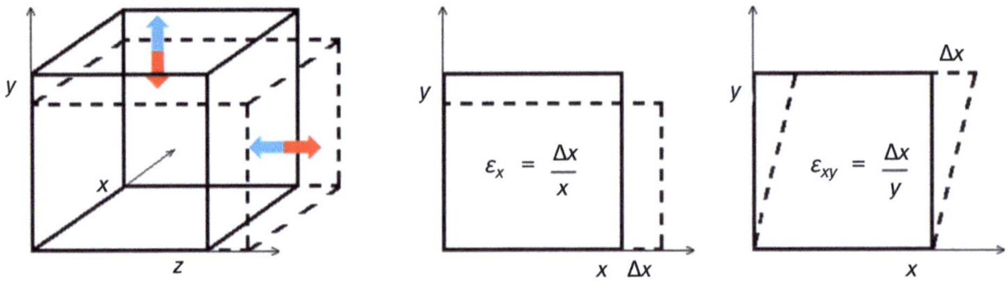

Figure 3.15: Longitudinal and shear deformation of a simple volume of material.

where the stress (σ) is a measure of the average force applied over an equivalent area (F/A), and the strain (ε) is the ratio of deformation with respect to the initial length (see Figure 3.15). Such relation can be generalized to three-dimensional space to relate all the components of the stress and strain tensors (generalized Hooke's law):

$$\sigma_{ij} = C'_{ijkl}\,\varepsilon_{kl}, \tag{3.12}$$

where C'_{ijkl} are the components of the fourth-order stiffness tensor of the considered material. Note that the C' tensor is symmetric ($\sigma_{ij} = \sigma_{ji}$ and thus $C'_{ijkl} = C'_{jikl}$, similarly for the strain tensor), and that the Voigt's notation [104] is typically adopted to simplify eq. (3.12):

$$\sigma_i = C'_{ij}\varepsilon_j, \tag{3.13}$$

with $i, j = 1, 2, \ldots, 6$. Therefore, for any linearly elastic material, the stiffness matrix can be written as

$$C' = \begin{bmatrix} C'_{11} & C'_{12} & C'_{13} & C'_{14} & C'_{15} & C'_{16} \\ C'_{21} & C'_{22} & C'_{23} & C'_{24} & C'_{25} & C'_{26} \\ C'_{31} & C'_{32} & C'_{33} & C'_{34} & C'_{35} & C'_{36} \\ C'_{41} & C'_{42} & C'_{43} & C'_{44} & C'_{45} & C'_{46} \\ C'_{51} & C'_{52} & C'_{53} & C'_{54} & C'_{55} & C'_{56} \\ C'_{61} & C'_{62} & C'_{63} & C'_{64} & C'_{65} & C'_{66} \end{bmatrix} \tag{3.14}$$

MD simulations can be used to apply a deformation $\varepsilon = [\varepsilon_{11}\ \varepsilon_{22}\ \varepsilon_{33}\ 2\varepsilon_{23}\ 2\varepsilon_{13}\ 2\varepsilon_{12}] = [\varepsilon_1\ \varepsilon_2\ \varepsilon_3\ \varepsilon_4\ \varepsilon_5\ \varepsilon_6]$ to a representative unit cell of the material under investigation, and then to measure the respective residual stress in each direction [105]. Given these values, eq. (3.13) allows then to obtain the 6 × 6 elastic stiffness matrix (C') of the simulated material.

The expression of C' is simpler in materials characterized by symmetry, since some of its coefficients are equal to zero. For example, the elastic stiffness tensor of orthotropic materials can be reduced to

$$C' = \begin{bmatrix} C'_{11} & C'_{12} & C'_{13} & 0 & 0 & 0 \\ C'_{21} & C'_{22} & C'_{23} & 0 & 0 & 0 \\ C'_{31} & C'_{32} & C'_{33} & 0 & 0 & 0 \\ 0 & 0 & 0 & C'_{44} & 0 & 0 \\ 0 & 0 & 0 & 0 & C'_{55} & 0 \\ 0 & 0 & 0 & 0 & 0 & C'_{66} \end{bmatrix} \tag{3.15}$$

namely to nine different stiffness coefficients ($C'_{12} = C'_{21}$, $C'_{13} = C'_{31}$, $C'_{23} = C'_{32}$). Instead, the elastic properties of homogeneous isotropic linear elastic materials are determined by two moduli only, namely λ_m and μ_m:

$$\mu_m = \frac{1}{3}(C'_{44} + C'_{55} + C'_{66}) \tag{3.16}$$

and

$$\lambda_m = \frac{1}{3}(C'_{11} + C'_{22} + C'_{33}) - \frac{2}{3}(C'_{44} + C'_{55} + C'_{66}), \tag{3.17}$$

and thus

$$\lambda_m + 2\mu_m = \frac{1}{3}(C'_{11} + C'_{22} + C'_{33}).$$

In this case, the elastic stiffness tensor simplifies to [103]

$$C' = \begin{bmatrix} \lambda_m + 2\mu_m & \lambda_m & \lambda_m & 0 & 0 & 0 \\ \lambda_m & \lambda_m + 2\mu_m & \lambda_m & 0 & 0 & 0 \\ \lambda_m & \lambda_m & \lambda_m + 2\mu_m & 0 & 0 & 0 \\ 0 & 0 & 0 & \mu_m & 0 & 0 \\ 0 & 0 & 0 & 0 & \mu_m & 0 \\ 0 & 0 & 0 & 0 & 0 & \mu_m \end{bmatrix} \tag{3.18}$$

Once λ_m and μ_m are determined, the following relations for homogeneous isotropic linear elastic materials hold:

$$G_m = \mu_m \tag{3.19}$$

$$K_m = \lambda_m + \frac{2}{3}\mu_m \tag{3.20}$$

$$\nu = \frac{\lambda_m}{2(\lambda_m + \mu_m)} \tag{3.21}$$

$$E_m = \mu_m \frac{3\lambda_m + 2\mu_m}{\lambda_m + \mu_m}, \tag{3.22}$$

where G_m is the shear modulus, which regards forces parallel to one surface of the material; K_m is the bulk modulus, which regards the isobaric compression of the material; ν is the Poisson's ratio, which relates the expansion of the material in one direction while it is compressed in another; E_m represents the Young's modulus, which is a measure of the linear elastic stiffness of the material.

MD can be also used to mimic the experimental tensile test of nanomaterials. Typically, the system is first equilibrated; then, a small strain is applied along one longitudinal direction of the computational box with a ramped function of time (see Figure 3.16), and the resulting strain measured [106, 107].

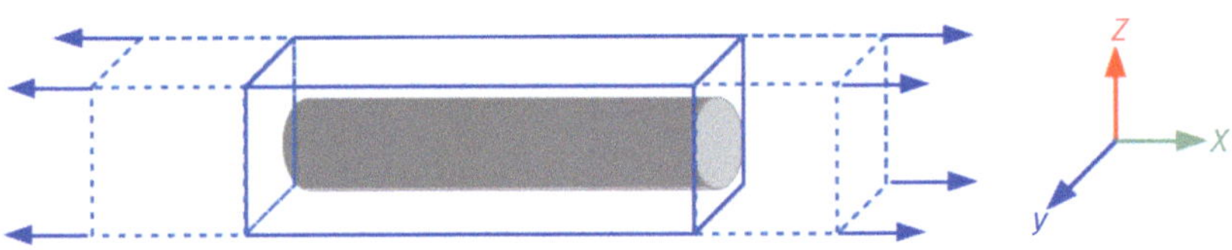

Figure 3.16: Longitudinal deformation of a simulation box.

Electrical properties

Polymeric materials are typically characterized by very low (or no) electrical conductivity; however, the insertion of carbon fillers – such as graphene sheets, CNTs or carbon fibers (CFs) – can improve the electrical properties of the resulting polymer nanocomposites.

A limited volume fraction of CFs has been observed to significantly improve electrical conductivity of polymer nanocomposites [108, 109]. This is due to the formation of continuous conductive networks of CFs in the polymer matrix, thus allowing a transition from nonconductive to conductive electrical properties. The percolation threshold is the critical volume fraction of conductive fillers in the nanocomposite above which conductive pathways in the system always occur. This threshold determines the transition from electrical insulating to conducting behavior, and it is therefore the crucial figure of merit to understand the electrical features of nanocomposites. Furthermore, researchers have reported that the conductive behavior of nanocomposites depends on the spatial distribution of CFs as well. On the one hand, high volume fractions of fibers poorly dispersed in the polymer matrix may lead to fiber agglomeration, which possibly results in electrical conductivity enhancement (but degraded mechanical properties). On the other hand, spatially uniform distributions of fibers at low volume fractions are unfavorable for the electrical conductance of nanocomposites.

Molecular simulation is an efficient tool to study the electrical properties and percolation mechanism in nanocomposites, since it allows determining whether CFs are forming conductive networks within the polymeric matrix or not. According to the electric tunneling theory, two carbon fillers can be considered as electrically connected to each other if their shortest distance is less than a critical gap, which is known as tunneling distance. The electrical resistance offered by the thin (i.e., less than the tunneling distance) insulating polymer matrix between the conductive fibers is therefore known as tunneling resistance.

The generalized formula for electric tunnel effect through a potential barrier between electrodes separated by a thin insulating film was derived by Simmons in 1963 [110]. Hence, if the thickness of insulating film between the conductive fibers is treated as uniform, the curvature of fibers is neglected, the potential barrier is taken as rectangular and the image forces are included, the current density (j, considered in A/cm^2) flowing through the thin film of polymer matrix between the fibers is given as [110]

$$j = \frac{6.2\times 10^{10}}{\Delta s^2}[\phi \exp(-1.025\Delta s\phi^{1/2}) \\ -(\phi + U)\exp(-1.025\Delta s(\phi + U)^{1/2})], \tag{3.23}$$

where

$$\phi = \phi_0 - (U/2s)(s_1 + s_2) - \frac{5.75}{\kappa\Delta s}\ln\frac{s_2(s - s_1)}{s_1(s - s_2)}, \tag{3.24}$$

s is the thickness of the insulating polymer film (in Å), $\Delta s = (s_2 - s_1)$ is the difference of the limits of barrier at Fermi level (in Å), κ is the dielectric constant of insulating polymer film, U is the voltage across the insulating polymer film (in V) and ϕ_0 is the height of potential barrier (or work function) (in V). If $U < \phi_0$, then

$$s_1 = 6/\kappa\phi_0 \\ s_2 = s[1 - 46/(3\phi_0\kappa s + 20 - 2U\kappa s)] + 6/\kappa\phi_0. \tag{3.25}$$

Moreover, the voltage across the insulating polymer film can be calculated as

$$U = e/C = \frac{es}{A_c\kappa\varepsilon_0}, \tag{3.26}$$

where e, C, A_c and ε_0 are electron charge, capacitance, area of the insulating film in contact and permittivity of free space, respectively. The electric tunneling resistance can be finally computed as

$$R_{tunnel} = \frac{U}{j\,A_c}. \tag{3.27}$$

Li et al. [111] performed Monte Carlo simulations to understand the effect of tunneling resistance between CNTs on the electrical conductivity of the whole polymer nanocomposite. Their results indicate that the tunneling resistance increases with the thickness of insulating polymer film (see Figure 3.17). They also reported that the conductivity of nanocomposites drops below 10^{-12} S/m, namely the nanocomposite behaves as an electrical insulator, when the tunneling resistance is larger than 10^{19} Ω. The thickness of polymeric film corresponding to this Ω value is found to be 1.8 nm, which is then considered as the maximum tunneling distance.

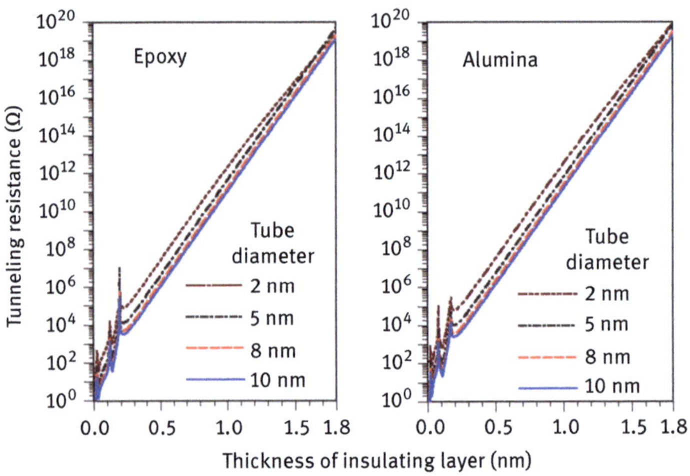

Figure 3.17: Tunneling resistance as a function of the thickness of insulating polymer film between carbon nanotube fillers in a nanocomposite. Different diameters of nanotubes and matrix materials are reported. Reprinted from Ref. [111] with the permission of AIP Publishing.

Hu et al. [112] studied the effect of CNT aspect ratio on the electrical properties of nanocomposites by means of three-dimensional statistical percolation modeling. They reported that the higher aspect ratio of CNTs leads to larger aggregates of fillers in the nanocomposite, which ease the creation of paths with high electrical conductivity. As a result, both lower percolation threshold and higher electrical conductivity of nanocomposites are reported for larger aspect ratio of fillers [112]. Furthermore, the electrical percolation threshold is found to be significantly influenced by the intertube van der Waals interaction and tunneling resistance for nanocomposites with low CNT aspect ratio (i.e., length/diameter ratio <200), whereas negligible effects are observed with high aspect ratio [113].

Researchers agree that the tunneling resistance, which shows values several orders of magnitude larger than the intrinsic resistance of carbon fillers, plays a dominating role in controlling the electrical behavior of nanocomposites [111–114]. Thus, to enhance computational efficiency and simulate larger domains, the intrinsic resistance of nanofillers is often neglected in simulations of nanocomposites with well-dispersed fillers [114].

3.2.2 Mesoscopic models

Thermal properties

The most popular approaches to study effective thermal properties of composites at mesoscale are off-lattice Monte Carlo, dissipative particle dynamics (DPD) and lattice Boltzmann method (LBM).

Duong and coworkers [115] developed an off-lattice Monte Carlo simulation approach to investigate the effective thermal conductivity (λ_{eff}) of CNT-reinforced nanocomposites, taking into account also the thermal boundary resistance (R_k) at CNT–matrix interface. In their model, heat transfer comes from the random motion of a large number of discrete thermal walkers. The proposed model has been widely utilized to investigate the impact of fibers distribution, morphology, volume fraction as well as R_k value at fiber–matrix interface on the λ_{eff} of nanocomposites [116–119], nanofluids [120, 121] and aerogels [122] based on CNTs, with a good capability to reproduce experiments. Kui et al. [123] modified Duong's model to evaluate the thermal properties of composite materials reinforced by graphene sheets. Later, Gong et al. [124–127] extended the initial model to evaluate the thermal conductivity of multiphase composite materials as well. As an example, they studied the λ_{eff} of a polymer composite consisting of polyether ether ketone (PEEK) matrix reinforced by SWCNT fibers and tungsten disulfide (WS_2) nanoparticles [125]. They explored the impact of interfacial thermal resistance at CNT–PEEK and WS_2–PEEK interfaces, CNT morphology (i.e., diameter and aspect ratio of SWCNTs) and SWCNT orientation (i.e., random, parallel or perpendicular to heat flux) on the effective thermal conductivity of composites. According to their findings, λ_{eff} increases with SWCNT concentration, while it decreases with larger Kapitza resistances at the different interfaces. Regarding SWCNT orientation, fibers parallel and perpendicular to the heat flux direction show the best and the worst effect on the overall thermal performance, respectively. They also reported that SWCNTs with higher aspect ratio and smaller diameter generate higher thermal conductivity enhancements at given mass fraction of fillers.

In DPD, the simulated particles (beads) represent whole molecules or fluid regions [128]. For instance, Zhou et al. [129] coupled DPD and smoothed particle hydrodynamics (SPH) to study, at mesoscopic level, the λ_{eff} of polymer composites reinforced by CNTs. This numerical model was used to investigate the impact of different parameters, such as filler dispersion, volume fraction, length and matrix characteristics, on λ_{eff} of nanocomposites. Their results show that λ_{eff} changes quadratically with the volume fraction of fillers, with both random and aligned distributions of fibers.

LBM is another method that can be used to study the thermal properties of composite materials at mesoscopic level. The most significant advantage of LBM is the easy implementation of different interparticle interactions and complex boundary conditions [130]. Wang and coworkers [131] developed a three-dimensional LBM model to calculate the thermal conductivity of composites reinforced by CFs, in good agreement with experimental evidence. Chiavazzo and Asinari [132], instead, computed through LBM the λ_{eff} of composites made of polypropylene polymer (matrix) and graphite particles (fillers). Fang et al. proposed a multiple relaxation time LBM to predict numerically the λ_{eff} of anisotropic heterogeneous materials, such as polymers reinforced by braiding yarns [133]. The model has been afterward applied to compute longitudinal and perpendicular λ_{eff} of computational domains with different volume fractions of fillers, considering the interfacial thermal resistance between components as well.

Mechanical properties

Atomistic simulation methods are well established for computing the interfacial properties of polymer composites at the molecular level (nanoscale). However, the considerable computational resources required limit the maximum length- and timescale of feasible simulations [134, 135]. While sophisticated strategies for speeding up complex MD simulations of systems that undergo significant configuration changes have been recently suggested on the basis of the equation-free method [136, 137], coarse-grained (CG) models have been specifically developed to simulate polymers [138–140] and carbon fillers (CNTs and graphene) [141, 142].

CG MD overcomes the complexity of atomistic simulations by reducing the degrees of freedom of particle–particle interactions, so that larger length- and timescales (from nanoscale to mesoscale) can be afforded [138, 143]. For the development of CG molecular models, chemically connected atoms are grouped together to form superatoms or CG beads that contain sufficient information to reproduce the chemical features of the molecular structure (see, for instance, Figure 3.18) [139]. Moreover, several methods have been proposed to parametrize the force field among CG beads starting from their respective atomistic details [134, 138, 139]. Arash et al. [143, 144] developed a CG model of CNT/polymer systems, with the aim to compute the mechanical behavior of polymer nanocomposites. Their results confirm that CNT reinforcement significantly enhances the Young's modulus of the composite, and that CG models provide comparable accuracy and lower computational cost than atomistic simulations. Later, Mousavi et al. [145] introduced nonbonded interactions between polymer chains, CNTs and polymer matrix in the CG model to investigate the influence of cross-linking, weight fraction and distribution of CNTs on the elastic properties of polymer nanocomposites.

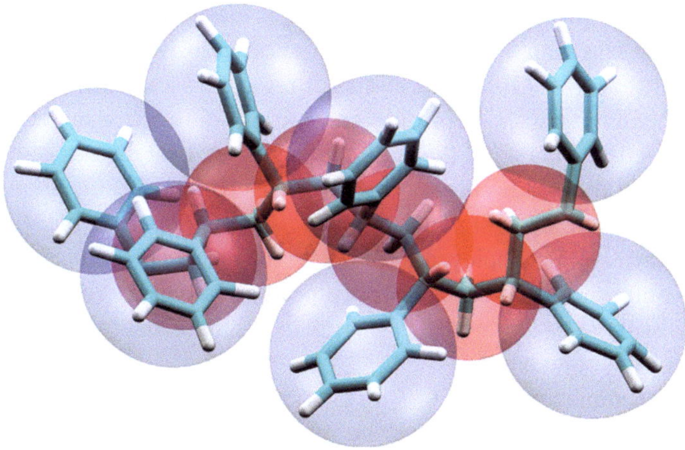

Figure 3.18: Coarse-grained model of polystyrene (red and gray beads) from atomistic details (white and cyan sticks). Reprinted from Ref. [139] with permission of American Chemical Society.

The effect of dispersion, alignment and morphology of nanofillers on the properties of polymer composites can be studied by DPD as well. Zhou et al. [146] studied the dispersion and alignment of CNTs using DPD simulations. Numerical experiments demonstrated that CNTs tend to align themselves in the polymer matrix with increasing nanotube length and volume fraction, especially in well-dispersed systems. Instead, a DPD study on graphene/PMMA (Poly(methyl methacrylate)) composite showed that, while pristine graphene and PMMA are immiscible, high concentrations of graphene functionalizations lead to better filler dispersions in the PMMA matrix [147]. Furthermore, when the volume fraction of fillers is larger, graphene needs to be functionalized more to reach good dispersion. Similar results were obtained by Lin et al. [148], who reported better dispersions of covalently functionalized graphene fillers in PMMA matrix compared to nonfunctionalized ones.

Electrical properties

The RVE concept is also applicable to predict the mesoscale electrical properties of nanocomposites. The electrical conductivity of carbon-based polymer nanocomposites mainly depends on [149–151]:

- the intrinsic electrical conductance of CFs;
- the normal transport through conductive networks of carbon fillers in contact with each other within the polymeric matrix;
- the hopping transport due to the tunneling of electrons between CFs separated by polymer matrix of thickness less than the tunneling distance.

To compute the electrical conductivity of nanocomposites at mesoscale, CFs should be first randomly distributed within a RVE of polymer nanocomposite. These simulations typically assume that fibers have low electrical resistance and thus act as conductors, whereas polymeric matrix is considered as a highly resistive material with low electrical conduction. Therefore, the most conductive path for a given composite is the one minimizing the relative distance between CFs [151, 152]. Different algorithms are available to find the shortest path in a given configuration (e.g., Dijkstra algorithm [153]), which can be thus employed to calculate the electrical conductivity of the simulated nanocomposites at mesoscale. This procedure is typically repeated over different random distributions of fibers in the matrix to reduce statistical fluctuations in terms of composite conductivity.

In a cuboid RVE with dimensions (L_x, L_y, L_z) containing randomly distributed fibers in a polymer matrix [154], each fiber can be modeled as a line segment in spherical coordinates (see Figure 3.19):

$$\begin{aligned} x_i^e &= x_i^s + l_i \sin\theta_i \cos\varphi_i, \\ y_i^e &= y_i^s + l_i \sin\theta_i \sin\varphi_i, \\ z_i^e &= z_i^s + l_i \cos\theta_i, \end{aligned} \tag{3.28}$$

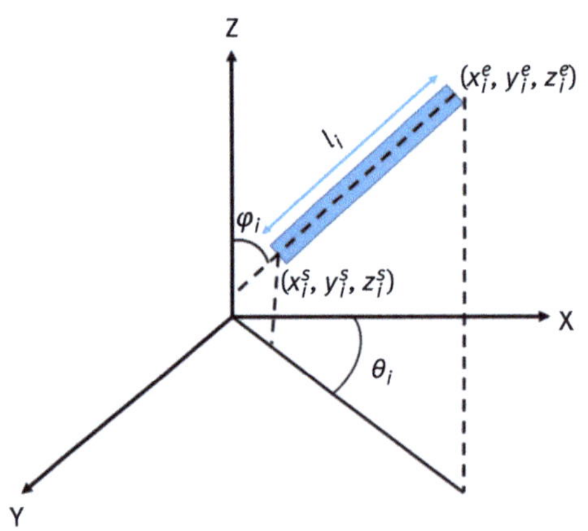

Figure 3.19: Fiber modeled as a line segment in spherical coordinates.

where i is the index of the ith fiber, (x_i^s, y_i^s, z_i^s) and (x_i^e, y_i^e, z_i^e) are the start and end point coordinates of the fiber. The length, polar and azimuthal angle of the ith fiber are denoted as l_i, θ_i and φ_i, respectively.

The start point coordinates, the polar angles and the azimuthal angles of each fiber can be generated randomly as

$$\begin{aligned} &x_i^s = L_x\,\xi, \quad y_i^s = L_y\,\xi, \quad z_i^s = L_z\,\xi \\ &\theta_i = 2\pi\,\xi, \quad \varphi_i = \cos^{-1}(2\xi - 1) \end{aligned} \tag{3.29}$$

where ξ is an uniformly generated random number in the range [0, 1]. To better represent experimental variability, the simulated length of fibers can follow a Weibull distribution [155]. It should be noted that the end points of fibers might initially lie outside the RVE cuboid: in that case, periodic boundary conditions are applied to relocate them inside the cuboid.

The electrical conductivity of nanocomposites can be estimated once the series of resistances in the percolating network is computed, including both intrinsic resistance of fibers (R_{ij}) and their contact resistance ($R_{contact}$). According to the Drude model, the R_{ij} of a fiber (i, j) with l_{ij} length, D diameter and σ_e^{fiber} intrinsic electrical conductivity (see Figure 3.20) can be calculated as [154]

$$R_{ij} = \frac{4l_{ij}}{\sigma_e^{fiber} \pi D^2}. \tag{3.30}$$

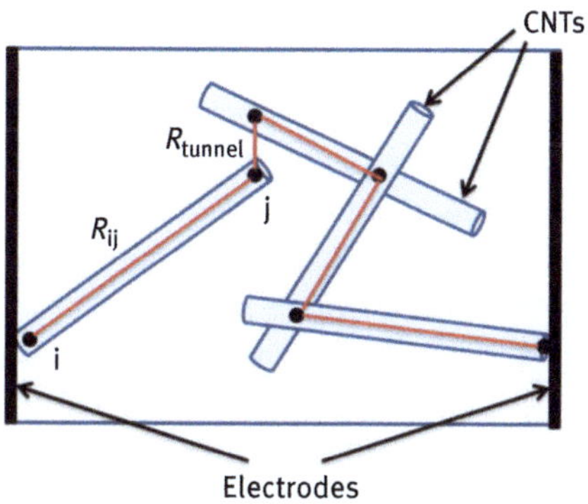

Figure 3.20: Network of electrical resistors in the RVE of a nanocomposite reinforced by nanofillers, for example, CNTs. Reprinted from Ref. [151] with permission from Elsevier.

Considering a marginal effect of temperature, the contact resistance $R_{contact}$ between two fibers can be instead estimated as [150]

$$R_{contact} = \frac{h}{2e^2 M \tau_p}, \tag{3.31}$$

being h the Planck constant, e the electron charge, M the number of conduction channels and τ_p the transmission probability for the electron to tunnel through the polymer layer between fibers. According to Bao et al. [150], the transmission probability τ_p can be estimated by the Wentzel–Kramers–Brillouin approximation [156] as follows:

$$\tau_p = \begin{cases} \exp\left(-\frac{d_{vdW}}{d_{tunnel}}\right) & 0 \le d \le D + d_{vdW} \\ \exp\left(-\frac{d-D}{d_{tunnel}}\right) & D + d_{vdW} \le d \le D + d_{cut} \end{cases} \tag{3.32}$$

where d is the minimum distance between adjacent fibers, d_{vdW} is the van der Waals separation distance and d_{tunnel} is the tunneling length. The latter can be described as [150]

$$d_{tunnel} = \frac{\hbar}{\sqrt{8m_e\Delta E}}, \tag{3.33}$$

where $\hbar = h/2\pi$ is the reduced Planck constant, m_e the mass of electron and ΔE the energy barrier. Note that in eq. (3.32), tunneling effects are neglected if the thickness of polymer matrix between contiguous fibers is larger than d_{cut}, namely, the considered cut-off distance.

Bao et al. [150] studied the effect of CNT alignment in polymer matrix on the critical percolation threshold, and their results revealed that the maximum electrical conductivity is obtained in nanocomposites with partially aligned CNTs. Similarly, Monte Carlo simulations by Zeng et al. [157] reported a clear dependence of the electrical percolation threshold on CNT length, waviness, distribution anisotropy and volume fraction. Improved three-dimensional percolating network theories can also be used to understand how the deformation of CNTs at crossed nanotube junctions influences the electrical conductivity of nanocomposites [158]. In detail, the study by Gong and colleagues [158] suggests that CNT deformation at crossed nanotube junctions determines a large increase in the intrinsic resistance of the nanotube while only a limited decrease in the CNT–CNT contact resistance, therefore, leading to enhanced overall resistance at the junction. Furthermore, numerical simulations based on the Dijkstra algorithm showed that increased contents of MWCNTs in polymer nanocomposites result in higher electrical conductivities, in good quantitative agreement with experiments (see Figure 3.21)

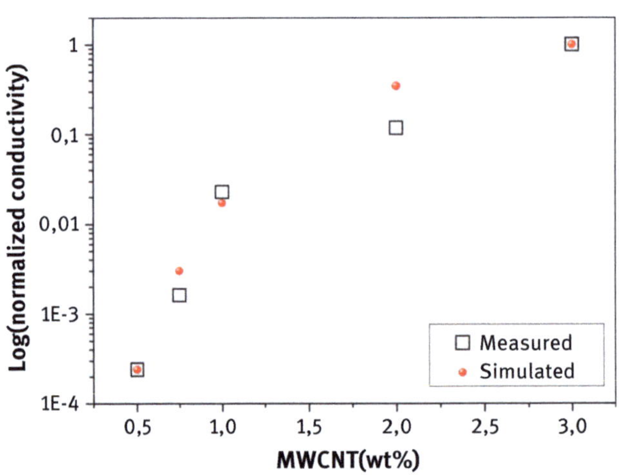

Figure 3.21: Simulation and experimental results of electrical conductivity of CNT-based polymer composites as a function of weight percentage of MWCNTs. Reprinted from Ref. [152] with permission from Elsevier.

[152]. Finally, Rahatekar et al. [159] performed DPD simulations to explore the influence of aspect ratio and fiber alignment on the electrical percolation threshold of nanocomposites.

3.2.3 Continuum models

Thermal properties

The effective thermal conductivity (λ_{eff}) of polymer nanocomposites at macroscale can be predicted by two approaches, namely analytical or simulation ones. Analytical models, for example, effective medium approximation, generally provide quick evaluations of the properties of composites, but cannot take into account neither the interactions between adjacent inclusions nor the actual morphology of composite. Further details on the analytical models for effective thermal conductivity of nanocomposites can be found in a recent review article by Zhai et al. [160].

Due to the current improvement of computational tools and numerical algorithms, finite element method (FEM) has been increasingly utilized to compute the thermal conductivity of composite materials at continuum level. FEM calculates the λ_{eff} of nanocomposites by solving numerically the Fourier's law for conduction at steady state, and this procedure is – in principle – capable to consider different morphologies (e.g., fiber distribution) of the composite material by proper meshing strategies. FEM simulations of composites are typically limited to an RVE of the system. The RVE is built in such a manner that the smallest constituent that can influence the first-order macroscopic behavior of the system is consistent with the simulated length scale; then, the results obtained from the RVE are extended to develop a full-scale model. For example, Figure 3.22 depicts 3D cubic RVEs of composites with different inclusions, namely cylindrical, platelet-like and spherical fillers. The major drawback of FEM in comparison with analytical approaches remains the high computational cost, as well as the modeling complexity given by systems including fibers with high volume fraction or aspect ratio.

Ramani and Vaidynathan [162] carried out an automated finite element analysis to compute λ_{eff} of composite materials. The employed FEM model allowed to study the effect of various parameters, such as fiber aspect ratio, volume fraction, dispersion, orientation and fiber–matrix interfacial thermal resistance on the resulting thermal behavior of composites. Numerical results were compared to analytical models and experiments, finding a good agreement especially at high volume fractions. Ahmed and Masud [163] employed FEM to study the effect of geometrical parameters on the λ_{eff} of polymer composites reinforced by MWCNTs. Their study revealed that the λ of nanocomposites increases with the aspect ratio of nanotubes, whereas interface thermal resistances do not have a prime contribution in lowering the thermal conductivity at the macroscopic scale. Instead, Li et al. [164] quantified

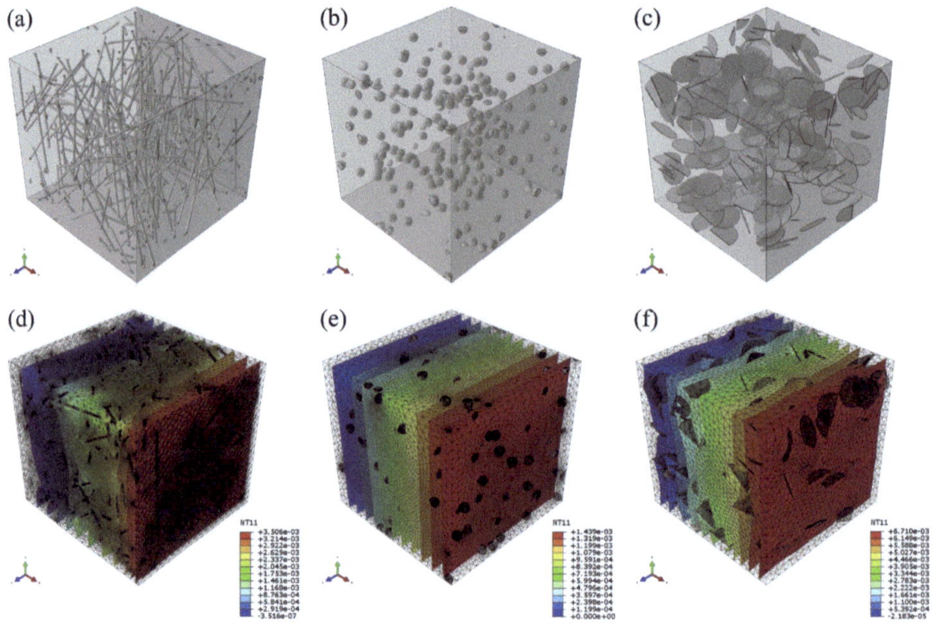

Figure 3.22: Temperature fields in 3D representative volume elements of different composites simulated by FEM method. Reprinted from Ref. [161] with permission from Elsevier.

the influence of SWCNT–matrix interfacial thermal resistance (R_{CNT-m}) on λ_{eff} of SWCNT/polymer composites. Their results showed that the λ_{eff} of composite can be lower than the one of pristine polymer matrix if R_{CNT-m} is greater than a critical value.

Mechanical properties

The FEM can be used to understand the macroscale mechanical properties of nanocomposites, for example, the elastic moduli. At molecular level, the RVE of a generic nanocomposite is a heterogeneous medium consisting of polymer matrix, carbon nanofibers and interface region. However, at continuum level, such heterogeneous RVE should be homogenized to compute the effective material properties through micromechanical analyses (see Figure 3.23). Notice that the homogenization procedure of RVEs to calculate the effective elastic moduli of nanocomposites is equivalent to average the mechanical properties of heterogeneous nanocomposites [165].

The constituents of the RVE can then be assumed as isotropically symmetric within the nanocomposite [166], and the constitutive relation given by the generalized Hooke's law:

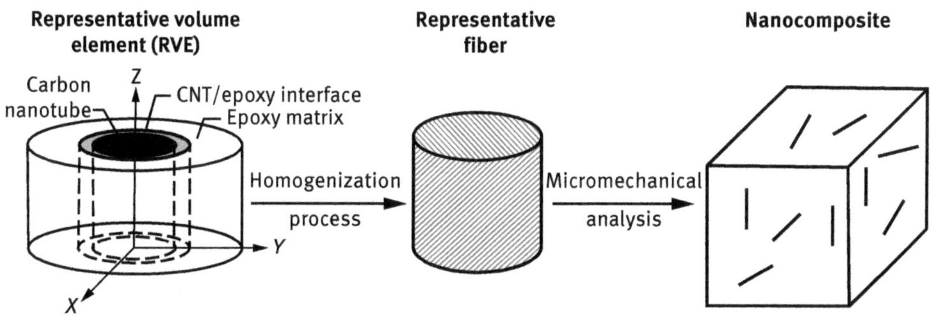

Figure 3.23: Homogenization of a representative volume element of nanocomposite, which includes the nanofiller (CNT), the CNT/epoxy interface and part of the surrounding epoxy matrix into a representative, homogeneous fiber of the composite to be considered for micromechanical analysis. Adapted from Ref. [166] with permission from Elsevier.

$$\sigma_{ij} = C'_{ijkl}\,\varepsilon_{kl}, \qquad i, j, k, l = 1, 2, 3 \tag{3.34}$$

being C'_{ijkl} the elastic stiffness tensor of the RVE, σ_{ij} and ε_{kl} the stress and strain tensors, respectively. The effective stiffness coefficient $(C'^{\,e}_{ijkl})$ of the homogenized multi-constituent nanocomposite can thus be written as

$$\bar{\sigma}_{ij} = C'^{\,e}_{ijkl}\,\bar{\varepsilon}_{kl}, \tag{3.35}$$

being $\bar{\sigma}_{ij}$ and $\bar{\varepsilon}_{kl}$ the stress and strain tensors averaged over the RVE volume (V_{RVE}), respectively, that is:

$$\bar{\sigma}_{ij} = \frac{1}{V_{RVE}} \int_{V_{RVE}} \sigma_{ij} \mathrm{d}V, \tag{3.36}$$

$$\bar{\varepsilon}_{kl} = \frac{1}{V_{RVE}} \int_{V_{RVE}} \varepsilon_{kl} \mathrm{d}V. \tag{3.37}$$

The effective elastic moduli of nanocomposite are finally calculated using the averaged stress and strain tensors as follows:

$$C'^{\,e}_{ijkl} = \frac{\bar{\sigma}_{ij}}{\bar{\varepsilon}_{kl}}, \quad i = j = k = l \quad \text{(Young's modulus)} \tag{3.38}$$

$$C'^{\,e}_{ijkl} = \frac{\bar{\sigma}_{ij}}{2\bar{\varepsilon}_{kl}}, \quad (i = k) \neq (j = l) \quad \text{(Shear modulus)} \tag{3.39}$$

Instead, the Poisson's ratio of nanocomposite is computed as

$$\nu_{ij} = -\frac{\bar{\varepsilon}_{jj}}{\bar{\varepsilon}_{ii}}. \tag{3.40}$$

Odegard et al. [167] proposed a method to transform the discrete molecular structures of nanotube fillers, polymer matrix and CNT/polymer interface at nanoscale into an equivalent continuum model of composite. Liu and Chen [168] computed the effective mechanical properties of CNT-reinforced composites using a nanoscale RVE and the FEM. They showed that the stiffness of composite increases several times with volume fractions of CNT nanofillers up to 5%. Instead, an equivalent continuum modeling study by Huang and Rodrigue [169] found that both aspect ratio and length of CNT fillers have a significant influence on the mechanical properties of polypropylene-based composites. In a similar study, Kumar and Srivastava [170] reported that the elastic stiffness of polymer matrix was improved by carbon-based filler reinforcements. This study also concluded that graphene reinforcements provided the best in-plane stiffness properties of nanocomposite, whereas CNTs the best out-of-plane ones. Finally, a continuum simulation using multiscale homogenization procedure was applied by Golestanian and Gahruei [171] to understand the effect of CNT waviness on the elastic properties of CNT-reinforced polymer composites. The simulation study revealed that wavier CNTs reduce elastic modulus of nanocomposite along the longitudinal direction, while they increase elastic modulus along the transverse one.

Electrical properties

The effective electrical properties of nanocomposites at the micro- and macroscale are substantially affected by phenomena at molecular level, such as tunneling mechanism and formation of conductive networks.

In case of a 1D conductor with l length and A_c cross section, the effective electrical conductivity of composite (σ_e) can be computed by measuring the electric current (I) flowing through the computational domain, given a certain electric potential (U), that is

$$\sigma_e = \frac{Il}{A_c U}. \tag{3.41}$$

Instead, in case of 3D anisotropic materials, a more general expression should be considered:

$$\mathbf{j} = \boldsymbol{\sigma}_\mathbf{e} \mathbf{E}, \tag{3.42}$$

where $\mathbf{j}$ being the current density vector, $\mathbf{E} = -\nabla U$ the electric field vector under electrostatic conditions and $\boldsymbol{\sigma}_\mathbf{e}$ the effective conductivity tensor. Typically, FEM analysis with micromechanical corrections is adopted to compute effective electrical properties of nanocomposites at continuum level as, for example, depicted in Figure 3.24.

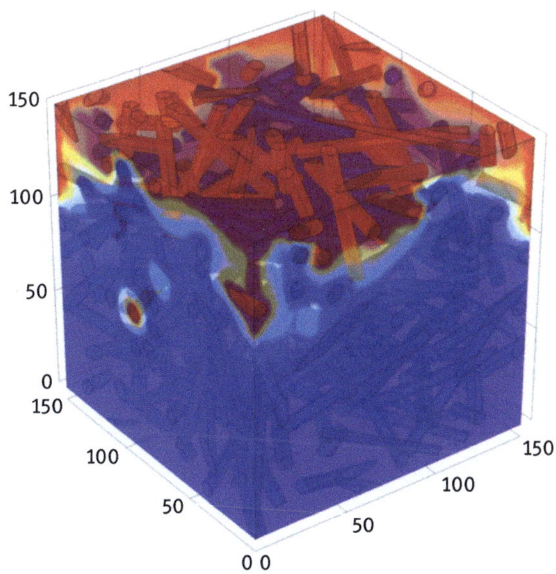

Figure 3.24: Electrical potential within a nanocomposite evaluated by FEM analysis. Reprinted from Ref. [172] with permission from Elsevier.

Seidel and Lagoudas [173] developed a micromechanical model to estimate the electrical properties of CNT–polymer nanocomposites. They reported that the large increment in the σ_e of MWCNT-based nanocomposites at low nanotube concentrations is mainly due to formation of conductive networks, whereas in SWCNT-based nanocomposites, both electron hopping and conductive network formation play a relevant role [173, 174]. The macroscale piezoresistive properties of CNT–polymer composites along both transverse and axial directions were studied instead by Ren and Seidel [37], considering both electric tunneling and inherent piezoresistivity of CNT in the selected RVEs. In that study, the macroscale piezoresistive response of different mesoscale CNT dispersion scenarios in the nanocomposite was predicted using computational micromechanic techniques based on the finite element analysis. The authors found that the electric tunneling mechanism dominates the overall piezoresistive properties of CNT–polymer composites along both transverse and axial directions, whereas the inherent piezoresistivity contribution is more evident along the axial direction rather than along the transverse one [37]. The influence of electric tunneling was found to be negligible in composites with well-dispersed distributions of CNTs, where the intertube distance was larger than the tunneling distance; conversely, agglomerated CNT dispersions eased electric tunneling mechanism.

3.3 Perspectives

The idea to combine different materials to develop an engineered material having unique capabilities, distinct from their constituent materials, is very fascinating, and it has been well known by long time. Nowadays, the possibilities offered by materials processing and, sometimes, nanotechnology have revamped the emphasis on composite materials. In particular, carbon-based composites have already proved to provide remarkable improvements in terms of stiffness, durability, strength and lightweight characteristics in some industrial sectors, for example, automotive.

In spite of the previous successes, some industrial challenges remain for composite materials. According to a recent commentary by *Materials Today* [175], mass production of composites is a key aspect that must still be developed before composites will prove to be useful for many new applications. The lack of mass production/automation techniques sometimes reveals the lack of knowledge of the composite manufacturing process, which forces to rely on try and error in developing new products [175]. Clearly, this approach, which is still dominant in many applications, cannot be properly standardized. Hence, the lack of standardization, design standards and good technical support/backup from suppliers tends to reduce the composite acceptance in an industrial context [175]. Composite materials without a warranty are not well accepted by engineers, who have to deal with high-fidelity standards in the design and production of complex products. Things become even worse in case of recycling CFs, as envisioned by circular economy. End-of-life fiber-reinforced composites are particularly challenging for implementing circular economy, even though some large-scale demonstration has been already funded by the European Commission [176]. All the above factors significantly increase the price of carbon-based composites with regard to their competitors, namely steel, concrete and thermoplastics, and this induces a loss of market share [175]. Price reduction can hardly be achieved by combining raw materials, because this intrinsically requires some additional processing costs. However, one promising alternative option consists in increasing the value chain by adding new functionalities for composite materials. This leads to the so-called smart materials (see Table 3.1 and references therein). Materials with self-sensing and structural health monitoring capabilities, devices with fast triggering shape memory, thermally driven actuators and high-performance all-solid-state supercapacitors are good examples of this strategy. On the other hand, this strategy leads necessarily to even more scientific challenges, because high precision and reproducibility of the arrangement of CFs into the composites are required.

The previous discussion reveals that the exploitation of carbon-based composites requires addressing many scientific challenges, which could benefit from systematic and state-of-the-art materials modeling. Essentially, modeling should address two main issues: (i) elucidating the atomistic details of interface between CF and polymer and (ii) taking into account properly the extreme variability in the CF orientation within the composite.

Table 3.1: Modeling methods for computing thermophysical properties of carbon-based nanocomposites of interest as smart materials. PDMS is polydimethylsiloxane; PVDF is polyvinylidene fluoride.

Material	Modeling method	Properties	Applications as smart material
CNT/PDMS composite	FEM	Structural, mechanical and thermal properties	Thermally driven actuator [181]
CNT/polymer composite	Multiphysics modeling	Electromechanical properties	Strain sensor [182]
CNT/polymer	Percolation network model	Electrical conductivity	High accuracy sensor [183, 184]
PVDF/carbon fiber composite	Multiphysics micromechanics and FEM	Mechanical and piezoelectric properties	Self-sensing and structural health monitoring [39]
CNT sponge/shape memory polymer nanocomposite	FEM	Electrical properties	Fast triggering shape memory material [185]
Graphene oxide (GO)/ reduced GO/water	Quantum/MD simulations	Capacitance	High-performance all-solid-state supercapacitors [186]

The first issue is very well known and is also largely explored from the experimental point of view. For example, CF-reinforced thermoplastic composites gained recent interest due to ease of production and recycling compared to that of thermoset composites, but surface treatments are used to increase the concentration of surface functional groups and hence the fiber–polymer interfacial adhesion [177]. Similarly, different methods of preparation for thermoset composites rely on functionalizing the interface between various nanofillers and epoxy, which is typically characterized by several experimental techniques [178]. Remarkably, nowadays both experimental and modeling techniques are used in order to investigate the relationships between interfacial characteristics and composite properties, for example, with regard to the advantages and disadvantages of covalent and noncovalent functionalization of CFs [179].

The second issue where materials modeling can provide some useful insights is due to the extreme variability in the microscopic structure of composites. Different arrangements of CFs within the composite, shape and orientation of the CF aggregates and dynamics response of the microstructure due to the applied loads are just examples of phenomena that require modeling at scales much larger than that of individual fiber. This may appear simply as a geometrical problem, but it leads immediately to (i) a significant increase in the computational demand and to (ii) a problem of model reduction in postprocessing the modeling results. The second problem is particularly severe, because it may hinder the opportunity to use materials modeling for increasing the understanding of composites under realistic conditions. In fact, understanding requires squeezing a large amount of information in very simple design guidelines. Moreover, this second problem has an indirect effect on the first problem about affordable computational demand as well. In fact, the possibility to develop simple interpolation models, which can summarize the outcome of more complex models, is essential for any coupling and/or linking strategy. Concerning the latter point, analytical simplified models, which can describe the transfer of energy, mechanical stresses and electricity between neighboring fibers, are highly desirable for developing reliable predictions about percolating networks within the composites. Nowadays, machine learning techniques and artificial intelligence offer solutions for developing analytical interpolating functions, which can be tuned on underlying detailed models and are suitable as input for up-scaled models for predicting macroscopic properties of the materials. These analytical functions keep some insights into the investigated phenomena and, at the same time, are very fast and transferable in up-scaled models, which can describe large portions of the material under investigation. This is a clear advantage with regard to previous techniques based on neural networks. Hence, we envision a huge impact of such techniques in dealing with realistic simulations of carbon-based composites for engineering applications.

Finally, last but not least, carbon-based composites still raise some concerns in terms of their toxicity, particularly when nanofibers are involved. These concerns are about the manufacturing processes of composites, the release of nanoparticles during operation and the end-of-life waste management. Clearly, the fact that some nanofibers, for example, CNTs, are similar in shape and size to asbestos raises further concern [180]. Even though materials modeling may help in assessing some guidelines and in clarifying some scenarios, quantitative and predictive computational nanotoxicology is still far from being mainstream. However, it represents a very interesting approach, which is currently under development in many different fields for assessing preliminarily the impact of new nanotechnologies, and we envision that also carbon-based composites will benefit from it.

Acknowledgments: This work partially received funding from the European Union's Horizon 2020 research and innovation programs, through MODCOMP (grant agreement no. 685844) and SMARTFAN (grant agreement no. 760779) projects.

References

[1] van der Zwaag, S., Self Healing Materials: An Alternative Approach to 20 Centuries of Materials Science, Springer Netherlands: Springer Series in Materials Science, 2007.

[2] Robinson, A. L., Metallurgy: Extraordinary alloys that remember their past. Science 1976, 191, 4230, 934–936.

[3] Jani, J. M., Leary, M., Subic, A., and Gibson, M. A., A review of shape memory alloy research, applications and opportunities. Mater. Des 2014, 56, 1078–1113.
[4] Calkins, F. T., and Mabe, J. H., Shape memory alloy based morphing aerostructures. ASME. J. Mech. Des 2010, 132, 11, 111012.
[5] Baer, G., Wilson, T. S., Matthews, D. L., and Maitland, D. J., Shape memory behavior of thermally stimulated polyurethane for medical applications. J. Appl. Polym. Sci 2007, 103, 6, 3882–3892.
[6] Kim, S., Hawkes, E., Choy, K., Joldaz, M., Foleyz, J., and Wood, R., "Micro artificial muscle fiber using NiTi spring for soft robotics," in 2009 IEEE/RSJ International Conference on Intelligent Robots and Systems, pp 2228–2234, Oct 2009.
[7] Song, G., Ma, N., and Li, H.-N., Applications of shape memory alloys in civil structures. Eng. Struct 2006, 28, 9, 1266–1274.
[8] Gök, M. O., Bilir, M. Z., and Gürcüm, B. H., "Shape-memory applications in textile design," Procedia - Social and Behavioral Sciences, vol. 195, pp. 2160–2169, 2015. World Conference on Technology, Innovation and Entrepreneurship.
[9] Olander, A., An electrochemical investigation of solid cadmium-gold alloys. J. Am. Chem. Soc 1932, 54, 10, 3819–3833.
[10] Buehler, W. J., Gilfrich, J. V., and Wiley, R. C., Effect of low-temperature phase changes on the mechanical properties of alloys near composition TiNi. J. Appl. Phys 1963, 34, 5, 1475–1477.
[11] Otsuka, K., and Ren, X., Physical metallurgy of Ti–Ni-based shape memory alloys. Prog. Mater Sci 2005, 50, 5, 511–678.
[12] van Humbeeck, J., Shape memory materials: State of the art and requirements for future applications. J. Phys. IV France 1997, 07, C5, C5–3–C5–12.
[13] Ratna, D., and Karger-Kocsis, J., Recent advances in shape memory polymers and composites: a review. J. Mater. Sci 2008, 43, 254–269.
[14] Gong, X., Liu, L., Liu, Y., and Leng, J., An electrical-heating and self-sensing shape memory polymer composite incorporated with carbon fiber felt. Smart Mater. Struct 2016, 25, 035036.
[15] Liu, Y., Zhao, J., Zhao, L., Li, W., Zhang, H., Yu, X., and Zhang, Z., High performance shape memory epoxy/carbon nanotube nanocomposites. ACS Appl. Mater. Interfaces 2016, 8, 1, 311–320.
[16] Zhang, Z.-X., Dou, J.-X., He, J.-H., Xiao, C.-X., Shen, L.-Y., Yang, J.-H., Wang, Y., and Zhou, Z.-W., Electrically/infrared actuated shape memory composites based on a bio-based polyester blend and graphene nanoplatelets and their excellent self-driven ability. J. Mater. Chem. C 2017, 5, 4145–4158.
[17] Wang, X., Sparkman, J., and Gou, J., Electrical actuation and shape memory behavior of polyurethane composites incorporated with printed carbon nanotube layers. Compos. Sci. Technol 2017, 141, 8–15.
[18] Yang, E., Xu, Z., Chur, L. K., Behroozfar, A., Baniasadi, M., Moreno, S., Huang, J., Gilligan, J., and Minary-Jolandan, M., Nanofibrous smart fabrics from twisted yarns of electrospun piezopolymer. ACS Appl. Mater. Interfaces 2017, 9, 28, 24220–24229.
[19] Yuan, X., Zhu, S., Li, X., Chen, C., Zhou, K., and Zhang, D., Mechanical performance of piezoelectric fiber composites and electroelastic field concentration near the electrode edges. Mater. Des 2017, 128, Supplement C, 71–79.
[20] Chen, Q., Sun, Y., Qin, L., and Wang, Q. M., Piezoelectric fiber-composite-based cantilever sensor for electric-field-induced strain measurement in soft electroactive polymer. IEEE Trans. Ultrason. Ferroelectr. Freq. Control 2013, 60, 2142–2153.

[21] Lin, X., Huang, S., Zhou, K., and Zhang, D., The influence of structural parameters on the actuation performance of piezoelectric fiber composites. Mater. Des 2016, 107, Supplement C, 123–129.
[22] Yang, Y., Tang, L., and Li, H., Vibration energy harvesting using macrofiber composites. Smart Mater. Struct 2009, 18, 11, 115025.
[23] Zeng, W., Tao, X.-M., Chen, S., Shang, S., Chan, H. L. W., and Choy, S. H., Highly durable all-fiber nanogenerator for mechanical energy harvesting. Energy Environ. Sci 2013, 6, 2631–2638.
[24] Zhong, J., Zhang, Y., Zhong, Q., Hu, Q., Hu, B., Wang, Z. L., and Zhou, J., Fiber-based generator for wearable electronics and mobile medication. ACS Nano 2014, 8, 6, 6273–6280.
[25] Lu, X., Qu, H., and Skorobogatiy, M., Piezoelectric micro- and nanostructured fibers fabricated from thermoplastic nanocomposites using a fiber drawing technique: Comparative study and potential applications. ACS Nano 2017, 11, 2, 2103–2114.
[26] Back cover. Energy Environ. Sci 2013, 6, 2778.
[27] Wang, J., Gao, Y., Zhang, J., and Tian, H., Invisible photochromism and optical anti-counterfeiting based on d-a type inverse diarylethene. J. Mater. Chem. C 2017, 5, 4571–4577.
[28] Ye, J.-T., Wang, L., Wang, H.-Q., Chen, -Z.-Z., Qiu, Y.-Q., and Xie, H.-M., Spirooxazine molecular switches with nonlinear optical responses as selective cation sensors. RSC Adv 2017, 7, 642–650.
[29] Renuka, K. D., Lekshmi, C. L., Joseph, K., and Mahesh, S., Sustainable electronic materials: Reversible phototuning of conductance in a noncovalent assembly of MWCNT and bioresource-derived photochromic molecule. ACS Appl. Mater. Interfaces 2017, 9, 2, 1167–1172.
[30] Wei, J., Jiao, X., Wang, T., and Chen, D., Electrospun photochromic hybrid membranes for flexible rewritable media. ACS Applied Materials & Interfaces 2016, 8, 43, 29713–29720.
[31] William J, S., “Doll or the like,” Oct. 27 1953. US Patent 2656647 A.
[32] Ercole, F., Davis, T. P., and Evans, R. A., Photo-responsive systems and biomaterials: photochromic polymers, light-triggered self-assembly, surface modification, fluorescence modulation and beyond. Polym. Chem 2010, 1, 37–54.
[33] Manouras, T., and Vamvakaki, M., Field responsive materials: photo-, electro-, magnetic- and ultrasound-sensitive polymers. Polym. Chem 2017, 8, 74–96.
[34] Fasano, M., Chiavazzo, E., and Asinari, P., Water transport control in carbon nanotube arrays. Nanoscale research letters 2014, 9, 1, 559.
[35] Gizzatov, A., Key, J., Aryal, S., Ananta, J., Cervadoro, A., Palange, A. L., Fasano, M., Stigliano, C., Zhong, M., Di, M. D., Guven, A., Chiavazzo, E., Asinari, P., Liu, X., Ferrari, M., Wilson, L. J., and Decuzzi, P., Hierarchically structured magnetic nanoconstructs with enhanced relaxivity and cooperative tumor accumulation. Adv. Funct. Mater 2014, 24, 29, 4584–4594.
[36] Chiavazzo, E., Fasano, M., Asinari, P., and Decuzzi, P., Scaling behaviour for the water transport in nanoconfined geometries. Nat. Commun 2014, 5, 3565.
[37] Ren, X., and Seidel, G. D., Computational micromechanics modeling of piezoresistivity in carbon nanotube–polymer nanocomposites. Compos. Interfaces 2013, 20, 9, 693–720.
[38] Ren, X., Burton, J., Seidel, G. D., and Lafdi, K., Computational multiscale modeling and characterization of piezoresistivity in fuzzy fiber reinforced polymer composites. Int. J. Solids Struct 2015, 54, 121–134.
[39] Greminger, M., and Haghiashtiani, G., Multiscale modeling of PVDF matrix carbon fiber composites. Modell. Simul. Mater. Sci. Eng 2017, 25, 4, 045007.
[40] Toth, R., Santese, F., Pereira, S. P., Nieto, D. R., Pricl, S., Fermeglia, M., and Posocco, P., Size and shape matter! A multiscale molecular simulation approach to polymer nanocomposites. Journal of Materials Chemistry 2012, 22, 12, 5398–5409.

[41] Posocco, P., Pricl, S., and Fermeglia, M., Multiscale modeling approach for polymeric nanocomposites. Modeling and Prediction of Polymer Nanocomposite Properties 2013, 95–128.
[42] Fermeglia, M., Posocco, P. and Pricl, S., Nano tools for macro problems: multiscale molecular modeling of nanostructured polymer systems. Composite Interfaces 2013, 20, 6, 379–394.
[43] Subramanian, N., Rai, A., and Chattopadhyay, A., Atomistically informed stochastic multiscale model to predict the behavior of carbon nanotube-enhanced nanocomposites. Carbon 2015, 94, 661–672.
[44] Laurini, E., Posocco, P., Fermeglia, M., and Pricl, S., Modena nanotools: An integrated multiscale simulation workflow to predict thermophysical properties of thermoplastic polyurethanes. Journal of Computational Science 2016, 15, 24–33.
[45] Khromov, K. Y., Knizhnik, A., Potapkin, B., and Kenny, J., Multiscale modeling of electrical conductivity of carbon nanotubes based polymer nanocomposites. Journal of Applied Physics 2017, 121, 22, 225102.
[46] Laurini, E., Marson, D., Fermeglia, M., and Pricl, S., Multimodel approach for accurate determination of industry-driven properties for polymer nanocomposite materials. Journal of computational science 2018, 26, 28–38.
[47] Li, Y., and Seidel, G., Multiscale modeling of the interface effects in CNT-epoxy nanocomposites. Computational Materials Science 2018, 153, 363–381.
[48] De Baas, A., "What makes a material function? let me compute the ways . . .," Modeling in FP7 NMP Programme, European Commission, vol. 6th edition, 2017.
[49] The CEN Workshop Agreement, "CEN/WS MODA - Materials modelling - terminology, classification and metadata," CWA, vol. 17284, 2018. ftp://ftp.cencenelec.eu/CEN/Sectors/TCandWorkshops/Workshops/WS%20MODA/CWA_17284.pdf.
[50] Ghedini, E., Hashibon, A., Friis, J., Goldbeck, G., Schmitz, G., and de Baas, A., "EMMO the European Materials Modelling Ontology," in EMMC Workshop on Interoperability in Materials Modelling, 2017. https://emmc.info/wp-content/uploads/2017/12/EMMC_IntOp2017-Cambridge_Ghedini_Bologna.pdf.
[51] Sáenz Ezquerro, C., Laspalas, M., Chiminelli, A., Serrano, F., and Valero, C., Interface characterization of epoxy resin nanocomposites: A molecular dynamics approach. Fibers 2018, 6, 3, 54.
[52] Humphrey, W., Dalke, A., and Schulten, K., VMD: Visual molecular dynamics. J. Mol. Graphics 1996, 14, 1, 33–38.
[53] Hanwell, M. D., Curtis, D. E., Lonie, D. C., Vandermeersch, T., Zurek, E., and Hutchison, G. R., Avogadro: an advanced semantic chemical editor, visualization, and analysis platform. J. Cheminf. Aug 2012, 4, 17.
[54] Fasano, M., Crisafulli, A., Cardellini, A., Bergamasco, L., Chiavazzo, E., and Asinari, P., Thermally triggered nanorocket from double-walled carbon nanotube in water. Molecular Simulation 2019, 45, 4–5, 417–424.
[55] Schulz, M., and Frisch, H. L., Monte Carlo studies of interpenetrating polymer network formation. J. Chem. Phys 1994, 101, 11, 10008–10022.
[56] Doherty, D., Holmes, B., Leung, P., and Ross, R., Polymerization molecular dynamics simulations. i. cross-linked atomistic models for poly (methacrylate) networks. Comput. Theor. Polym. Sci 1998, 8, 1–2, 169–178.
[57] Yarovsky, I., and Evans, E., Computer simulation of structure and properties of crosslinked polymers: application to epoxy resins. Polymer 2002, 43, 3, 963–969.
[58] Li, C., and Strachan, A., Molecular simulations of crosslinking process of thermosetting polymers. Polymer 2010, 51, 25, 6058–6070.

[59] Jang, C., Lacy, T. E., Gwaltney, S. R., Toghiani, H., and Pittman, C. U. Jr, Relative reactivity volume criterion for cross-linking: Application to vinyl ester resin molecular dynamics simulations. Macromolecules 2012, 45, 11, 4876–4885.
[60] Varshney, V., Patnaik, S. S., Roy, A. K., and Farmer, B. L., A molecular dynamics study of epoxy-based networks: Cross-linking procedure and prediction of molecular and material properties. Macromolecules 2008, 41, 18, 6837–6842.
[61] Lin, P.-H., and Khare, R., Molecular simulation of cross-linked epoxy and epoxy-POSS nanocomposite. Macromolecules 2009, 42, 12, 4319–4327.
[62] Rottach, D. R., Curro, J. G., Budzien, J., Grest, G. S., Svaneborg, C., and Everaers, R., Molecular dynamics simulations of polymer networks undergoing sequential cross-linking and scission reactions. Macromolecules 2007, 40, 1, 131–139.
[63] Wu, C., and Xu, W., Atomistic molecular modelling of crosslinked epoxy resin. Polymer 2006, 47, 16, 6004–6009.
[64] Jang, C., Sirk, T. W., Andzelm, J. W., and Abrams, C. F., Comparison of crosslinking algorithms in molecular dynamics simulation of thermosetting polymers. Macromol. Theory Simul 2015, 24, 3, 260–270.
[65] Maiti, A., Mahan, G., and Pantelides, S., Dynamical simulations of nonequilibrium processes—heat flow and the Kapitza resistance across grain boundaries. Solid State Commun 1997, 102, 7, 517–521.
[66] Kubo, R., Toda, M., and Hashitsume, N., Statistical physics II: nonequilibrium statistical mechanics, Vol. 31. Springer Science & Business Media, 2012.
[67] Evans, D. J., and Morriss, G. P., 4 - The Green–Kubo relations, In: Statistical Mechanics of Nonequilibrium Liquids (D. J. Evans and G. P. Morrisseds.), 77–93, Academic Press, 1990.
[68] Kumar, A., Sundararaghavan, V., and Browning, A., Study of temperature dependence of thermal conductivity in cross-linked epoxies using molecular dynamics simulations with long range interactions. Modell. Simul. Mater. Sci. Eng 2014, 22, 2, 025013.
[69] Sirk, T. W., Karim, M., Khare, K. S., Lenhart, J. L., Andzelm, J. W., and Khare, R., Bi-modal polymer networks: composition-dependent trends in thermal, volumetric and structural properties from molecular dynamics simulation. Polymer 2015, 58, 199–208.
[70] Fasanella, N. A., and Sundararaghavan, V., Atomistic modeling of thermal conductivity of epoxy nanotube composites. JOM 2016, 68, 5, 1396–1410.
[71] Anderson, C. V., and Tamma, K. K., An overview of advances in heat conduction models and approaches for prediction of thermal conductivity in thin dielectric films. Int. J. Numer. Methods Heat Fluid Flow 2004, 14, 1, 12–65.
[72] Müller-Plathe, F., A simple nonequilibrium molecular dynamics method for calculating the thermal conductivity. J. Chem. Phys 1997, 106, 14, 6082–6085.
[73] Varshney, V., Patnaik, S. S., Roy, A. K., and Farmer, B. L., Heat transport in epoxy networks: A molecular dynamics study. Polymer 2009, 50, 14, 3378–3385.
[74] Mortazavi, B., Benzerara, O., Meyer, H., Bardon, J., and Ahzi, S., Combined molecular dynamics-finite element multiscale modeling of thermal conduction in graphene epoxy nanocomposites. Carbon 2013, 60, 356–365.
[75] Alaghemandi, M., Müller-Plathe, F., and Böhm, M. C., Thermal conductivity of carbon nanotube—polyamide-6,6 nanocomposites: Reverse nonequilibrium molecular dynamics simulations. J. Chem. Phys 2011, 135, 18, 11B606.
[76] Shen, Z., Bateman, S., Wu, D. Y., McMahon, P., Dell'Olio, M., and Gotama, J., The effects of carbon nanotubes on mechanical and thermal properties of woven glass fibre reinforced polyamide-6 nanocomposites. Compos. Sci. Technol 2009, 69, 2, 239–244.

[77] Du, F., Guthy, C., Kashiwagi, T., Fischer, J. E., and Winey, K. I., An infiltration method for preparing single-wall nanotube/epoxy composites with improved thermal conductivity. J. Polym. Sci., Part B: Polym. Phys 2006, 44, 10, 1513–1519.
[78] Tascini, A. S., Armstrong, J., Chiavazzo, E., Fasano, M., Asinari, P., and Bresme, F., Thermal transport across nanoparticle-fluid interfaces: the interplay of interfacial curvature and nanoparticle-fluid interactions. Phys. Chem. Chem. Phys 2017, 19, 3244–3253.
[79] Fasano, M., Bigdeli, M. B., Sereshk, M. R. V., Chiavazzo, E., and Asinari, P., Thermal transmittance of carbon nanotube networks: Guidelines for novel thermal storage systems and polymeric material of thermal interest. Renewable Sustainable Energy Review 2015, 41, 1028–1036.
[80] Bigdeli, M. B., and Fasano, M., Thermal transmittance in graphene based networks for polymer matrix composites. Int. J. Therm. Sci 2017, 117, 98–105.
[81] Shenogin, S., Xue, L., Ozisik, R., Keblinski, P., and Cahill, D. G., Role of thermal boundary resistance on the heat flow in carbon-nanotube composites. J. Appl. Phys 2004, 95, 12, 8136–8144.
[82] Lervik, A., Bresme, F., and Kjelstrup, S., Heat transfer in soft nanoscale interfaces: the influence of interface curvature. Soft Matter 2009, 5, 2407–2414.
[83] Varshney, V., Roy, A. K., Michalak, T. J., Lee, J., and Farmer, B. L., Effect of curing and functionalization on the interface thermal conductance in carbon nanotube–epoxy composites. Jom 2013, 65, 2, 140–146.
[84] Shenogin, S., Bodapati, A., Xue, L., Ozisik, R., and Keblinski, P., Effect of chemical functionalization on thermal transport of carbon nanotube composites. Appl. Phys. Lett 2004, 85, 12, 2229–2231.
[85] Clancy, T. C., and Gates, T. S., Modeling of interfacial modification effects on thermal conductivity of carbon nanotube composites. Polymer 2006, 47, 16, 5990–5996.
[86] Liu, C., and Fan, S., Effects of chemical modifications on the thermal conductivity of carbon nanotube composites. Appl. Phys. Lett 2005, 86, 12, 123106.
[87] Shen, X., Wang, Z., Wu, Y., Liu, X., and Kim, J.-K., Effect of functionalization on thermal conductivities of graphene/epoxy composites. Carbon 2016, 108, 412–422.
[88] Luo, T., and Lloyd, J. R., Enhancement of thermal energy transport across graphene/graphite and polymer interfaces: a molecular dynamics study. Adv. Funct. Mater 2012, 22, 12, 2495–2502.
[89] Wang, Y., Yang, C., Pei, Q.-X., and Zhang, Y., Some aspects of thermal transport across the interface between graphene and epoxy in nanocomposites. ACS Appl. Mater. Interfaces 2016, 8, 12, 8272–8279.
[90] Zheng, Q., Xue, Q., Yan, K., Gao, X., Li, Q., and Hao, L., Effect of chemisorption on the interfacial bonding characteristics of carbon nanotube–polymer composites. Polymer 2008, 49, 3, 800–808.
[91] Varshney, V., Patnaik, S. S., Roy, A. K., and Farmer, B. L., Modeling of thermal conductance at transverse cnt- cnt interfaces. J. Phys. Chem. C 2010, 114, 39, 16223–16228.
[92] Clancy, T. C., Frankland, S.-J. V., Hinkley, J. A., and Gates, T. S., Multiscale modeling of thermal conductivity of polymer/carbon nanocomposites. Int. J. Therm. Sci 2010, 49, 9, 1555–1560.
[93] Varshney, V., Roy, A. K., and Baur, J. W., Modeling the role of bulk and surface characteristics of carbon fiber on thermal conductance across the carbon-fiber/matrix interface. ACS Appl. Mater. Interfaces 2015, 7, 48, 26674–26683.
[94] Huang, H., Chen, L., Varshney, V., Roy, A. K., and Kumar, S., Investigation of phonon transport and thermal boundary conductance at the interface of functionalized swcnt and poly (ether-ketone). Journal of Applied Physics 2016, 120, 9, 095102.

[95] Wang, Y., Zhan, H., Xiang, Y., Yang, C., Wang, C. M., and Zhang, Y., Effect of covalent functionalization on thermal transport across graphene–polymer interfaces. J. Phys. Chem. C 2015, 119, 22, 12731–12738.
[96] Kim, B., Choi, J., Yang, S., Yu, S., and Cho, M., Multiscale modeling of interphase in crosslinked epoxy nanocomposites. Composites Part B: Engineering 2017, 120, 128–142.
[97] Kuang, Y., and Huang, B., Effects of covalent functionalization on the thermal transport in carbon nanotube/polymer composites: a multi-scale investigation. Polymer 2015, 56, 563–571.
[98] Zhang, J., Jiang, C., Jiang, D., and Peng, H.-X., Nano-engineering thermal transport performance of carbon nanotube networks with polymer intercalation: a molecular dynamics study. Phys. Chem. Chem. Phys 2014, 16, 9, 4378–4385.
[99] Hida, S., Hori, T., Shiga, T., Elliott, J., and Shiomi, J., Thermal resistance and phonon scattering at the interface between carbon nanotube and amorphous polyethylene. Int. J. Heat Mass Transfer 2013, 67, 1024–1029.
[100] Carlborg, C. F., Shiomi, J., and Maruyama, S., Thermal boundary resistance between single-walled carbon nanotubes and surrounding matrices. Phys. Rev. B 2008, 78, 20, 205406.
[101] Chawla, R., Mechanical and tribological properties of graphene reinforced natural rubber composites: A molecular dynamics study. MRS Advances 2018, 3, 10, 525–530.
[102] Weerasinghe, A., Lu, C.-T., Maroudas, D., and Ramasubramaniam, A., Multiscale shear-lag analysis of stiffness enhancement in polymer–graphene nanocomposites. ACS applied materials & interfaces 2017, 9, 27, 23092–23098.
[103] Theodorou, D. N., and Suter, U. W., Atomistic modeling of mechanical properties of polymeric glasses. Macromolecules jan 1986, 19, 139–154.
[104] Weiner, J. H., Statistical mechanics of elasticity, New York: Wiley, 1983.
[105] Li, Y., Wang, S., and Wang, Q., A molecular dynamics simulation study on enhancement of mechanical and tribological properties of polymer composites by introduction of graphene. Carbon 2017, 111, 538–545.
[106] Zaminpayma, E., Molecular dynamics simulation of mechanical properties and interaction energy of polythiophene/polyethylene/poly (pphenylenevinylene) and CNTs composites. Polymer Composites 2014, 35, 11, 2261–2268.
[107] Sul, J.-H., Prusty, B. G., and Kelly, D. W., Application of molecular dynamics to evaluate the design performance of low aspect ratio carbon nanotubes in fibre reinforced polymer resin. Composites Part A: Applied Science and Manufacturing 2014, 65, 64–72.
[108] Coleman, J. N., Curran, S., Dalton, A. B., Davey, A. P., McCarthy, B., Blau, W., and Barklie, R. C., Percolation-dominated conductivity in a conjugated-polymer-carbon-nanotube composite. Phys. Rev. B Sep 1998, 58, R7492–R7495.
[109] Ahmad, K., Pan, W., and Shi, S.-L., Electrical conductivity and dielectric properties of multiwalled carbon nanotube and alumina composites. Appl. Phys. Lett 2006, 89, 13, 133122.
[110] Simmons, J. G., Generalized formula for the electric tunnel effect between similar electrodes separated by a thin insulating film. J. Appl. Phys 1963, 34, 6, 1793–1803.
[111] Li, C., Thostenson, E. T., and Chou, T.-W., Dominant role of tunneling resistance in the electrical conductivity of carbon nanotube–based composites. Appl. Phys. Lett 2007, 91, 22, 223114.
[112] Hu, N., Masuda, Z., Yan, C., Yamamoto, G., Fukunaga, H., and Hashida, T., The electrical properties of polymer nanocomposites with carbon nanotube fillers. Nanotechnology 2008, 19, 21, 215701.
[113] Lu, W., Chou, T.-W., and Thostenson, E. T., A three-dimensional model of electrical percolation thresholds in carbon nanotube-based composites. Appl. Phys. Lett 2010, 96, 22, 223106.

[114] Yu, Y., Song, G., and Sun, L., Determinant role of tunneling resistance in electrical conductivity of polymer composites reinforced by well dispersed carbon nanotubes. J. Appl. Phys 2010, 108, 8, 084319.

[115] Duong, H. M., Papavassiliou, D. V., Lee, L. L., and Mullen, K. J., Random walks in nanotube composites: Improved algorithms and the role of thermal boundary resistance. Appl. Phys. Lett 2005, 87, 1, 013101.

[116] Duong, H. M., Papavassiliou, D. V., Mullen, K. J., and Maruyama, S., Computational modeling of the thermal conductivity of single-walled carbon nanotube–polymer composites. Nanotechnology 2008, 19, 6, 065702.

[117] Duong, H. M., Yamamoto, N., Papavassiliou, D. V., Maruyama, S., and Wardle, B. L., Inter-carbon nanotube contact in thermal transport of controlled-morphology polymer nanocomposites. Nanotechnology 2009, 20, 15, 155702.

[118] Bui, K., Grady, B. P., and Papavassiliou, D. V., Heat transfer in high volume fraction CNT nanocomposites: Effects of inter-nanotube thermal resistance. Chem. Phys. Lett 2011, 508, 4, 248–251.

[119] Bui, K., Grady, B. P., Saha, M. C., and Papavassiliou, D. V., Effect of carbon nanotube persistence length on heat transfer in nanocomposites: A simulation approach. Appl. Phys. Lett 2013, 102, 20, 203116.

[120] Duong, H. M., Papavassiliou, D. V., Mullen, K. J., Wardle, B. L., and Maruyama, S., Calculated thermal properties of single-walled carbon nanotube suspensions. J. Phys. Chem. C 2008, 112, 50, 19860–19865.

[121] Duong, H. M., Papavassiliou, D. V., Mullen, K. J., Wardle, B. L., and Maruyama, S., A numerical study on the effective thermal conductivity of biological fluids containing single-walled carbon nanotubes. Int. J. Heat Mass Transfer 2009, 52, 23, 5591–5597.

[122] Gong, F., Tam, Y. S., Nguyen, S. T., and Duong, H. M., Prediction of thermal resistances and heat conduction of carbon nanotube aerogels in various permeated gases. Chem. Phys. Lett 2015, 627, 116–120.

[123] Bui, K., Duong, H. M., Striolo, A., and Papavassiliou, D. V., Effective heat transfer properties of graphene sheet nanocomposites and comparison to carbon nanotube nanocomposites. J. Phys. Chem. C 2011, 115, 10, 3872–3880.

[124] Gong, F., Papavassiliou, D. V., and Duong, H. M., Off-lattice Monte Carlo simulation of heat transfer through carbon nanotube multiphase systems taking into account thermal boundary resistances. Numerical Heat Transfer, Part A 2014, 65, 11, 1023–1043.

[125] Gong, F., Bui, K., Papavassiliou, D. V., and Duong, H. M., Thermal transport phenomena and limitations in heterogeneous polymer composites containing carbon nanotubes and inorganic nanoparticles. Carbon 2014, 78, 305–316.

[126] Gong, F., Hongyan, Z., Papavassiliou, D. V., Bui, K., Lim, C., and Duong, H. M., Mesoscopic modeling of cancer photothermal therapy using single-walled carbon nanotubes and near infrared radiation: insights through an off-lattice Monte Carlo approach. Nanotechnology 2014, 25, 20, 205101.

[127] Gong, F., Duong, H. M., and Papavassiliou, D. V., Inter-carbon nanotube contact and thermal resistances in heat transport of three-phase composites. J. Phys. Chem. C 2015, 119, 14, 7614–7620.

[128] Gooneie, A., Schuschnigg, S., and Holzer, C., A review of multiscale computational methods in polymeric materials. Polymers 2017, 9, 1, 16.

[129] Zhou, B., Luo, W., Yang, J., Duan, X., Wen, Y., Zhou, H., Chen, R., and Shan, B., Thermal conductivity of aligned CNT/polymer composites using mesoscopic simulation. Composites Part A: Applied Science and Manufacturing 2016, 90, 410–416.

[130] Raabe, D., Overview of the Lattice Boltzmann method for nano-and microscale fluid dynamics in materials science and engineering. Modell. Simul. Mater. Sci. Eng 2004, 12, 6, R13.
[131] Wang, M., Kang, Q., and Pan, N., Thermal conductivity enhancement of carbon fiber composites. Appl. Therm. Eng 2009, 29, 2, 418–421.
[132] Chiavazzo, E., and Asinari, P., Reconstruction and modeling of 3d percolation networks of carbon fillers in a polymer matrix. Int. J. Therm. Sci 2010, 49, 12, 2272–2281.
[133] Fang, W.-Z., Chen, L., Gou, -J.-J., and Tao, W.-Q., Predictions of effective thermal conductivities for three-dimensional four-directional braided composites using the Lattice Boltzmann method. International Journal of Heat and Mass Transfer 2016, 92, 120–130.
[134] Reith, D., Meyer, H., and Müller-Plathe, F., Mapping atomistic to coarse-grained polymer models using automatic simplex optimization to fit structural properties. Macromolecules 2001, 34, 7, 2335–2345.
[135] Rzepiela, A. J., Louhivuori, M., Peter, C., and Marrink, S. J., Hybrid simulations: combining atomistic and coarse-grained force fields using virtual sites. Phys. Chem. Chem. Phys 2011, 13, 10437–10448.
[136] Chiavazzo, E., Covino, R., Coifman, R. R., Gear, C. W., Georgiou, A. S., Hummer, G., and Kevrekidis, I. G., Intrinsic map dynamics exploration for uncharted effective free-energy landscapes. Proceedings of the National Academy of Sciences 2017, 114, 28, E5494–E5503.
[137] Georgiou, A. S., Bello-Rivas, J. M., Gear, C. W., Wu, H.-T., Chiavazzo, E., and Kevrekidis, I. G., An exploration algorithm for stochastic simulators driven by energy gradients. Entropy 2017, 19, 7, 294.
[138] Dirk, R., Mathias, P., and Florian, M., Deriving effective mesoscale potentials from atomistic simulations. J. Comput. Chem 2003, 24, 13, 1624–1636.
[139] Fritz, D., Harmandaris, V. A., Kremer, K., and van der Vegt, N. F. A., Coarse-grained polymer melts based on isolated atomistic chains: Simulation of polystyrene of different tacticities. Macromolecules 2009, 42, 19, 7579–7588.
[140] Hsu, D. D., Xia, W., Arturo, S. G., and Keten, S., Systematic method for thermomechanically consistent coarse-graining: A universal model for methacrylate-based polymers. J. Chem. Theory Comput 2014, 10, 6, 2514–2527.
[141] Zhao, J., Jiang, J.-W., Wang, L., Guo, W., and Rabczuk, T., Coarse-grained potentials of single-walled carbon nanotubes. J. Mech. Phys. Solids 2014, 71, 197–218.
[142] Ruiz, L., Xia, W., Meng, Z., and Keten, S., A coarse-grained model for the mechanical behavior of multi-layer graphene. Carbon 2015, 82, 103–115.
[143] Arash, B., Park, H. S., and Rabczuk, T., Mechanical properties of carbon nanotube reinforced polymer nanocomposites: A coarse-grained model. Composites Part B: Engineering 2015, 80, 92–100.
[144] Arash, B., Park, H. S., and Rabczuk, T., Tensile fracture behavior of short carbon nanotube reinforced polymer composites: A coarse-grained model. Compos. Struct 2015, 134, 981–988.
[145] Mousavi, A. A., Arash, B., Zhuang, X., and Rabczuk, T., A coarse-grained model for the elastic properties of cross linked short carbon nanotube/polymer composites. Composites Part B: Engineering 2016, 95, 404–411.
[146] Zhou, B., Luo, W., Yang, J., Duan, X., Wen, Y., Zhou, H., Chen, R., and Shan, B., Simulation of dispersion and alignment of carbon nanotubes in polymer flow using dissipative particle dynamics. Comput. Mater. Sci 2017, 126, 35–42.
[147] Ju, S.-P., Wang, Y.-C., Huang, G.-J., and Chang, J.-W., Miscibility of graphene and poly(methyl methacrylate) (PMMA): molecular dynamics and dissipative particle dynamics simulations. RSC Adv 2013, 3, 8298–8307.

[148] Lin, F., Yang, C., Zeng, Q., and Xiang, Y., Morphological and mechanical properties of graphene-reinforced PMMA nanocomposites using a multiscale analysis. Comput. Mater. Sci 2018, 150, 107–120.
[149] Kilbride, B. E., Coleman, J. N., Fraysse, J., Fournet, P., Cadek, M., Drury, A., Hutzler, S., Roth, S., and Blau, W. J., Experimental observation of scaling laws for alternating current and direct current conductivity in polymer-carbon nanotube composite thin films. J. Appl. Phys 2002, 92, 7, 4024–4030.
[150] Bao, W., Meguid, S., Zhu, Z., Pan, Y., and Weng, G., A novel approach to predict the electrical conductivity of multifunctional nanocomposites. Mech. Mater 2012, 46, Supplement C, 129–138.
[151] Grabowski, K., Zbyrad, P., Uhl, T., Staszewski, W. J., and Packo, P., Multiscale electro-mechanical modeling of carbon nanotube composites. Comput. Mater. Sci 2017, 135C, 169–180.
[152] Castellino, M., Rovere, M., Shahzad, M. I., and Tagliaferro, A., Conductivity in carbon nanotube polymer composites: A comparison between model and experiment. Composites Part A 2016, 87, Supplement C, 237–242.
[153] Dijkstra, E., A note on two problems in connexion with graphs. Numer. Math 1959, 1, 1, 269–271.
[154] Bao, W. S., Meguid, S. A., Zhu, Z. H., and Weng, G. J., Tunneling resistance and its effect on the electrical conductivity of carbon nanotube nanocomposites. J. Appl. Phys 2012, 111, 9, 093726.
[155] Wang, S., Liang, Z., Wang, B., and Zhang, C., Statistical characterization of single-wall carbon nanotube length distribution. Nanotechnology 2006, 17, 3, 634.
[156] Bohm, D., Quantum Theory, Dover Publications: Dover Books on Physics, 1989.
[157] Zeng, X., Xu, X., Shenai, P. M., Kovalev, E., Baudot, C., Mathews, N., and Zhao, Y., Characteristics of the electrical percolation in carbon nanotubes/polymer nanocomposites. J. Phys. Chem. C 2011, 115, 44, 21685–21690.
[158] Gong, S., Zhu, Z. H., and Haddad, E. I., Modeling electrical conductivity of nanocomposites by considering carbon nanotube deformation at nanotube junctions. J. Appl. Phys 2013, 114, 7, 074303.
[159] Rahatekar, S. S., Hamm, M., Shaffer, M. S. P., and Elliott, J. A., Mesoscale modeling of electrical percolation in fiber-filled systems. J. Chem. Phys 2005, 123, 13, 134702.
[160] Zhai, S., Zhang, P., Xian, Y., Zeng, J., and Shi, B., Effective thermal conductivity of polymer composites: Theoretical models and simulation models. Int. J. Heat Mass Transfer 2018, 117, 358–374.
[161] He, B., Mortazavi, B., Zhuang, X., and Rabczuk, T., Modeling Kapitza resistance of two-phase composite material. Compos. Struct 2016, 152, Supplement C, 939–946.
[162] Ramani, K., and Vaidyanathan, A., Finite element analysis of effective thermal conductivity of filled polymeric composites. J. Compos. Mater 1995, 29, 13, 1725–1740.
[163] Ahmed, S., and Masud, A., "Evaluation of effective thermal conductivity of multiwalled carbon nanotube reinforced polymer composites using finite element method and continuum model," Procedia Eng., vol. 90, pp.129–135, 2014. 10th International Conference on Mechanical Engineering, ICME 2013.
[164] Li, X., Fan, X., Zhu, Y., Li, J., Adams, J. M., Shen, S., and Li, H., Computational modeling and evaluation of the thermal behavior of randomly distributed single-walled carbon nanotube/polymer composites. Comput. Mater. Sci 2012, 63, 207–213.
[165] Srivastava, A., and Kumar, D., A continuum model to study interphase effects on elastic properties of CNT/GS-nanocomposite. Mater. Res. Express 2017, 4, 2, 025036.
[166] Wernik, J., and Meguid, S., Multiscale micromechanical modeling of the constitutive response of carbon nanotube-reinforced structural adhesives. Int. J. Solids Struct 2014, 51, 14, 2575–2589.

[167] Odegard, G., Gates, T., Wise, K., Park, C., and Siochi, E., Constitutive modeling of nanotube–reinforced polymer composites. Compos. Sci. Technol 2003, 63, 11, 1671–1687.
[168] Liu, Y., and Chen, X., Evaluations of the effective material properties of carbon nanotube-based composites using a nanoscale representative volume element. Mech. Mater 2003, 35, 1, 69–81.
[169] Huang, J., and Rodrigue, D., Equivalent continuum models of carbon nanotube reinforced polypropylene composites. Materials & Design 2013, 50, 936–945.
[170] Kumar, D., and Srivastava, A., Elastic properties of CNT-and graphene-reinforced nanocomposites using RVE. Steel and Composite Structures 08 2016, 21, 1085–1103.
[171] Golestanian, H., and Gahruei, M. H., Effective mechanical properties of nanocomposites reinforced with wavy carbon nanotubes. Mater. Sci. Technol 2013, 29, 8, 913–920.
[172] Nilsson, F., Krückel, J., Schubert, D. W., Chen, F., Unge, M., Gedde, U. W., and Hedenqvist, M. S., Simulating the effective electric conductivity of polymer composites with high aspect ratio fillers. Compos. Sci. Technol 2016, 132, 16–23.
[173] Seidel, G. D., and Lagoudas, D. C., A micromechanics model for the electrical conductivity of nanotube-polymer nanocomposites. J. Compos. Mater 2009, 43, 9, 917–941.
[174] Feng, C., and Jiang, L., Micromechanics modeling of the electrical conductivity of carbon nanotube (CNT)–polymer nanocomposites. Compos. A Appl. Sci. Manuf 2013, 47, Supplement C, 143–149.
[175] Jacob, A., "What is the biggest challenge facing the composites industry today?," Plastics magazine, Materials Today, 2013. https://www.materialstoday.com/composite-industry/comment/what-is-the-biggest-challenge-facing-the-composite/.
[176] FiberEUse, "Large scale demonstration of new circular economy value chains based on the reuse of end-of-life fiber reinforced composites," CORDIS Europe, vol. 730323, 2017–2021. https://cordis.europa.eu/project/rcn/210178_en.html.
[177] Yao, -S.-S., Jin, F.-L., Rhee, K. Y., Hui, D., and Park, S.-J., Recent advances in carbon-fiber-reinforced thermoplastic composites: A review. Composites Part B: Engineering 2017, 142, 241–250.
[178] Gantayat, S., Rout, D., and Swain, S. K., Carbon nanomaterial–reinforced epoxy composites: A review. Polymer-Plastics Technology and Engineering 2018, 57, 1, 1–16.
[179] Chen, J., Yan, L., Song, W., and Xu, D., Interfacial characteristics of carbon nanotube-polymer composites: A review. Composites Part A: Applied Science and Manufacturing 2018, 114, 149–169.
[180] Chernova, T., Murphy, F. A., Galavotti, S., Sun, X.-M., Powley, I. R., Grosso, S., Schinwald, A., Zacarias-Cabeza, J., Dudek, K. M., Dinsdale, D., Quesne, J. L., Bennett, J., Nakas, A., Greaves, P., Poland, C. A., Donaldson, K., Bushell, M., Willis, A. E., and MacFarlane, M., Long-fiber carbon nanotubes replicate asbestos-induced mesothelioma with disruption of the tumor suppressor gene Cdkn2a (ink4a/Arf). Current Biology 2017, 27, 21, .e6, 3302–3314.
[181] Zhang, W., Chen, L., Zhang, J., and Huang, Z., Design and optimization of carbon nanotube/polymer actuator by using finite element analysis. Chinese Physics B 2017, 26, 4, 048801.
[182] Grabowski, K., Blacha, I., Staszewski, W. J., Uhl, T., and Packo, P., Optimization of multi-scale modelling of CNT/polymer composite strain sensors, In: Health Monitoring of Structural and Biological Systems 2017, Vol. 10170, 101700R, International Society for Optics and Photonics, 2017.
[183] Gong, S., Wang, Y., Xiao, Z., Li, Z., Wang, Z. X., Lei, R. S., and Zhu, Z. H., Effect of temperature on the electrical property of epoxy composites with carbon nanotube. Compos. Sci. Technol 2017, 149, 48–54.

[184] Gong, S., Wu, D., Li, Y., Jin, M., Xiao, T., Wang, Y., Xiao, Z., Zhu, Z., and Li, Z., Temperature-independent piezoresistive sensors based on carbon nanotube/polymer nanocomposite. Carbon 2018, 137, 188–195.
[185] Zhou, G., Zhang, H., Xu, S., Gui, X., Wei, H., Leng, J., Koratkar, N., and Zhong, J., Fast triggering of shape memory polymers using an embedded carbon nanotube sponge network. Sci. Rep 2016, 6, 24148.
[186] Gao, Y., Wan, Y., Wei, B., and Xia, Z., Capacitive enhancement mechanisms and design principles of high-performance graphene oxide-based all-solid-state supercapacitors. Adv. Funct. Mater 2018, 28, 17, 1706721.

Marta Martins, Joana F. Guedes, Celeste Pereira, Stefan Forero, Volker Liedtke, Jiří Zelenka, Jan Václavík, Ernst Pfeiffer, Laurent Pambaguian, Advenit Makaya, Nuno Rocha

4 Carbon nanotube-based materials for space applications

4.1 Introduction

Carbon nanomaterials embedded in carbon fiber-reinforced composite materials can improve specific performance characteristics such as electrical conductivity, resistance to mechanical fatigue and crack propagation, structural damping, Passive Intermodulation (PIM) reduction, Electromagnetic compatibility (EMC) shielding, remachining and coating capabilities. Some of these improvements were investigated and demonstrated in the ESA projects "NACO-1" and "NACO-2" performed between 2007 and 2013. This chapter outlines the work performed during these activities for the development of carbon fiber-reinforced polymer (CFRP) composites integrating carbon nanotube (CNT) buckypapers (BP) to be used in space applications.

The ranking of space applications potentially benefitting CNT technology has been reviewed throughout these activities. The final activity is aimed at finally selecting an application with highest potential for performance improvements. For that, an optical CFRP mirror with comparable characteristics of conventional Zerodur-based mirrors featuring the inclusion of 3D nanotube structures was selected and manufactured.

Although CFRP composites are being used for reflectors in Space applications, due to its outstanding mechanical and thermal properties at low mass, optical mirroring properties are limited by the surface properties of the CFRP. One should note that each composite layer is composed of 1D and/or 2D patterns of continuous carbon fibers, each one of them having about 10 μm of diameter, which are combined in tows of thousands of carbon fibers leading to a tow a few millimeters large. This configuration creates an undesired composite surface roughness, in a phenomenon usually known as print-through effect (Figure 4.1).

CFRP surface roughness limits its reflective properties. To afford further mirroring characteristics, polishing and metal coatings strategies can be followed. However, the

---

Marta Martins, Joana F. Guedes, Celeste Pereira, Nuno Rocha, INEGI – Institute of Science and Innovation in Mechanical and Industrial Engineering, Porto, Portugal
Stefan Forero, FutureCarbon, Bayreuth, Germany
Volker Liedtke, AAC – Aerospace & Advanced Composites, Wiener Neustadt, Austria
Jiří Zelenka, TOSEDA s.r.o., Staré Čívice, Czech Republic
Jan Václavík, Institute of Plasma Physics of CAS, Prague, Czech Republic
Ernst Pfeiffer, HPS GmbH, München, Germany
Laurent Pambaguian, Advenit Makaya, ESA ESTEC, Noordwijk, Netherlands

https://doi.org/10.1515/9783110479133-004

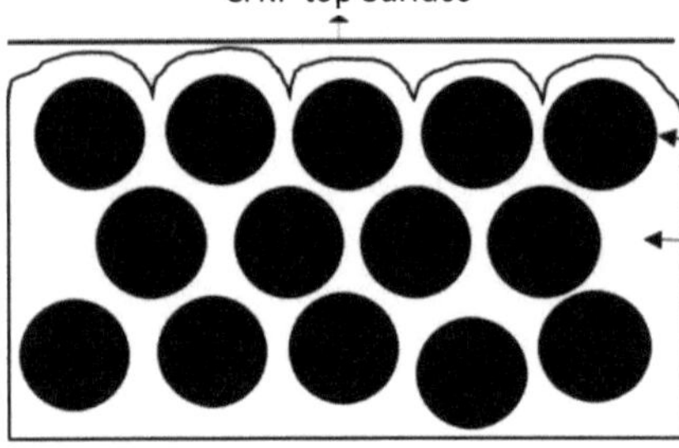

Figure 4.1: Representative image of print-through effect in a CFRP.

former is inherently limited by the continuous form of the carbon fibers, whereas the latter is also limited by the print-through characteristics of the CFRP. To overcome this limitation, inclusion of carbon nanotubes (CNT) at the top surface of the composites has been proposed as a solution to reduce its print-through behavior. In addition, the nearly isotropic nature of the CNT top layer is expected to ease polishing and metal coating processing. A material configuration in which a top layer of resin rich in CNT is added in the CFRP lay-up allows the reduction of the print-through effect on the manufactured CFRP and improves the reflective properties of the top surface.

The potential of CNTs is considered as tremendous; their intrinsic properties make them very attractive for Space applications where the characteristics of the materials often represent a limit to the development of technologies. Much of the development efforts have been focused on retaining CNTs large aspect ratio, achieving good interfacial stress transfer (surface modification), good dispersion and alignment in the polymer matrix during processing, all for maximizing the nanotubes effect. As a result, a growing number of companies already supply CNT-based masterbatches or compounds to be used in plastic/composite processing. Nevertheless, these materials are expensive, limited in the type of dispersing resin systems and on the CNT weight loading, and very difficult to incorporate in the conventional processes due to the very high viscosities.

In fact, one of the main difficulties that have been associated with the manufacturing of CFRP materials containing nanotubes is the poor processability of the resin systems modified with CNT, due to the extremely high viscosities. It is practically impossible to obtain a surface with quality when using high viscosity systems.

CNT BP consist of a 3D structure (skeleton) of CNTs either being single-walled (SWNT) or multiwalled or carbon nanofibers (CNF), which show a unique combination of inherent CNT/CNF properties while maintaining their lengths in a mat-like or 3D structure. These structures can be an alternative to overcome dispersion issues usually associated with the use of CNTs in resin masterbatches. In addition, higher CNT contents can also be attained by the use of three-dimensional CNT structures; therefore, CNT in the form of BP is a promising choice for boosting CFRP properties.

4.2 Preparation of epoxy nanocomposites based on high CNT content buckypapers

The main objective of the first study was to develop and characterize highly loaded CNT-based nanocomposites, by infiltrating nanotube skeletons with either organic or inorganic matrix. We herein present the main results obtained with an organic thermoset (epoxy) resin.

A new approach was used to produce these CNT nanocomposites, where the CNT skeletons or papers were infiltrated with a low-viscosity epoxy resin in a closed mold. The CNT were first produced in thin papers called BP to form networks of elongated nanotubes. The impregnation was performed by infiltrating the resin along all the 3D CNT structure. The level of impregnation of the final structure was assessed by physical measurements and observations under scanning electron microscopy (SEM). The results showed that the proposed processing approach is capable of producing well-dispersed nanocomposites with high CNT loading (more than 12 wt% in resin weight), which are important for developing high-performance structures based on CNT with good thermal and electrical conductivity.

4.2.1 Materials

CNT thin papers, with 80 mm of diameter, functionalized (-COOH) and solvent (hexane or ethanol, EtOH) were produced by Future Carbon, GmbH. The properties of the nanotubes used for the preparation of the skeletons are described in Table 4.1.

Table 4.1: Properties of the CNTs used for the development of buckypapers.

Property	
Diameter (nm)	10–20
Length (μm)	20–40
Purity (min wt% C)	>98%
Number of walls	5–15

The epoxy resin used to infiltrate the CNT skeleton was SR 8100 with SD 8824 hardener from SICOMIN. This system has low viscosity at room temperature (110 mPa s), facilitating the resin flow and impregnation of the skeletons. The gel time of about 1 h at 40 °C is suitable for a low-pressure infiltration. The recommended cure and postcure cycles for this system are 24 h at room temperature and 8 h at 60 °C, respectively.

4.2.2 Sample manufacturing

The impregnation of the CNT BP was performed at room temperature using a closed mold, as represented in Figure 4.2. For this mold, two spacers are available: one of 0.5 mm used for one layer composites and the other of 1 mm used for two layer composites. Three runs were performed using CNT papers with different chemical treatments, namely hexane or ethanol.

Figure 4.2: (Left) CNT skeleton positioned in the closed mold, before resin infiltration; (right) resin trap, placed before the vacuum pump, used to store eventual excess resin from the infiltration process.

The procedure used for the infiltration process can be roughly divided into the following steps:

1. Prepare the mold with the application of release agent
2. Put spacer in place
3. Place the CNT BP in the bottom mold
4. Close the mold
5. Connect the resin entry and the vacuum pump outlet
6. Turn on vacuum pump and verify if mold is completely sealed
7. Prepare the resin
8. Infiltrate the resin

The vacuum level used during the infiltration process was of 0.2 bar; the resin was injected in the mold at room temperature and the mold itself was kept at room temperature. The skeletons were allowed to cure 24 h at room temperature inside the mold, followed by demolding and postcure cycle in an electrical oven.

4.2.3 Characterization

CNT papers and composite samples were visual inspected, measured and weighted before and after infiltration. Fractured surfaces of samples were observed under a high-resolution SEM for the level of impregnation and surface quality assessment, as well as for measurements of the nanotube and resin layer thicknesses.

4.2.4 Results and discussion

4.2.4.1 Physical characterization

Figure 4.3 shows an SEM image of the BP surface (a), appearance of a CNT paper before (b) and after (c) infiltration. CNT papers have good strength and flexibility to allow handling like carbon fiber mat. The SEM image (a) shows that CNT possess diameters in the range of 10–20 nm and lengths in the range of 20–40 µm. CNT are produced solely by catalytic chemical vapor deposition as described elsewhere [1]. The number of walls is between 5 and 15.

Table 4.2 summarizes the main physical characteristics of the CNT papers used and of the nanocomposites obtained, namely the thickness and mass values. The CNT BP papers used in the manufacturing of samples 1 and 2 are hexane treated and the one used for sample 3 is ethanol treated. The thickness of the CNT papers is uniform, ranging from 350 to 375 µm.

After resin infiltration, the CNT BP thicknesses increase more than doubling its value. This increase is, however, very close to all samples, in the range of 405–430 µm. The mass of the obtained nanocomposites ranges between 3.11 and 3.98 g and the CNT weight content calculated directly (from the variation of mass before and after infiltration) is between 11.6% and 15.4% [2]. The higher CNT content was obtained for sample 1, with hexane-treated nanotubes. From these results we cannot conclude about a relation between solvent treatment and CNT content in the composite. Sample 2 is the one that has higher mass and consequently lower CNT content. This is due to the higher amount of resin that infiltrated the paper. The infiltration process shall be controlled in order to achieve highly loaded CNT composites, the reproducibility of the process and thus be able to conclude on the effect of the paper solvent treatment on composite properties.

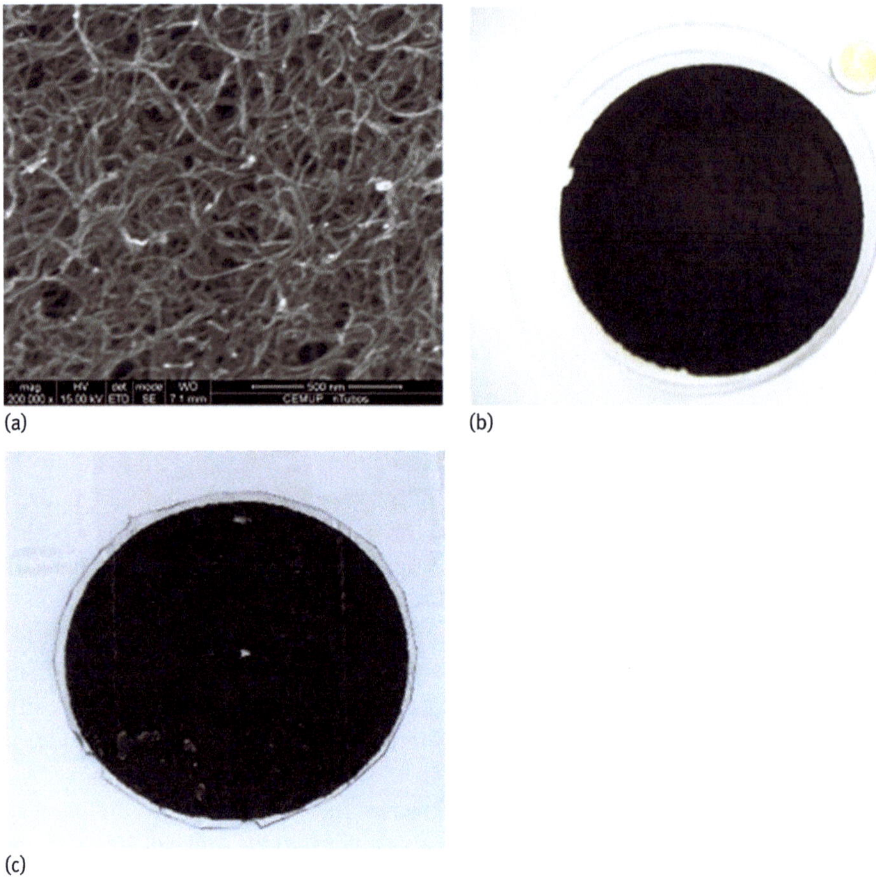

Figure 4.3: SEM image of the buckypaper surface (a) and the aspect of the buckypaper before (b) and after infiltration (c).

Table 4.2: Summary of the CNT papers and nanocomposite characteristics.

Sample	CNT paper			Nanocomposite		
	Solvent	Thickness (mm)	Mass (g)	Thickness (mm)	Mass (g)	CNT wt%
1	Hexane	0.350	0.4798	0.760	3.11	15.4
2	Hexane	0.370	0.4616	0.803	3.98	11.6
3	Ethanol	0.375	0.4576	0.780	3.27	14.0

4.2.4.2 Structure observations

In Figure 4.4, two microscopic pictures are presented, which are obtained under two magnifications and from different locations along the thickness of the sample 1. It is

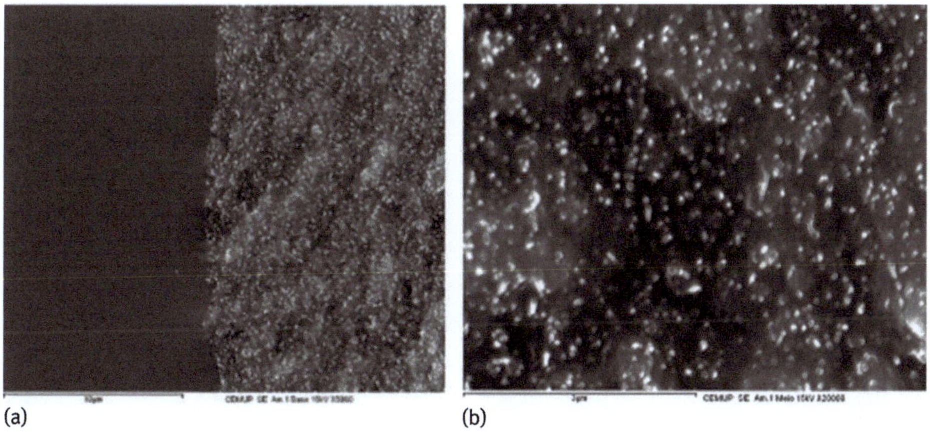

Figure 4.4: SEM pictures of sample 1 (a) in the bottom side of the composite and (b) in the middle of the composite.

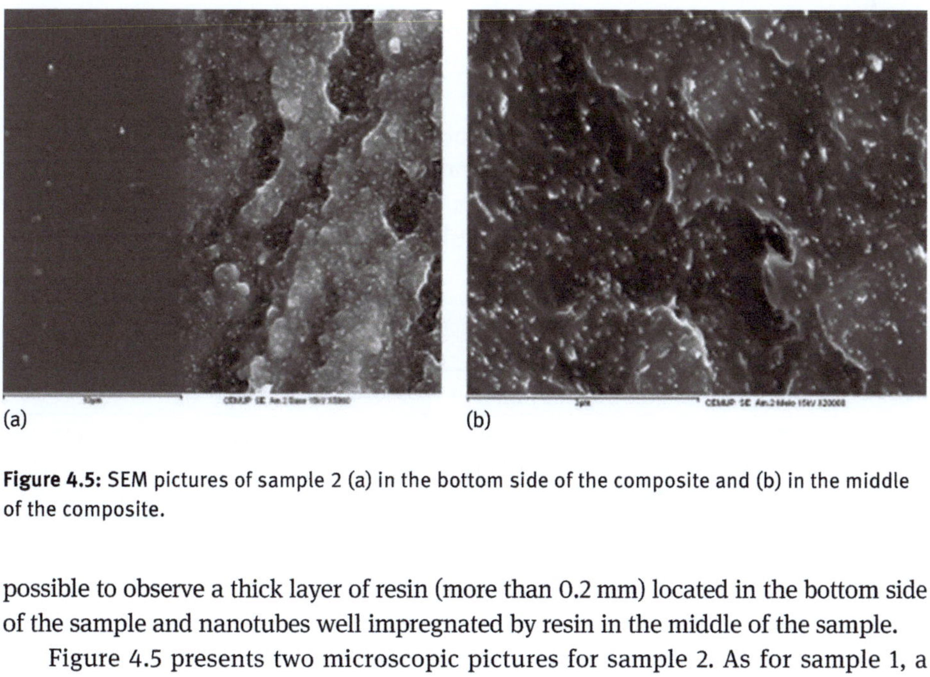

Figure 4.5: SEM pictures of sample 2 (a) in the bottom side of the composite and (b) in the middle of the composite.

possible to observe a thick layer of resin (more than 0.2 mm) located in the bottom side of the sample and nanotubes well impregnated by resin in the middle of the sample.

Figure 4.5 presents two microscopic pictures for sample 2. As for sample 1, a rich resin layer is observed in the bottom side (lower than 0.2 mm) and upper side (lower than 0.100 mm) of the sample, and a good matrix infiltration in the middle is witnessed.

Figure 4.6 shows the pictures obtained for the third run sample. It is observed that the BP does not maintain the initial form (appearing slightly curved in the images) and their thickness is variable along the diameter. Once again, a thick layer of

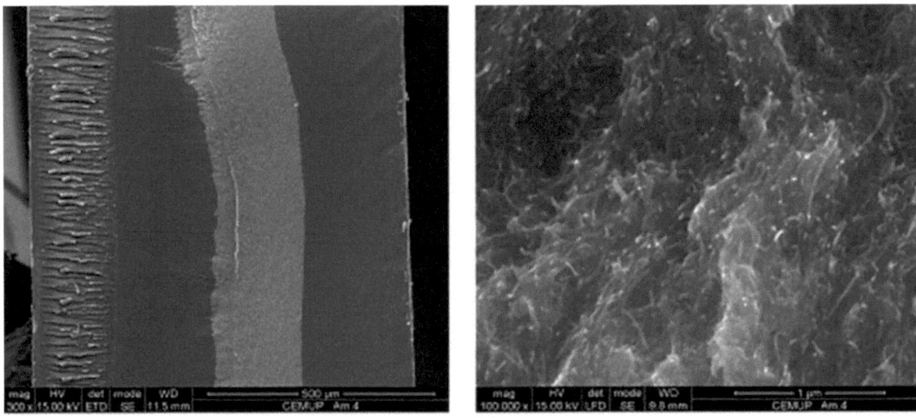

(a) (b)

Figure 4.6: Composite SEM pictures of the third run sample: (a) along the thickness of the composite and (b) in the middle of the composite.

resin is covering the paper. In fact, it is possible to see that the thickness of the resin layer is much higher then the CNT BP one. Once more, a good matrix infiltration in the middle is observed.

Table 4.3 summarizes the measurements of the top and bottom resin-rich layers and CNT paper thicknesses, after the infiltration process, measured from the SEM observations. It is observed that the thickness of the paper does not remain unchangeable, since it swells for samples 1 and 2 and shrinks considerably for sample 3.

Table 4.3: Resume of the characteristics of CNT papers and CNT nanocomposites.

Sample	**Nanocomposite**			
	CNT thickness before resin infiltration (mm)	**CNT thickness after resin infiltration (mm)**	**Thickness of resin layers (top + bottom)**	**CNT wt% (eliminating resin-rich layers)**
1	0.350	0.428	0.366	34.8
2	0.370	0.462	0.282	17.4
3	0.375	0.220	0.624	47.4

Two possible explanations for this can be prompted: the first one is that the solvent treatment has an effect on the resin behavior – contraction or expansion – during the cure [3]; the second is that the solvent treatment increases or decreases the infiltration of the resin into the paper, either by changing the wettability or the permeability of the CNT papers [4]. P.E. Lopes and coauthors [5] studied the wettability (by contact angle measurements) of CNT BP with different treatments. However, the CNT treatments studied were heat, carboxyl or nonfunctionalized, which are different than the ones reported here.

A rough estimation of the possible CNT content that could be achieved, if the resin-rich layers were eliminated, calculated and presented in Table 4.3. The most surprising results are the differences between samples 1 and 2, which have the same CNT treatment. The second sample has the lower CNT content and confirms the high portion of resin in the composite. Once more, we conclude that the infiltration process should be well controlled, and a release film must be used in the surfaces of the CNT papers intended for reducing the thickness of top and bottom layers of resin and the permeability of the CNT papers accessed. The resin content is a critical aspect of this CNT composites, since the excess of resin can affect negatively some properties, namely the thermal and electrical conductivity as well the mechanical properties [6–8].

These extrapolated results of CNT content (presented in Table 4.3) are more in agreement with those reported in literature [9]. Wang and coworkers have prepared SWNT BP/epoxy resin nanocomposites by using very thin papers (0.1–0.5 mm) stacked together in a hot press machine. The resin was infiltrated in papers after being diluted with acetone due to its high viscosity (2,700 mPa.s). They produced nanocomposites with SWNT loadings as high as 39% with nanopore sizes in the BP in the range of 100–200 nm.

4.2.5 Conclusions

It was the purpose of this experimental work to produce CNT/epoxy resin composites by infiltrating a low-viscosity epoxy resin in membranes of nanotube networks called BP, supplied by Future Carbon. The CNT content achieved in the composites is between 11.6% and 15.4% in weight. In all the SEM observations of the composites produced, there is evidence that there is a good level of impregnation, yet the resin layer around the CNT BP is very thick. The process must be optimized in order to obtain composites without excess of resin in the surfaces, since the resin affects negatively some properties of the composites, namely the thermal and electrical conductivity as well the mechanical properties.

4.3 Manufacturing process scale-up and integration into CFRP composite structures

The main purpose of project NACO2 was the scale-up processing of CNT BP and their integration on CFRP composites. In order to find the most suitable manufacturing method, a number of tests were planned taking into account critical aspects of the manufacturing parameters (such as mold materials, vacuum and pressure levels during curing and resin impregnation temperature) [9]. Based on this, several manufacturing runs were experimentally performed, aiming to obtain high-quality CNT–CFRP composites. From these runs, it was concluded that the use of a steel mold coated with Teflon release films, combined with manual resin impregnation at room temperature and cure in the autoclave under vacuum and pressure lead to improved materials. At the end of the project, a CNT–skeleton composite demonstrator was obtained; herein we present the manufacturing details of that demonstrator.

4.3.1 Materials

Oxidized CNT skeletons in paper containing binder were used. The epoxy resin system used was: Araldite LY556, Aradur HY906 and Accelerator DY070, manufactured by HUNTSMAN, Switzerland. A Style 450/T300 Plain Weave 140 g/m^2 carbon fiber fabric of Cramer Fabrics, Germany, was used for the manufacturing of the demonstrator. Table 4.4 presents the characteristics and suppliers of the raw materials used for manufacturing the polymer demonstrator.

Table 4.4: Resume of the materials used in CNT–polymer demonstrator.

Materials	Main characteristics	Supplier
CNT skeleton	Oxidized CNTs, paper form, with binder	Future Carbon
Resin	Epoxy system (LY556 + HY906 + DY070); pot-life: 50–55 h at 25 °C; viscosity: 1,900–2,100 mPa S	Huntsman
Carbon fiber	CF Style 450/T300 Plain Weave 140 g/m^2	CC Fabrics

The CNT skeleton was produced by tape casting and has a binder amount of 29% in weight. The thickness is about 70 µm, as measured by optical microscopy and the size was 107 × 11 cm^2.

4.3.2 Sample manufacturing

The demonstrator was produced by hand lay-up followed by vacuum bagging in the autoclave and using Teflon as release film. The resin was applied at room temperature between all the reinforcement layers, and the CNT BP was applied as the top layer. The “striped” surface of the CNT skeleton (BP side that was cast in the release film) was faced toward the CFRP material during the impregnation. No resin in the top layer (on the BP) was applied. The pressure used in the autoclave was of 3 bar, the vacuum level was of 150–250 mbar and the temperature program was 2 h at 120 °C + 2 h at 180 °C. Figure 4.7 shows the different stages of the demonstrator manufacturing.

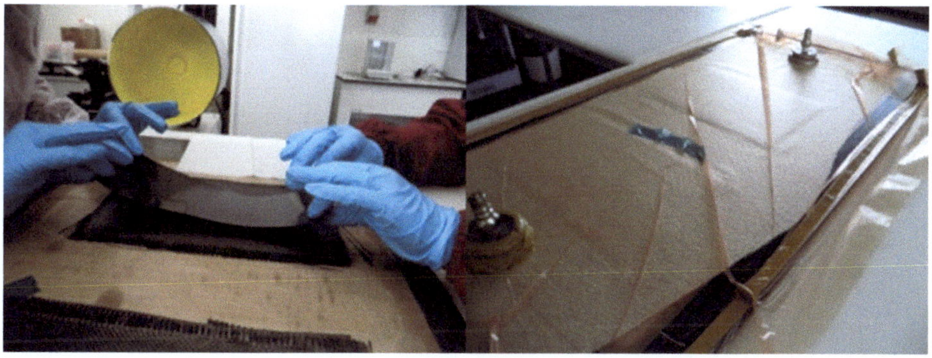

Figure 4.7: Different stages of the manufacturing process of the demonstrator.

4.3.3 Results and discussion

4.3.3.1 CNT skeleton handling

Since the CNT skeleton used had a very low thickness (70 µm), it was important that the material is very well conditioned. Also, due to its large size, it is crucial to be easy to manipulate so the composite manufacturing process is feasible. The CNT skeleton used in the demonstrator was packaged between two polymer substrate foils. The upper substrate foil was very easy to remove; however, the lower foil was completely "glued" to the BP. The cutting of the structure was easily performed with the use of supporting plates to maintain the CNT BP immobile and fully supported. Figure 4.8 shows two pictures of the conditioning and handling of the skeleton.

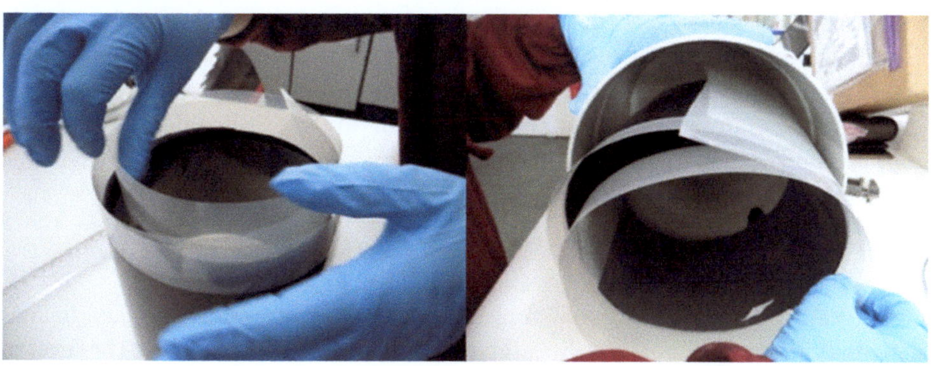

Figure 4.8: Conditioning, handling and aspect of BP used for manufacturing the demonstrator.

4.3.3.2 CNT skeleton preparation

As mentioned earlier, the BP was "glued" to the supporting foil. The procedure for removing the CNT paper, used in the demonstrator, from the casting substrate is given as follows:

1. cutting of the edges with the aid of supporting plates;
2. face the supporting film upward;
3. release a film tip with the aid of a tweezer;
4. separate the film from the BP using a metallic roller, while pulling the film upward and
5. continue until both materials are completely separated.

Figure 4.9 shows the sequence of procedures performed to separate the CNT skeleton from the supporting foil. It can be seen that the face of the BP that was "glued" to the polymer foil showed a different aspect, where a "striped" pattern is clearly visible. This stripped surface was faced toward the CFRP material during the impregnation run.

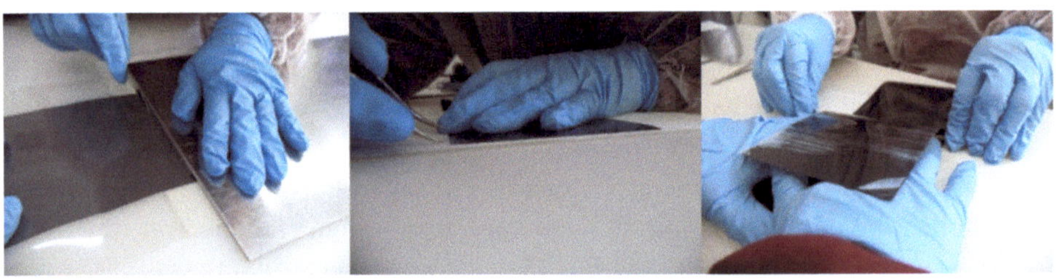

Figure 4.9: Procedure used for cutting and removing the supporting film from the buckypaper.

4.3.3.3 Visual inspections

Figure 4.10 shows the aspect of the whole sample after curing cycle and demolding, as well as samples already cut and ready for characterization tests. Some cracks in the BP were visible in the corners of the samples, after the cutting procedure. This can be due to either (i) lack of adhesion of the BP to the CFRP and/or (ii) the poor impregnation of the BP.

4.3.3.4 Structure observations

Figure 4.11 shows the resulting SEM pictures of the demonstrator. The SEM images show that, in some extent, the BP is separated from the CFRP laminate. The level of impregnation of the fibers seems good, and only a small resin layer (50–90 μm) is

Figure 4.10: Polymer demonstrator photographs.

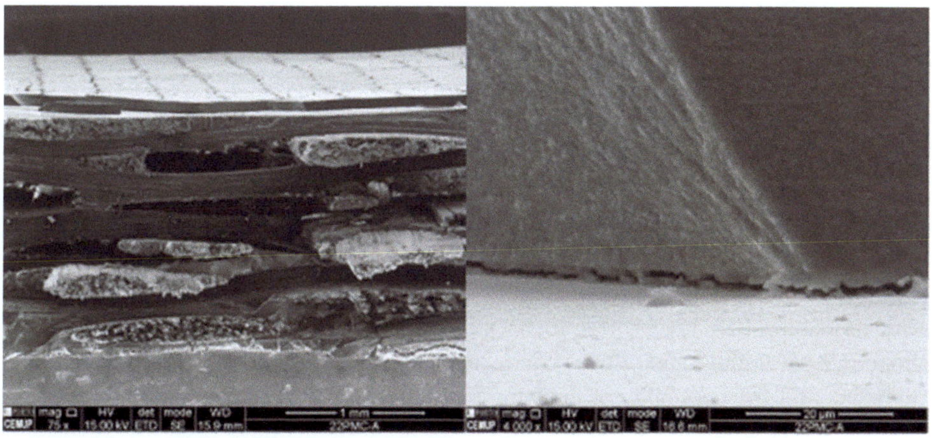

Figure 4.11: Polymer demonstrator photographs.

visible on the bottom surface. As previously mentioned, it is visible that the BP is not bonded to the laminate.

4.3.4 Conclusions

A small demonstrator was manufactured with the main following objectives:
1. to evaluate the quality of packaging;
2. to assess the hand ability of large and thin BP;
3. to determine the effect of the autoclave cycle on the BP's binder;
4. to verify the level of wettability and quality of impregnation of the BP of drop of resin and
5. to demonstrate that the manufacturing of large composites with large and very thin CNT layers is feasible.

The main conclusions from this activity can be summarized as follows [10, 11]:
- The CNT paper packaged between two polymer substrate foils and rolled inside a metallic container is effectively protected.
- BP can be easily manipulated when confined between the substrate foils. However, the removal of the lower substrate requires additional caution.
- The most significant change on the BP after autoclave cycle is the appearance of a layer of polymer (binder) in the surfaces of the skeleton.
- The resin spreads quite easily in the surface (nonstriped) and is able to enter into the material.
- The manufacturing of large composites with large and very thin CNT layers is feasible. However, some improvements or additional studies must be done, since the BP and the laminate were poorly bonded.

4.4 Technology demonstration in CFRP mirrors for space applications

The aim of NATAP activity was to identify, develop and test two selected demonstrators with functional CNT layers. Due to the high variety of options for adapting material properties, an initial trade-off was conducted to identify various space products, where the effect and the possibility of implementation would be maximized. Based on previous development activities, a large number of applications have been investigated in detail with respect to the likelihood to be used in the near future and with potential for improvement of weight, cost, performance and resistance to environmental impacts. From these, an optical mirror with comparable characteristics of conventional Zerodur-based mirrors was selected [12]. The realized demonstrator (28 cm diameter) consists of a sandwich structure, composed of CFRP honeycomb. CFRP face sheet and an infiltrated CNT BP, which serves as a "grindable surface" for highest accuracy. This surface was finally coated with a metal layer (by vapor deposition of aluminum coating and top protective SiO/SiO_2 coating layer) for optical reflectivity in the appropriate wavelength.

4.4.1 Materials

FutureCarbon was in charge of providing a dense CNT-based paper or nonwoven structure to be used as top carbon layer on the concave (optical) side of the mirror surface. From the manufacturing point of view, the CNT paper was the first layer on the mold surface. The CNT papers were manufactured using a filtration process, which resulted in 4–5 wt% of –COOH groups on the nanotube surface, aiming to provide covalent bonding with the resin system. CNT papers with a thickness of approx. 200–300 µm were manufactured, with diameters in the range of 31–32 cm [13].

TOSEDA supplied cyanate ester-based prepregs, using a pitch carbon fiber fabric (UD) [Granoc PF(S)-YSH 50A-75 (Mitsui, JP)]. The prepreg sheets used have Fibre Volume Fraction (FVF) of about 50% and were manufactured using a solvent-based process [14].

The honeycomb used was UCF-126-3/8-2.0 RS3 with YSH50 fibers and cyanate ester resin, supplied by Ultracore (Table 4.5).

Table 4.5: Resume of the materials used [15].

Materials	Main characteristics	Supplier
CNT skeleton	–COO groups, 6,96–7,42 mg/cm², 31–32 cm diameter	FutureCarbon
Resin	CE Primaset DT4000 (included as prepreg matrix)	Lonza
Carbon fiber	Granoc PF(S)-YSH50A-75 75 g/m² (included as prepreg reinforcement)	Nippon
Honeycomb	4 pieces 100 × 200 mm², UCF-126-3/8-2.0 RS3	Ultracore
Adhesive film	Redux 312L	Hexcel
Prepregs	Average carbon fabric volume concentration in prepreg 51 vol%	Toseda

The mold used for manufacturing the CNT–CFRP mirror was made from Invar, with a nickel surface treatment. The active surface of the mold, however, was machined with sharp tolerances to give the mirror the best surface accuracy possible before the polishing operations.

The prepregs manufacturing process includes the following steps:
- Impregnation of Granoc fabric with cyanate ester resin diluted to 35% in acetone
- Air-drying of the impregnated prepregs to remove the solvent (acetone concentration < 3%)
- Cutting the air dried pre-preg sheets into the requested size followed
- Marking one corner of the prepreg sheets to keep the same orientation of the carbon fabric fibers (warp and weft)
- Placement of the prepregs between PP foils for protection against bonding during storage and transport.

4.4.2 Sample manufacturing

After several manufacturing methodologies used (mainly deriving from the presence of cracks on the CNT surface, which are presented below), the procedure for manufacturing the mirror demonstrator was defined as follows:
1. Cut the CNT BP using a clean sharp tool, with the aid of metal plates, if necessary (Figure 4.12).
2. Dry the CNT BP during 16 h at 110 °C before stacking the materials.

3. If necessary, cut the prepregs in the appropriate dimensions (the warp and weft directions of the prepregs are identified).
4. Register the weight, area and thickness of the materials.
5. Place the Teflon release film in the mold and hold the materials by applying an adhesive tape in the corners (Figure 4.13).
6. Prepare the vacuum bag.
7. Place the CNT-BP in the Teflon release film inside the vacuum bag (Figure 4.14).
8. Lay up the two first prepregs layers, taking into account fabric warp and weft directions (Figure 4.15).
9. Close the vacuum bag and apply vacuum during 10 min with a vacuum level of 800 mbar with a Teflon release film between the prepreg layers and the vacuum bag (Figure 4.16).
10. Repeat the prepreg lay-up procedure until completion.
11. Place the demolding release film as top layer, use breather to eventually absorb excess resin and close the vacuum bag.
12. Apply the vacuum pressure (Figure 4.17). If necessary, the materials can remain under vacuum overnight.
13. Place the materials in the autoclave with the proper curing schedule.

Figure 4.12: Cutting the CNT-BP with the proper size.

Figure 4.13: Teflon release film.

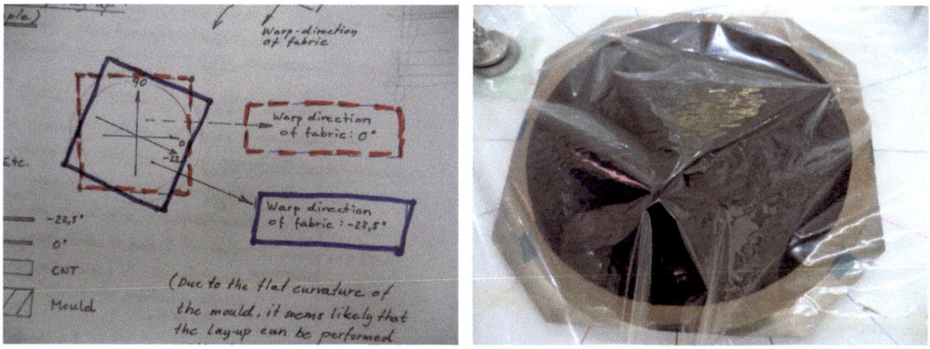

Figure 4.14: CNT-BP facing the mold.

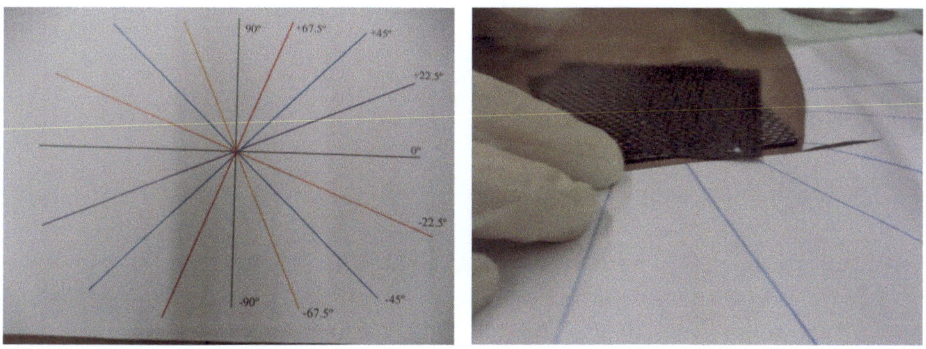

Figure 4.15: Prepreg orientations.

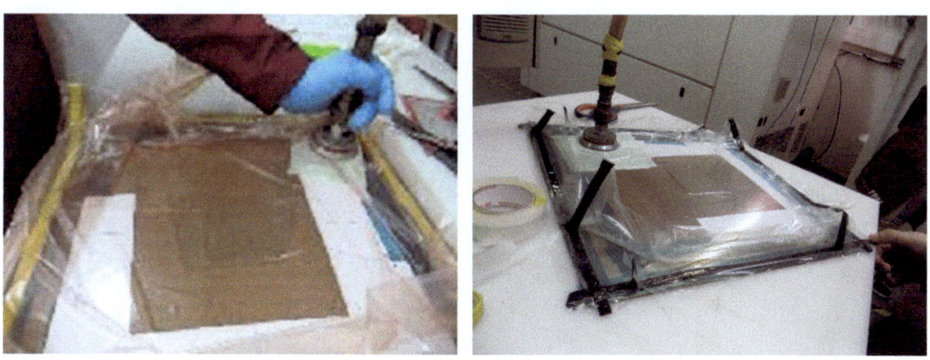

Figure 4.16: Vacuum application.

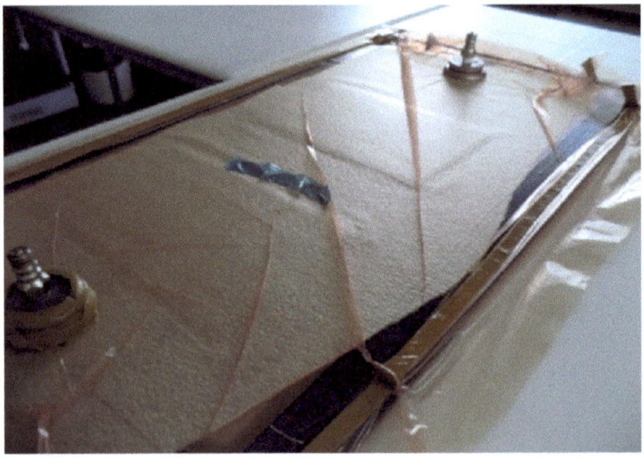

Figure 4.17: Final vacuum-bagged sample, before curing in the autoclave.

4.4.3 Results and discussion

4.4.3.1 Requirement definitions

For specifying the optical mirror properties, the nominal requirements for optical surfaces (not superpolished) were used as baseline. The usual base materials for classical optical mirrors are materials like Zerodur (lithium aluminosilicate glass-ceramic produced by Schott AG), Astrositall glass, ULE (ultralow expansion titaniasilicate glass), fused silica, SiC or nickel Canigen. Table 4.6 lists out the requirements [16]:

Table 4.6: CNT–CFRP optical mirror requirements.

Property	Requirement
Mass	At least 20% less than conventional mirrors
Reflectivity	R > 0.93
Surface form tolerance	RMS < 70 nm
Operating temperatures	−65 °C/+45 °C
Nonoperating temperatures	±120 °C
Outgassing	According to ECSS [17]
CME	24 ppm
CTE	$<1 \times 10^{-6}$
Thermal conductivity	>0.75 W/mK
Electrical conductivity	$>10^3$ S/m

4.4.3.2 Mirror design

The design of the mirror demonstrator was performed taking into account both the requirements and material properties. The active surface of the mirror is parabolic, circular symmetric, in order to ease the machining operations. The design (Figure 4.18) is based on a sandwich with a flat skin and a front skin with low curvature, rotationally symmetrical with a radius of curvature of 3,360 mm [18].

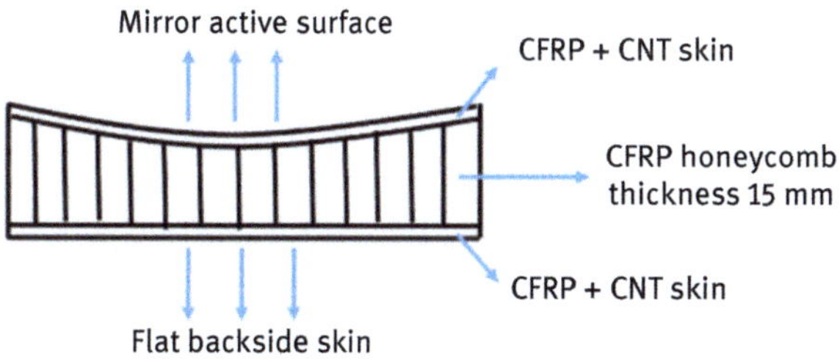

Figure 4.18: CNT–CFRP overall mirror design.

The detailed design also includes the following considerations:
- The reflector sandwich is fully symmetrical in thickness direction.
- Each skin is composed of 16 layers fabric prepreg with surface weight of 75 g/m^2 (without resin) together with a CNT layer.
- The honeycomb is made of CFRP using YSH50A carbon fibers.
- The CNT–CFRP skins are further metalized with vacuum deposit aluminum and protected with a SiO_2 layer.
- The adhesive film bonding the CFRP and the honeycomb is based on cyanate ester resin.
- The CTE of the combined skin + CNT + adhesive film shall be zero.
- The CTE of the CFRP honeycomb shall be zero.
- A design of the complete sandwich as close as possible to CTE zero is necessary.

In order to create a very accurate quasi-isotropic lay-up, the following lay-up of the CNT–CFRP layers is proposed. All orientations are based on warp fabric direction, explained in Figure 4.19.

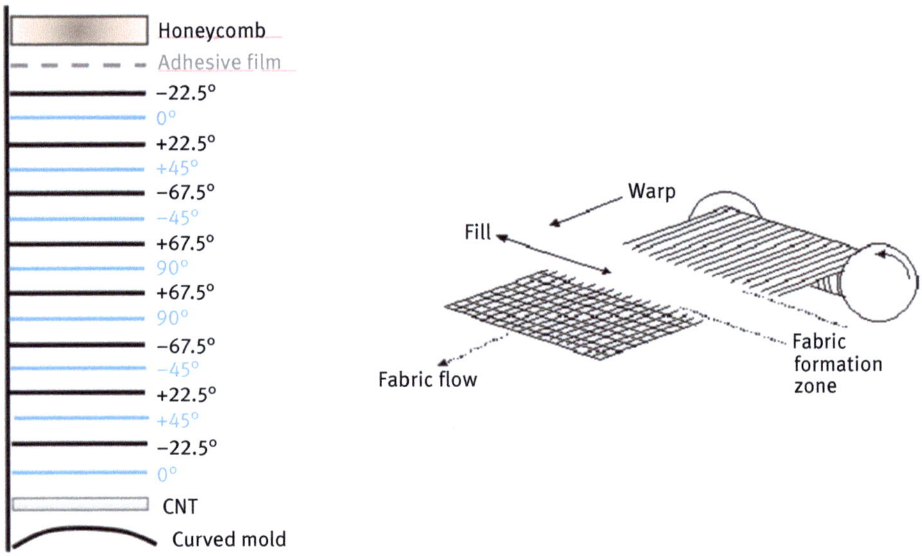

Figure 4.19: CNT–CFRP mirror design: detail of CNT–CFRP skins.

4.4.3.3 Manufacturing trials

NATAP sample manufacturing was performed in several phases. These phases and main objectives of each one are presented in Figure 4.20 [19].

From batch 1 and 2 sample manufacturing, the following conclusions were taken [20]:

- The preimpregnated fiber sheets and BP are easy to manipulate during the manufacturing process – including the production of curved samples – therefore, no major risks are attributed to draping for the expected geometries for the final applications.
- Prepregs prepared for batch 2 with 51% of FVF were found to be a good approach for manufacturing the demonstrator, since the resin amount present is enough to guarantee a good BP impregnation without the need to add additional resins. The CFRP layers were found to be well compacted, reducing the risk of having dry CFRP layers.
- Theoretical calculations also show that these new prepregs result in a final FVF of approximately 65%, which is a good balance between high fiber volume fraction for increased performance and enough resin to guarantee an homogenous structure of the CFRP structure (if previous prepregs with higher FVF were used, it was expected that composite was significantly dry – with 71–78 vol% of fiber).

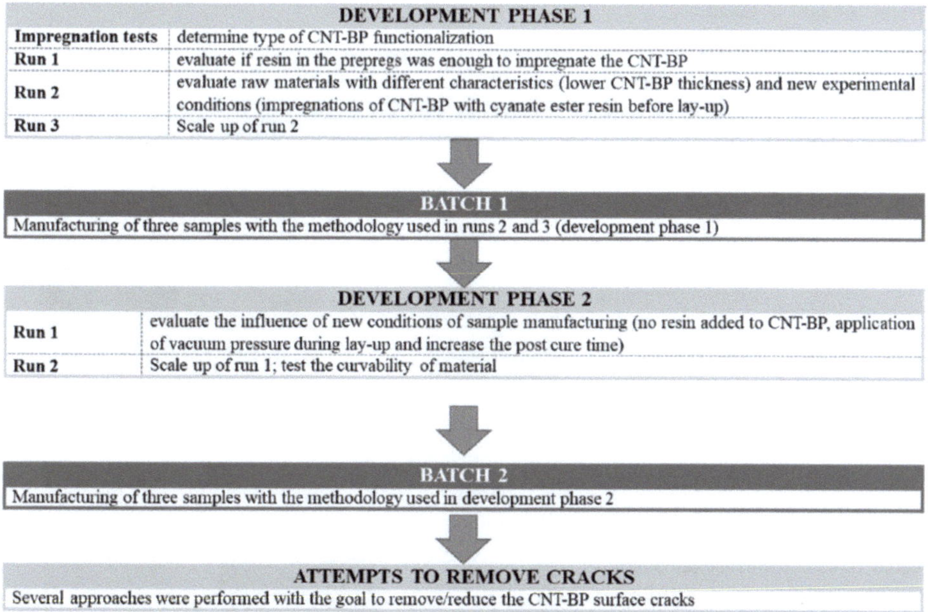

Figure 4.20: NATAP project phases and main objectives of each one.

- The vacuum pressure applied during lay-up is an important step to improve cohesion between layers.
- Curved samples can be manufactured without significant additional effort, however, they were found to be more difficult to demold. It is recommended to coat the curved mold with a Teflon release film with extended length (larger than the mold) to ease the demolding process;
- In batch 2, two main problems were detected: samples warping after cure cycle (only detected during demolding) and some marks on the surface. Warping can be attributed to the formation of tension forces formed during the cure cycle (at high temperatures and due to the geometry of the samples) and, therefore, the prepregs lay-up sequence may be reviewed for specific geometries;
- The existence of a surface pattern is should not be a problem, since the marks are expected to disappear after polishing and metalization processes.

Sample characterizations performed throughout the project allowed to detect some issues are shown below. The actions taken to avoid/eliminate those issues are also presented.

As referred before, some changes were made in both the characteristics of the raw materials used and experimental conditions during sample manufacturing. Figure 4.21 summarizes the experimental conditions, changes in the methodology used and problems encountered in the different stages of NATAP.

Project phase	Issue	Characterization	
Development phase 1: Run 1	Dry CNT structure, only bonded to CFRP at the ends	 Detail showing the adherence only in the edges of the CNT layer and the carbon fiber; the CNT structure is dry	
Batch 1	CNT BP thickness after impregnation was higher than expected (BP expands with resin addition)	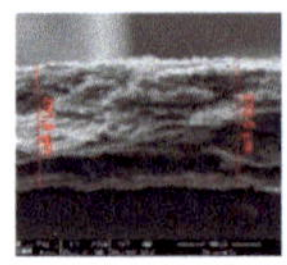 Cross-sectional view performed by INEGI (nonimpregnated CNT BP average thickness 300 μm)	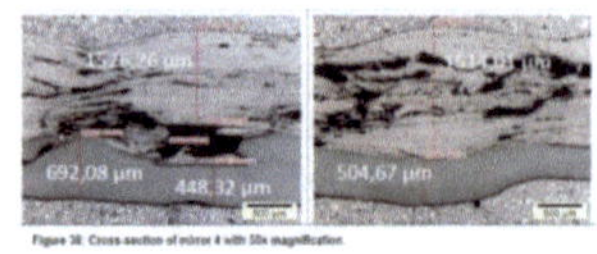 Cross-sectional view performed by AAC (BP-impregnated average thickness 430 μm)/ nonimpregnated CNT BP average thickness
Batch 1	Presence of cracks in the CNT BP structure	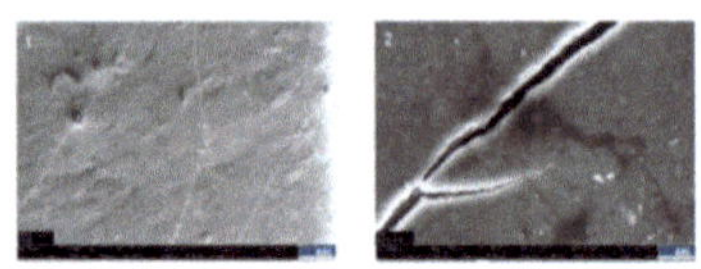 SEM images of coated surface mirror sample of batch 1 provided by AAC, showing the existence of cracks	
Batch 1	Glass transition temperature lower than expected	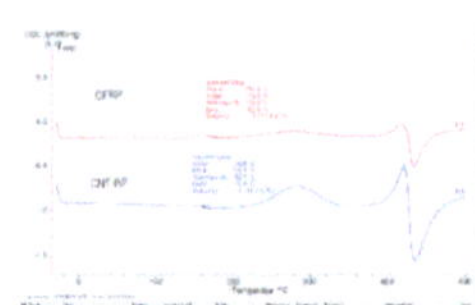 T_g measurements performed by AAC	
Development phase 2: Run 1	Lack of resin	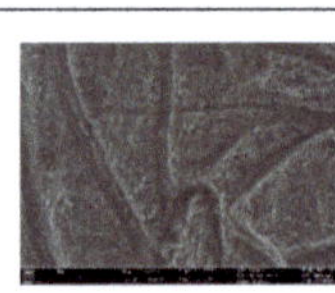 BP surface (before impregnation) – marks are visible	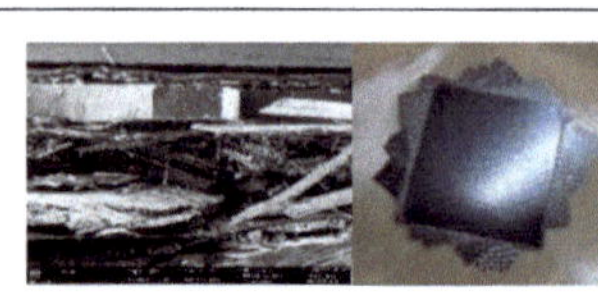 Cross-sectional SEM picture, showing that the structure is dry (low resin content); bright surface shows that the BP is well impregnated

(continued)

Project phase	Issue	Characterization
Batch 2	Cracks on CNT-BP structure	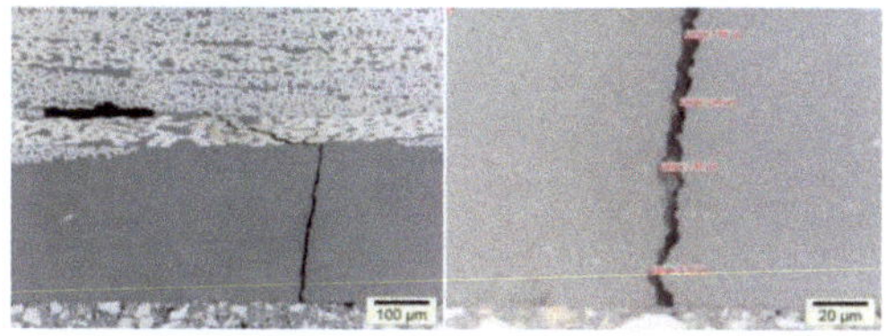 Cross-sectional view performed by AAC. Cracks are visible.

4.4.3.4 Mirror demonstrator manufacturing

The main steps taken in the demonstrator manufacturing are summarized as follows [15]:

1. Mold preparation
2. Back skin manufacturing
3. Front skin manufacturing
4. Bonding honeycomb to front skin
5. Honeycomb machining
6. Back skin bonding to honeycomb (final assembly)
7. Final machining

Back and front skins were prepared according to the stacking sequence previously presented. Figure 4.22 shows the aspect of both CNT–CFRP skins after demolding (Figures 4.23–4.25).

The first assembly step consisted in bonding the honeycomb to the front skin. The CNT–CFRP skin was placed in the mold and prepared for the adhesive and honeycomb settlement. Afterward, the assembly was placed in the oven under vacuum pressure and temperature cycles.

After the first assembly, the honeycomb was machined on the free side in order to have a flat surface, for further bonding with the back skin bonding. A flat disk installed in a defined position in the turning machine was used for the in-house machining process.

The final assembly consisted in bonding the back skin to the first assembly. The back skin, containing adhesive, was placed on the top of the machined honeycomb. The entire assembly was then placed in the oven under vacuum pressure and temperature cycles.

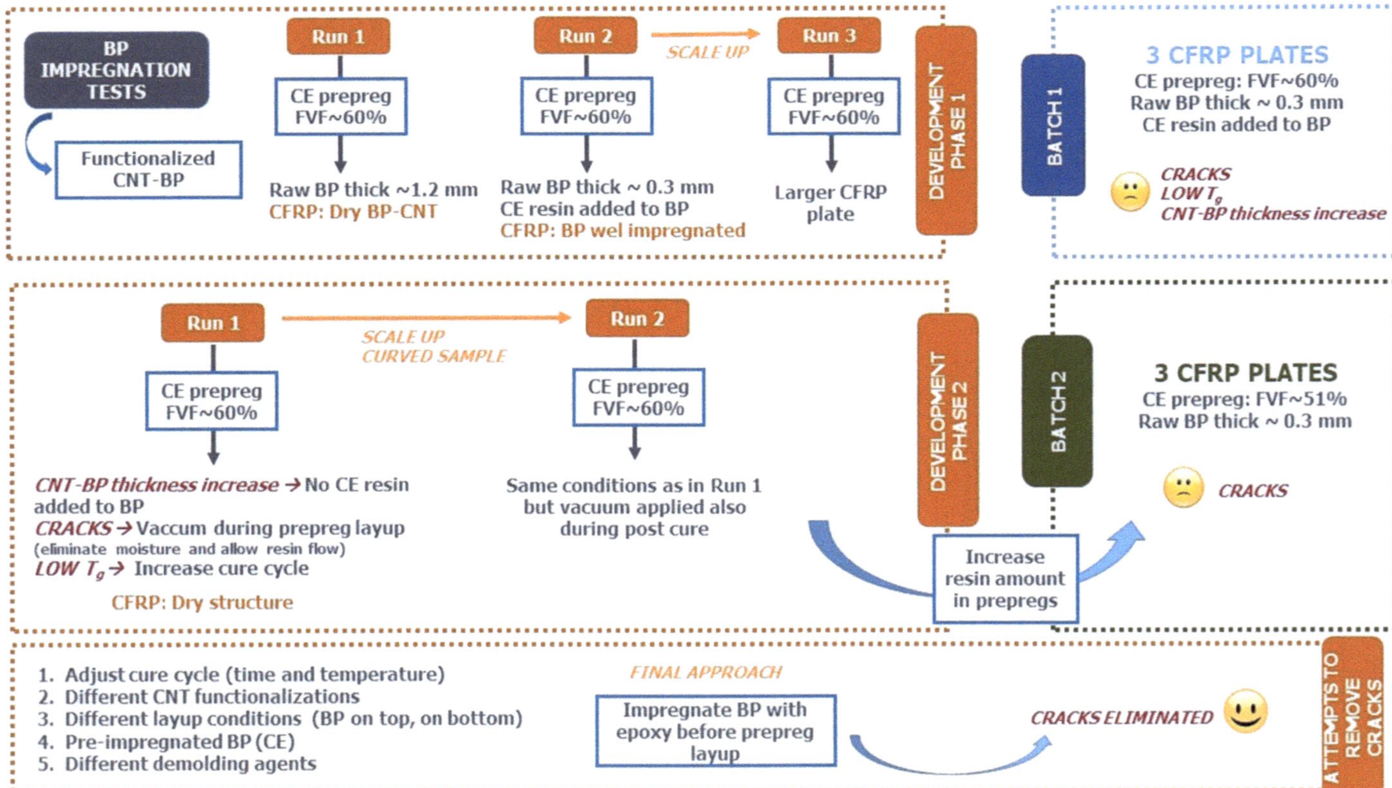

Figure 4.21: Conditions of each stage of sample manufacturing, problems and actions to solve them.

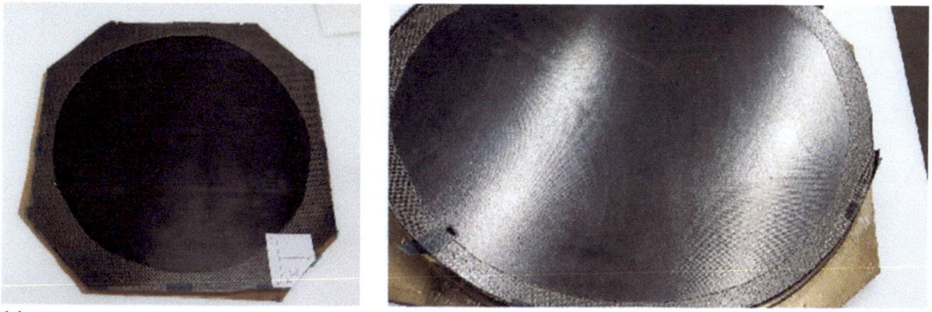

(a) (b)

Figure 4.22: (a) Mirror back skin and (b) mirror front skin.

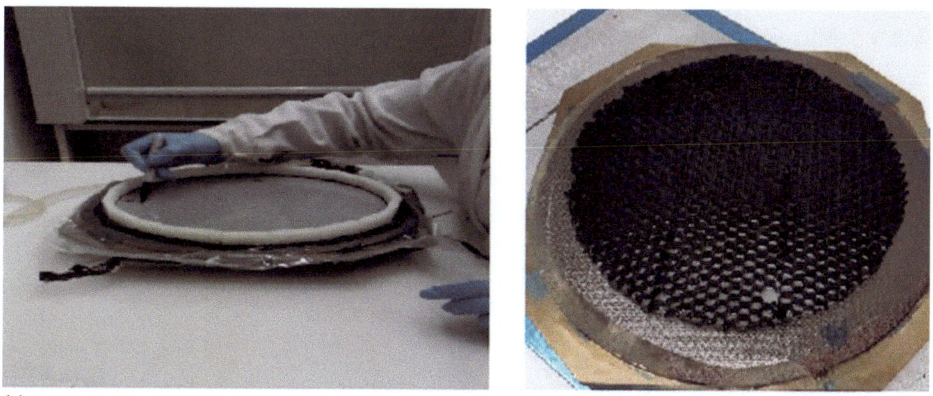

(a) (b)

Figure 4.23: (a) Application of adhesive and (b) view of the honeycomb bonded to front skin.

Figure 4.24: Honeycomb machining to obtain a flat surface.

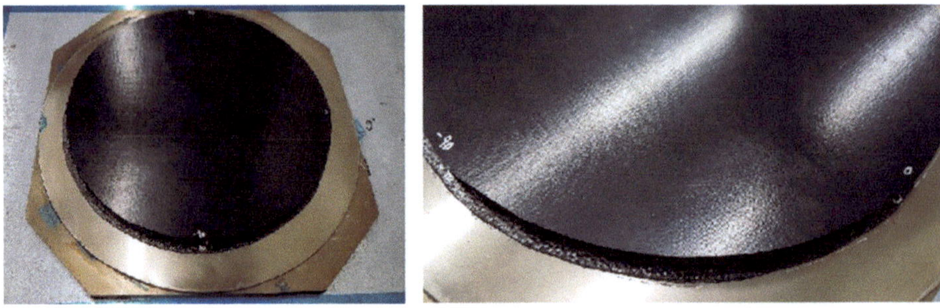

(a) (b)

Figure 4.25: (a) Assembly after curing with the back skin in the top. (b) Detail of the back surface, where no cracks are visible after all the assembly operations.

4.4.3.5 Mirror polishing and metalization

One of the project goals was to determine the most suitable polishing process conditions applicable on CNC polishing machines and allowing to achieve targeted microroughness Rq (Sq) of 10–15 nm on the optical mirror surface. The individual work steps are summarized as follows [21]:

4.4.3.6 Polishing

- Machining outside diameter
- Smoothing an optical surface
- Surface form correction
- Roughness correction

4.4.3.7 Metalization

- By vacuum deposition using e-beam evaporator unit.

The selected technologies were:
- Polishing by two-step process using CeO_2 polishing slurry and polyurethane pad
- Metalization by physical vapor deposition (PVD) technology (e-beam evaporator)

Figure 4.26 shows the surface appearance of a CNT–CFRP skin after the polishing procedure and the surface microroughness of that surface measured using an interferometer, showing that the surface root mean square height (Sq) value is 11.82 nm.

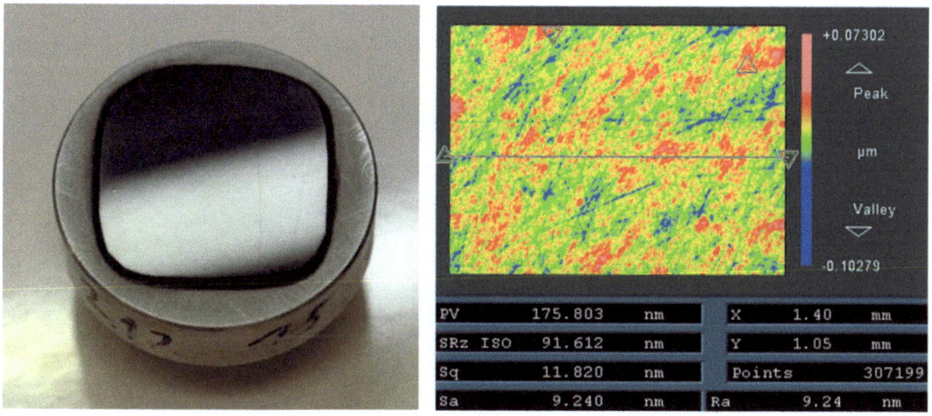

(a) (b)

Figure 4.26: (a) Surface appearance of CNT–CFRP skin after polishing and (b) interferometer results after polishing.

The polished CNT surface of the optical mirror was coated by combination of Al (100–120 nm) and SiO_2 (70 nm) protective layer without loss of adhesion. This combination is a standard type of coating for optical mirrors. SiO_2 greatly increases durability of the surface (protection of the Al coating against oxidation). Resulting mirror surface has good abrasion resistance, but the top CNT layer of the composite is soft (compared to ceramics and glasses); hence, the optical mirror surface can be damaged due to damage of the CNT layer.

The achieved RMS value of 412 nm, which included the outer border, does not reach the target value of 70 nm. A reduction of the effective area to exclude the edge effect enhances the RMS to ~230 nm, which still is not sufficient for the targeted quality.

Nevertheless, the quality is well passing the necessary surface quality for the backup option of the radiometer example MetOp, where 1,000 nm is targeted.

Further improvements on RMS quality will be possible when the print-through of the backing structure as well as the cracks in the CNT layer can be reduced or, respectively, avoided.

An enhancement of hardness of the top layer could improve the local roughness, which has been measured to be in the range of 10 nm [22].

4.4.3.8 Mechanical characterization

The mechanical properties have been determined at ambient temperature before thermal cycling (RT) and after thermal cycling (RT, TC), and at the maximal (120 °C) and minimal (–120 °C) nonoperational temperatures foreseen for the respective application. The summary of determined mechanical properties for the mirror is shown in Table 4.7. The measured density of the material was 1.493 ± 0.015 g/cm^3.

Table 4.7: Summary of mirror mechanical properties [19].

Temperature (°C)	3-Point bending strength (MPa)	Young's modulus (GPa)	4-Point bending strength (MPa)	Young's modulus (GPa)	Interlaminar shear strength (MPa)
−120	282.8 ± 12.9	24.9 ± 1.4	324.7 ± 16.7	27.7 ± 1.5	29.1 ± 1.6
RT	284.6 ± 9.9	31.9 ± 1.6	270.5 ± 5.5	29.1 ± 1.1	26.2 ± 1.4
+120	262.2 ± 5.9	24.9 ± 2.7	236.6 ± 6.3	32.4 ± 1.3	23.6 ± 1.1
RT, TC	251.2 ± 11.1	28.6 ± 1.6	254.5 ± 9.2	27.0 ± 0.5	27.1 ± 1.2

Summarizing all the above presented results, the mechanical properties of the materials are considered suitable for the application with sufficient safety margin.

4.4.3.9 Electrical characterization

A total of four measurement points were performed onto the antenna polished and metallized surface; three to be placed at the outer circumference of the antenna and one into the center (Table 4.8).

Table 4.8: Radial positions for electrical measurements.

Position	Code	Details
1	FS-R1 to FS-C	Front side radial-1 to front side–center
2	FS-R2 to FS-C	Front side radial-2 to front side–center
3	FS-R3 to FS-C	Front side radial-3 to front side–center
4	FS-R1 to FS-R2	Front side radial-1 to front side-2
5	FS-R1 to FS-R3	Front side radial-1 to front side-3
6	FS-R2 to FS-R3	Front side radial-2 to front side-3

The DC resistance using a Metrohit Multimeter was measured from each of the three radial positions (FS-R1 to FS-R3) to the center FS-C (Figures 4.27 and 4.28).

The results of this test are shown in Table 4.9. The standard deviation of the measurements before thermal cycling is very high. A reason for this can be a capacitor effect of the honeycomb structure under the surface.

The electrical conductivity after the thermal tests was much higher than before the thermal cycling. The reason for this can be the many small cracks of the metalized surface; therefore, the insulating silica top layer was probably fractured, exposing

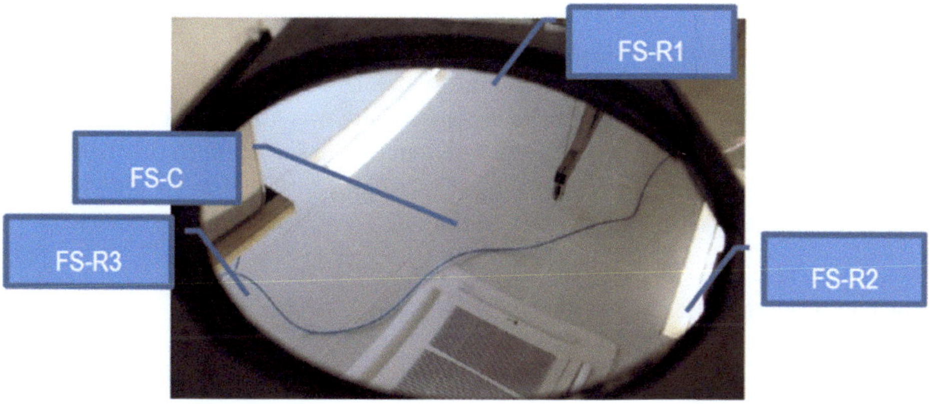

Figure 4.27: Electrical surface resistance test.

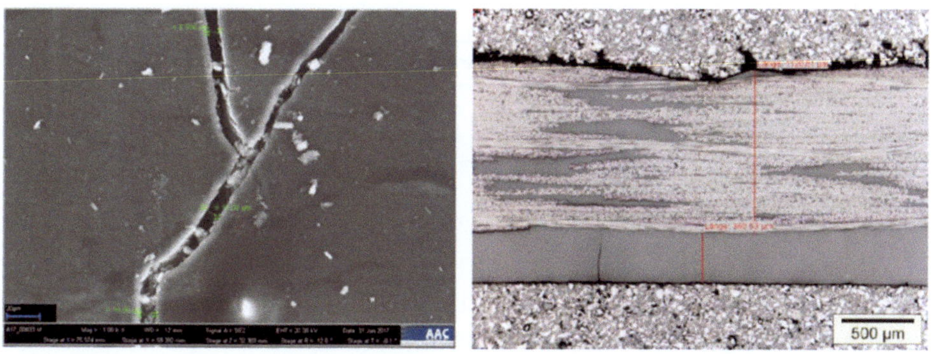

Figure 4.28: Surface crack in mirror sample after thermal cycling (left), Cross section of mirror sample (right).

Table 4.9: Summary of electrical measurement results.

Code	Before thermal cycling			After thermal cycling	
	Average (Ω)	Standard deviation	Conductivity (S/m)	Average (Ω)	Standard deviation
FS-R1 to FS-C	9,166.7	1040.8	9,92E-04	21.7	7.2
FS-R2 to FS-C	6,166.7	1040.8	1.47E-03	13.0	4.2
FS-R3 to FS-C	3,600.0	1216.6	2.53E-03	21.3	4.6
FS-R1 to FS-R2	2,066.7	378.6	2.42E-03	27.1	5.8
FS-R1 to FS-R3	2,000.0	435.9	2.50E-03	23.3	2.5
FS-R2 to FS-R3	2,566.7	907.4	1.95E-03	22.7	3.0

either the aluminum or the CNT layer, and we could measure the electrically conductive surface through the top layer.

The requirement of 10^3 S/m, that is, "metallic" electrical conductivity, could not be reached. The as-received surface conductivity was 4.28 S/m for the CNT side and 2.59 S/m for the CFRP side. Careful roughening the surface with Grit 1200 sandpaper improved the surface conductivity substantially to 4.28 S/m for the CNT side, and 16.18 S/m for the CFRP side where carbon fibers were apparently contacted.

4.4.3.10 Thermo-optical properties

Thermo-optical properties were tested at ZAE Würzburg, each for the CNT and for the CFRP side. The thermo-optical properties do not differ from CFRP to CNT side (Table 4.10).

Table 4.10: Summary of mirror absorbance values.

Sample	Visual absorbance	Solar absorbance	UV absorbance
CFRP face	0.92 ± 0.02	0.91 ± 0.02	0.93 ± 0.02
CNT face	0.94 ± 0.02	0.94 ± 0.02	0.94 ± 0.02

4.4.3.11 Structure observations

The mirror samples consisted of about 1.1–1.2 mm CFRP, and a CNT layer of about 350 µm. The surface shows cracks after thermal cycling.

Table 4.11 summarizes the main results achieved with the CNT–CFRP mirror, highlighting that mass, CTE and thermal conductivity properties achieved are in high-performance levels; however, specific heat, surface accuracy and reflectivity values are below the requirement definitions.

4.4.4 Conclusions

The selected demonstrator – a CFRP-based mirror – was manufactured using two CFRP skins based on cyanate ester prepregs, with BP on the top surface. The CNT structure was incorporated as top layer of the skin during prepregs lay-up and then impregnated with the resin from prepregs. These CFRP skins were assembled with a CFRP honeycomb for obtaining the final part, further polished and metalized. Mirror mass was significantly improved at very good thermoelastic distortion values, but the CNT surfaces have many cracks, which further propagated after thermal cycling.

Table 4.11: Optical mirror demonstrator – major results.

Property	Results	Performance level
Mass	More than 20% mass saving in relation to a conventional optical mirror can be achieved. The mirror demonstrator's mass was only 270 g.	++
CTE	-0.706×10^{-6}/K in X/Y-direction (requirement:<±1 × 10^{-6} /K)	++
Thermal conductivity	0.84 W/(m*K), better than required (>0.75 W/(m*K))	+
Specific heat	0.73 J/(kg*°C) acceptable in relation to Zerodur 0.85 J/(kg*°C)	–
Surface accuracy	0.135 µm has been reached, not far away from requirements (<0.1 µm)	–
Surface quality	Not reached – presence of many cracks	--
Reflectivity	0.7 (after cycling), required value (>0.93) not reached due to cracks in surface	--
Temperature range	−120 °C → +100 °C	

Several development steps were performed, aiming to eliminate or reduce the formation of cracks, such as the study of the influence of the lay-up, curing conditions and mold conditions, but none of the proposed approaches was effective. After all physical parameters being tested, the chemical influence of the resin was investigated. The cracks were eliminated by replacing the CTE for an epoxy resin in the CNT layer.

In fiber-reinforced composites, the CTE is affected primarily by the carbon fibers, while on the CNT layer, the resin CTE has the dominant effect. CTE disproportion can generate fatal stresses.

4.5 Conclusions

In this chapter, we presented a brief overview of the activities performed under three ESA projects, as well as the main results, targeting the use of CNT structures in the space sector. The first project was intended to prove the feasibility of CNT structures – BP – to be infiltrated with an organic matrix, potentially as an alternative to poorly dispersed and high viscosity nanotube-filled resin systems. Focus was given on assessing the morphology of the developed materials, by using high-resolution microscopic analyses. The results show the proposed processing approach as capable of producing well-dispersed nanocomposites with high CNT loading (more than 15 wt% CNT by composite weight), which are important for developing high-performance structures based on CNT with good thermal and electrical conductivity.

The main objective of this activity was to scale up the material developments and thereby increase the Technology readiness level (TRL) to 2 for CNT skeleton-based polymer composites. The developed materials are used for manufacturing and testing of demonstrators for typical space applications. Several experimental trials were performed, trying to optimize process conditions such as mold tools, curing conditions or the use of different methods for BP impregnation. In a first step, the manufacturing process for obtaining CNT–skeletons with well-defined and reproducible characteristics was followed. In a second step, the composite manufacturing processes were further developed on sample level and in parallel the dimensions of the CNT–skeletons were scaled up to 0.1 m × 1 m with thickness ranging from few tens of micrometers up to few centimeters. In the third step, the CNT-reinforced composite manufacturing was stabilized on scaled-up dimensions, suitable for manufacturing technological demonstrators. It was possible to obtain CFRP/CNT skeleton composites based on small papers in a controlled manner, with high CNT content, no defects and voids and without excess of resin in top, interlayer and bottom surfaces.

The final activity targeted a space application, and a demonstrator was manufactured and tested during the project. The selected application was an optical mirror, in which the CNT structures were intended to improve the surface characteristics of the material, facilitating the polishing and metalization steps. Extensive mechanical, thermal and optical characterizations were performed at sample and demonstrator level. The CNT–CFRP mirror resulted in more than 20% of mass savings, when compared to a conventional mirror, with a final mass of only 270 g. The measured surface accuracy was of 0.135 μm, not much distant of requirements (< 0.1 μm). The main issue of the optical mirror is related to the presence of several cracks on the CNT layer, limiting the surface quality and the reflectivity values obtained. Nevertheless, the principle of producing a highly accurate optical mirror based on a CFRP/CNT is rated as promising, with good thermoelastic distortion values. It is estimated that a TRL level between 3 and 4 has been reached.

Acknowledgments: The authors thank the financial support from the European Space Agency (ESA) under the frame of the following projects:
- Non-conventional Matrix/CNT Reinforced Composite for Applications in Space (NACO), Contract 20521/06/NL/SFe
- Non-conventional Matrix/CNT Reinforced Composite 2, Contract 4000104354/11/NL/RA
- Carbon Nanotube Technology and Material Engineering for Various Space Applications (NATAP), Contract 4000116757/16/NL/LvH/fg

References

[1] NACO-FCA-TN-06, Optimization Procedures and CNT Network Characteristics

[2] NACO-INE-TN, *Production of CNT-PMC at INEGI- first trials*

[3] Abdalla, M., Dean, D., Robinson, P. and Nyairo, E., "Cure behavior of epoxy/MWCNT nanocomposites: the effect of nanotube surface modification", Polymer 49 (2008), 3310–3317

[4] Wang, Z., Liang, Z., Wang, B., Zhang, C. and Kramer, L., "Processing and property investigation of single-walled carbon nanotube (SWNT) buckypaper/epoxy resin matrix nanocomposites"; Composites: Part A 35 (2004) 1225–1232

[5] Lopes, P.E. et al. Composite Structures 92 (2010) 1291–1298

[6] Gojny F., Wichmann, M.H.G., Fiedler, B., Bauhofer, W., Windle, A. H., Schulte, K. Polymer 47 (2006), 2036–2045

[7] Hernández-Pérez, A., Avilés, F., May-Pat, A., Valadez-González, A., Herrera-Franco, P. J. and Bartolo-Pérez, P., "Effective properties of multiwalled carbon nanotube/epoxy composites using two different tubes", Composites Science and Technology 68 (2008) 1422–1431

[8] Gonnet, P., Liang, Z., Choi, E. S., Kadambala, R. S., Zhang, C., Brooks, J. S., Kramer, L., "Thermal conductivity of magnetically aligned carbon nanotube buckypapers and nanocomposites", Current Applied Physics, 6 (2006), 119–122

[9] Non-Conventional Matrix/CNT Reinforced Composite 2, NACO2-INE-TN-240, *Polymer CNT-Composite 1 Development and Test*

[10] Non-Conventional Matrix/CNT Reinforced Composite 2, NACO2-HPS-RP-100, *Summary Report*

[11] Non-Conventional Matrix/CNT Reinforced Composite 2, NACO2-INE-TN-312, *Polymer Demonstrator Report*

[12] Carbon Nanotube Technology and Material Engineering for Various Space Applications (NATAP), NATAP-HPS-TN-110, *Selection of Space Applications and Associated Requirements*

[13] Carbon Nanotube Technology and Material Engineering for Various Space Applications (NATAP), NATAP-FC-TN-211 *Production of Composite Samples and Breadboards Batch 1 & 2*

[14] Carbon Nanotube Technology and Material Engineering for Various Space Applications (NATAP), NATAP-TOS-E-TN215, *Production of composite samples and breadboards: Batch 1 & 2*

[15] Carbon Nanotube Technology and Material Engineering for Various Space Applications (NATAP), NATAP-HPSP-TN-351, *Mirror Demonstrator Manufacturing*

[16] Carbon Nanotube Technology and Material Engineering for Various Space Applications (NATAP), NAT-KTO-RS-001, *Mirror Specifications*

[17] ECSS-Q-ST-70-02C, Thermal vacuum outgassing test for the screening of space materials, 15 November 2008

[18] Carbon Nanotube Technology and Material Engineering for Various Space Applications (NATAP), NATAP-HPS-TN-120, *Preliminary concept for demonstrator optical mirror*

[19] Carbon Nanotube Technology and Material Engineering for Various Space Applications (NATAP), NATAP-HPS-RP-200, *Final Report*

[20] Carbon Nanotube Technology and Material Engineering for Various Space Applications (NATAP), NATAP-INE-TN2.1.3, *Production of Material and Subelement Specimens 2*

[21] Carbon Nanotube Technology and Material Engineering for Various Space Applications (NATAP), NATAP-TOS-E-TN351, *Demonstrator 1 Manufacturing*

[22] Carbon Nanotube Technology and Material Engineering for Various Space Applications (NATAP), NATAP-AAC-TN-400, *Test Report Demonstrator Manufacturing*

Emmanuel N. Skountzos and Vlasis G. Mavrantzas

5 Molecular dynamics simulations of graphene-based polymer nanocomposites

5.1 Introduction

5.1.1 Carbon-based materials – the case of graphene

During the last decades, carbon has attracted a great deal of scientific and industrial attention due to the discovery of several allotropes (graphene, fullerene, carbon nanotubes [CNTs]), which are characterized by unprecedented physical and chemical properties such as high mechanical strength, extremely high electrical and thermal conductivity, high optical transparency and excellent gas barrier properties. Because of these unique features, carbon materials are widely used as nanofillers for the fabrication of composite materials with applications in several fields (biology, energy storage, transport and aviation, optoelectronics, pharmaceutics, medicine and many others). Fullerenes, for example, are used as electron acceptors for the fabrication of organic solar cells based on semiconducting polymers such as poly-3-hexyl thiophene, which has significantly improved the efficiency of the corresponding devices, reaching values comparable to those of inorganic solar cells. Graphene-based biosensors characterized by remarkable detection efficiency toward certain target molecules have also been manufactured by exploiting the excellent electrical and optical properties of graphene even when present at very low concentrations. The very high specific surface area of CNTs combined with their low electrical resistance and high charge transport capability have allowed the production of high-performance supercapacitors with increased energy storage, power-delivery capabilities and rather long-life cycle compared to conventional batteries.

From the broad family of the carbon-based nanoparticles (NPs), the one that has attracted the most significant attention is graphene, a one-atom-thick planar sheet of sp2-bonded carbon atoms densely packed in a honeycomb crystal lattice. Graphene is the basic building block for graphitic materials of all other dimensionalities (Figure 5.1), because it can be wrapped up into 0D fullerenes, rolled into 1D nanotubes or stacked into 3D graphite [1].

Emmanuel N. Skountzos, Department of Chemical Engineering, University of Patras, Patras, Greece; Foundation of Research and Technology Hellas, Institute of Chemical Engineering Sciences (FORTH/ICE-HT), Rio-Patras, Greece
Vlasis G. Mavrantzas, Department of Chemical Engineering, University of Patras, Patras, Greece; Foundation of Research and Technology Hellas, Institute of Chemical Engineering Sciences (FORTH/ICE-HT), Rio-Patras, Greece; Particle Technology Laboratory, Department of Mechanical and Process Engineering, ETH Zürich, Switzerland

https://doi.org/10.1515/9783110479133-005

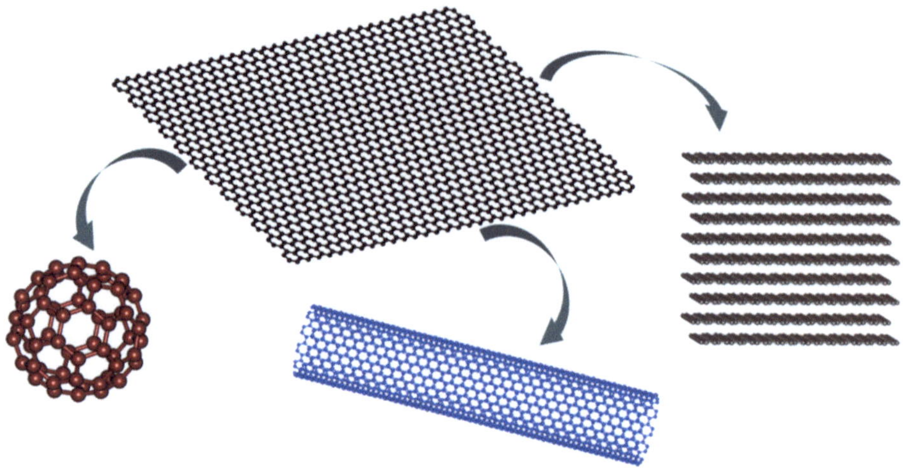

Figure 5.1: Graphene, the building block of all graphitic forms.

Until 2004, the production of single-layer (SL) graphene sheets (GS) was considered impossible, since theoretical calculations had predicted that GS would be completely unstable. The successful isolation of SL, defect-free GS using the "scotch tape method" [2] led to a rapid expansion of the field. The discovery was made by Prof. Sir Kostya Novoselov and Prof. Sir Andre Geim, who were awarded the Nobel Prize in Physics in 2010.

The impact of graphene on the scientific community has been enormous. Thousands of scientific articles (experimental and theoretical) are published every year and hundreds of these are heavily cited, while the paper describing the simple method to produce perfect GS has been characterized as "one of the most cited recent papers in the field of Physics" according to the ISI citation index. Apart from the great scientific interest in this exotic material, a significant impact on the largest technological industries of the world has also emerged after its discovery. Every year, thousands of patents are filed around the world on the design of new materials and new devices with important implications for practically all aspects of modern life (medicine, transportation, electronics, etc.), which explains why graphene has been characterized as the material of the future.

5.1.2 Graphene and graphene-based polymer nanocomposites

Numerous excellent reviews have been written in the last few years on the production, properties and applications of graphene [3–7] and graphene-based nanocomposites [8–13]; hence, here we will limit ourselves only to a brief overview of its unique structural and physicochemical properties, and for more details we refer the interested reader to the original articles.

Production of graphene

After a long and tenacious series of unsuccessful attempts to produce SL graphene, the publication of a simple method was presented in 2004 [2] known today as the "scotch tape method." By repeatedly cleaving a graphite crystal flake with an adhesive tape to its limit, and then transferring the thinned-down graphite onto an oxidized silicon wafer with the appropriate color, a 2D carbon lattice is produced.

In general, production methods of graphene fall into two main categories:
- Bottom-up methods
- Top-down methods

In bottom-up processes, graphene is synthesized by a variety of methods such as
1. Epitaxial growth on metal carbides [14–16]
2. Chemical vapor deposition (CVD) [17–18]
3. Unzipping CNTs [19–20]

In top-down processes, it is synthesized from a bulk material (e.g., graphite), which is broken down into smaller pieces using mechanical, chemical or other forms of energy. Typical examples of the top-down production processes include
1. the micromechanical exfoliation of graphite [2];
2. the direct sonication of graphite [21–22];
3. the chemical reduction of organically treated graphite oxide [23] and
4. the thermal exfoliation/reduction of graphite oxide [24].

In principle, with bottom-up approaches, large-sized, defect-free monolayer graphene is produced, a perfect material for the subsequent studies. The only drawback with these methods is that they afford only the production of tiny amounts of graphene. Top-down processes, on the other hand, are suitable for the large-scale production of graphene as (for example) required for the fabrication of graphene-based polymer nanocomposites.

Properties of graphene

The rapid adoption of graphene as the material of interest lies primarily in the excellent spectrum of properties (mechanical, electrical, thermal, optical, etc.) characterizing monolayer, few-layer graphene and graphene oxide (GO). The goal of many research efforts nowadays is to exploit these extraordinary properties for applications in nanotechnology by fabricating materials with improved mechanical, electrical, thermal and optical performance.

Carbon-based nanomaterials such as graphite, diamond and CNTs have their own record in terms of mechanical strength, hardness or Young's modulus. The newest member of the family, graphene, is no exception despite that its mechanical behavior has not been investigated as much as its electronic and optical properties. The reported stiffness of about 300–400 N/m (with a breaking strength of about 42 N/m) represents the intrinsic strength of a defect-free sheet [25] while estimates of the Young's modulus are on the order of 0.5–1.0 TPa [25]. Interestingly, and despite their defects, suspended GO sheets retain to a large extent their mechanical performance, characterized by a Young's modulus of 0.25 TPa [26]. These features, combined with the relatively low cost for the production of thin graphite and the ease of processes for blending GO into matrices [11], render these materials as ideal candidates for electrical and mechanical reinforcement [27, 28].

Regarding electrical properties of graphene, we mention that its electrical resistivity at room temperature is about 1 μΩ cm; as a result, graphene is about 35% less resistant than silver, the lowest resistivity material known today at room temperature. In semiconductors, a different measure is used to quantify electronic motion known as mobility. Mobility is often expressed as the conductivity of the material per electronic charge carrier. This implies that high mobility is advantageous also for chemical or biochemical sensing applications in which a charge signal (e.g., a molecule adsorbed on a device) is translated into an electrical signal, thanks to the changing conductivity of the device. Thermal vibrations of atoms set the upper limit to electron mobility in graphene, which is ~200,000 cm^2/V s [29–30] at room temperature, which should be compared to ~100,000 cm^2/V s [31] in CNTs.

Chen et al. [32–33] showed that although the room temperature limit of mobility in graphene can be as high as 200,000 cm^2/V s, in present-day samples the actual mobility is lower (around 10,000 cm^2/V s) leaving significant room for improvement. Because graphene is only one-atom thick, current samples must be supported by a substrate, typically silicon dioxide. Trapped electrical charges in silicon dioxide (a sort of atomic-scale dirt) can interact with electrons in graphene, which can cause a reduction in mobility.

In addition to being considered as a promising material for applications in chemical and biochemical sensing, its low resistivity and extremely thin nature render graphene a very promising material also for use in thin, mechanically tough, electrically conducting, transparent films that are needed in many applications in electronics (ranging from touch screens to photovoltaic cells).

Apart from its excellent mechanical and electrical properties, graphene exhibits amazing performance as a thermal conductor. Balandin et al. [34] reported values of thermal conductivity for SL graphene at room temperature in the range of 4,840–5,300 W/m K. These extremely high values of thermal conductivity suggest that graphene can outperform CNTs in heat conduction [35–36]. However, when graphene is in contact with a substrate, its thermal transport properties can be significantly affected. Seol et al. [37] showed experimentally that the value of thermal conductivity of monolayer graphene exfoliated on a silicon dioxide support is still as high as about 600 W/m K near room temperature, exceeding those of metals such as copper. It is lower, however, than that of suspended graphene because of phonons leaking across the graphene–support interface and strong interface scattering of flexural modes, which make a large contribution to thermal conductivity for suspended graphene. In general, the superb thermal conduction properties of graphene have established it as an excellent material for thermal management.

Production of graphene-based polymer nanocomposites

In recent years, a variety of processing methods have been proposed for dispersing GS into polymer matrices. Many of these procedures are like those used for other nanocomposite systems (e.g., CNT/polymer nanocomposites) [38], while others apply only to graphene-based polymer nanocomposites.

A crucial step in the production of any polymer nanocomposite is the dispersion of the nanofiller. A well-dispersed state ensures a maximized reinforced surface area, which directly affects all properties of the nanocomposite. Efforts are therefore focused on achieving a well-dispersed, homogeneous system by developing either covalent or noncovalent functionalization of the filler surface, an issue that will be discussed in some more detail in the following sections.

Most polymer/graphene composites are produced today with one of the following three strategies: (1) solvent processing, (2) in situ polymerization and (3) melt processing with each one of them having its own advantages and disadvantages. In the solvent processing method, GS are initially dispersed in a suitable solvent, a procedure typically assisted by ultrasonication [11, 39]. Then, the polymer is added in the solvent/graphene blend, and the solvent is finally removed by evaporation or distillation. It is a rather simple method, used widely to prepare polymer/graphene composites. Its most important drawback is that common organic solvents adsorb on GS stronger than most of the polymers.

In the in situ polymerization method, GS are mixed with the targeted monomers, and by adjusting parameters (such as temperature and pressure), the polymerization reaction proceeds [40–41]. The advantages of the method are twofold: (a) it provides a strong interaction between the polymer matrix and the surface of GS and (b) it leads to highly homogeneous dispersions. However, an increase in the viscosity of the blend is usually recorded as a side effect, which affects processability (especially at high GS loadings).

Melt processing is commercially the most attractive method to produce graphene-based polymer composites. It involves the direct inclusion of GS into the melted polymer using a twin-screw extruder by suitably adjusting parameters such as screw speed, temperature and time [42–43]. Drawbacks of the method include the low density of thermally exfoliated graphene that makes extruder feeding a troublesome task, and the lower degree of dispersion achieved compared to solvent blending. Reduced degree of dispersion typically results in poorer mechanical, electrical and thermal properties.

Properties of graphene-based polymer nanocomposites

In the modern literature, significant improvements in the mechanical, electrical, thermal and barrier properties of graphene-based polymer nanocomposites have been reported, as summarized in several reviews [8–10, 12, 44]. Fang et al. [40], for example, have reported an increase in the tensile strength by 70% and in the Young's modulus by 57% for graphene–polystyrene nanocomposites with polystyrene (PS) chains grafted onto GS by atomic transfer radical polymerization. The experimental studies of Ramanathan et al. [39] and Li and McKenna [45] have shown that behind the extraordinary mechanical properties and the increase in the glass transition temperature (T_g) of poly(methyl methacrylate) (PMMA) nanocomposites filled with functionalized graphene sheets (FGS) are the enhanced interfacial interactions with PMMA chains as driven by oxygen functionalities across the surface of graphene. FGS contain pendant epoxy, hydroxy and/or carboxy groups on their surface, which may form hydrogen bonds with the ester branches of PMMA. However, Liao and coworkers [46–47] have argued that an 80% increase in the PMMA modulus [39] at only 1 wt. % loading of the nanocomposite in FGS, and an increase in T_g by 29 °C [39] at only 0.05 wt.% loading seem unrealistically high. Thus, Liao et al. repeated the experiments carried out by Ramanathan et al. [39] and found an increase of only 25% in the Young's modulus measured [47] and no change in the T_g [46]. These very different observations were attributed to the experimental procedure followed by Ramanathan et al. [39]. In an earlier study [48], significant improvements in the Young's modulus and in ultimate tensile strength had been reported for poly(vinyl alcohol) (PVA) samples enhanced with GO sheets functionalized with PVA chains (PVA chains had been grafted onto the GO surface). The reported enhancement in the mechanical performance reached almost 60% for GO loadings below 0.3 vol.% [48]. For the same nanocomposite, Zhao et al. [49] have reported an improvement of ~150% in the tensile strength and an order of magnitude increase in the Young's modulus for only a 1.8 vol.% loading in graphene.

As already mentioned, graphene-based materials are very promising for the development of new devices with applications in electronics, owing to their high degree of electrical conductivity. In the past, several carbon-based NPs (i.e., carbon filler, carbon nanofibers, expanded graphite, etc.) have been exploited for the production of electrically conductive composites; however, the key advantage of graphene is that the insulator-to-conductor transition (known as the percolation threshold) can be achieved at significantly lower loadings. Production of electrically conductive polymers has been reported [50–51] upon successful GS dispersion in the host polymer matrix. Stankovich et al. [11, 52] determined the percolation threshold for a polystyrene solvent blended with GO to be 0.1 vol.%, perhaps the lowest percolation threshold ever reported. Eda and Chhowalla [53] studied the electrical properties of solution-processed, semiconducting thin films consisting of FGS as the filler and polystyrene as the host material and found that upon increasing the average size of FGS significantly enhanced carrier mobility and thus device performance. This study demonstrated how a commodity plastic can be used to develop low-cost, macroscale thin-film electronics.

Simulations

In addition to experimental efforts, theoretical and computational works have addressed several aspects of the structure–property–processing relationship in graphene or graphene-based nanocomposite materials. Several simulation techniques and approaches have been employed, extending from the quantum level to the atomistic to the mesoscopic and finally to the macroscopic. The findings of these theoretical studies have significantly improved our understanding of the microscopic mechanisms and interactions governing the macroscopically exhibited properties of these new classes of materials.

Using classical ab initio calculations, Van Lier et al. [54] and Liu et al. [55] reported values of graphene's Young's modulus equal to 1.11 and 1.05 TPa, respectively, which are in reasonable agreement with the experimentally measured ones [25]. It is worth mentioning that the computational values were reported in 2000 [54] and 2007 [55], respectively, while the experimental one in 2008 [25]. In a recent study that combined density functional theory calculations and classical molecular dynamics (MD) simulations, Kalosakas and coworkers [56] proposed a new force field specifically for graphene that takes into account only bond stretching (described by a Morse-style potential) and bond bending (described by a nonlinear function containing quadratic and cubic terms) interactions between carbon atoms. The new potential was employed in simulations with model graphene systems subject to uniaxial tension or to hydrostatic compression, yielding a value of 0.95 TPa [56] for graphene's Young's modulus, which is also consistent with the one measured experimentally [25]. The dependence of graphene's Young's modulus on temperature and size of GS was systematically studied by Jiang et al. [57], through MD deformation simulations of SL graphene using progressively larger GS, and an increase in the Young's modulus was observed with increasing GS size. The plateau value was reached for a GS size equal to 25 Å × 25 Å beyond which no further increase was recorded [57]. The MD deformation experiments were carried out at temperatures ranging from 100 to 600 K; a slight increase in the value of the Young's modulus (within the statistical error) was monitored for temperatures up to 500 K, followed by a rapid decrease at higher temperatures.

Ab initio studies of suspended GS are not restricted solely to the estimation of the mechanical properties. Excellent articles have been reported in the literature addressing atom–atom interactions between graphene and popular substrates (e.g., SiC [58] και MoS_2 [59]) that are extensively used for the fabrication of SL graphene. Also important are computational and simulation studies of the mechanical, thermal, barrier and electronic properties of polymers filled with GS. Issues addressed here include microscopic structure, chain conformation and local and terminal dynamics of polymer matrix chains in the presence of graphene. Earlier works focused on the study of the interfacial behavior of graphene-based polymer composites. Awasthi and coworkers [60] carried out atomistic MD simulations with the consistent valence force field to study nanoscale load transfer between polyethylene (PE) and GS and characterize the force-separation behavior between CNTs and a polymer matrix. Separation studies were conducted for opening and sliding modes, and cohesive zone parameters (such as the peak traction and the energy of separation for each mode) were evaluated as a first step toward the development of continuum length-scale micromechanical models for tracking the overall material response by incorporating information about the underlying interfacial interactions. MD simulations have also been employed by Li et al. [61] in their study on the effect of the shape of carbon-based NPs on the viscoelastic properties of a PE matrix. They found that it is the surface-to-volume ratio of the NPs that plays the most important role in the structural, dynamical and viscous properties [61]. More recent MD simulations [62–63] of PMMA/graphene nanocomposite showed strong adhesion of PMMA chains (especially of the side groups) on graphene, and considerably slower segmental and chain mobility in the interfacial area. The MD simulations suggest that local mass density, segmental dynamics and chain terminal relaxation all differ from the bulk behavior up to distances equal to several nanometers from the GS surface. Very similar results have been reported for a different matrix, PE [64], demonstrating large density inhomogeneities due to strong PE chain adsorption on the surfaces of GS, exactly as was reported in the case of PMMA/graphene nanocomposites [62]. Close to graphene, PE chains prefer to stand parallel to the graphene surface [64], and all polymer conformational and dynamic properties [64–65] are significantly affected: (a) the size of polymer chains (as measured by their radius of gyration) increases and (b) their dynamics change dramatically because their orientational relaxation time increases almost by one order of magnitude compared to the bulk value. In a very recent work [66], the effect of GS on the crystallization process of PE, polyvinylidene fluoride (PVDF) and PS oligomers was examined. It was reported that GS tend to act as nucleation sites for the crystallization of PE and PVDF but not for PS, which remains almost amorphous [66]. It was also reported that at high temperatures (e.g., close to 600 K), the crystalline structure of PE is destroyed, a result that is in accordance with the recent MD study of Gulde et al. [67].

Simulations have also addressed polymer nanocomposites enhanced not with pristine GS but with GO. GO is graphene-bearing epoxy, hydroxy and/or carboxy groups on its surface. When GO is used as a nanofiller of a polar (e.g., acrylic) polymer, the interactions between polymer atoms and graphene are intensified due to strong attractive forces that develop between polymer and GO oxygen atoms. Lv et al. [68] examined two different polymers as the host matrix, PMMA and PE, and concluded that with increasing concentration in carboxyl content, the interaction energy between polymer chains and GO sheets decreased (it became more attractive), followed by a significant increase in the value of the shear stress. However, an upper bound in the concentration was found, beyond which no further change in the values of these two properties was observed. This was explained as a saturation effect: high concentrations of functional groups strengthen the interactions between the surface functionalization groups themselves, thus no space is left for interactions with the surrounding polymer chains. Karatasos and Kritikos [69] have reported a strong increase (by 38 °C) of the T_g of GO-based poly(acrylic acid) (PAA) nanocomposites compared to pure polymer (PAA). This strong increase was explained by the strong adsorption of PAA chains onto the surface of GO sheets facilitated by the hydrogen bonds that develop between the hydroxyl groups of GO and the oxygen atoms of PAA branches. Earlier, Xue et al. [70] had studied the T_g of PMMA matrices enhanced either with pristine GS or with GS modified with -COOH and -NH_2 groups. By employing classical MD simulations, they observed a 30 °C increase in the T_g of the PMMA/GS system, whereas for the nanocomposites with the modified GS, the shift was higher (40 °C) [70]. More recently, Azimi et al. [71] showed that the dynamics of a polar (e.g., PVA) matrix is affected more by the presence of GO than by the presence of GS, whereas for an apolar polymer matrix (e.g., poly(propylene)) the effect of the two types of graphene is almost the same. That less polymer is adsorbed on GO than on pristine graphene can be explained by the roughness of GO particles due to OH- and -O- groups on their surface and agrees with a recent detailed MD study by Skountzos et al. [72]. However, the strength of interactions between oxygen atoms of the polar polymer and of GO particles is so strong that despite the smaller adsorbed amount on GO (in comparison to pristine GS), the dynamics of the PVA/GO system is considerably slower than the dynamics of the PS/GS nanocomposite.

In the last years, significant progress has been made in predicting the unique mechanical properties of graphene-based polymer nanocomposites through detailed atomistic-level simulations and understanding the underlying molecular mechanisms behind these properties [55, 72–76]. A typical example is the atomistic simulation work of Skountzos et al. [72] on the effect of pristine graphene and GO on the structure, conformation and mechanical properties of a syndiotactic PMMA (sPMMA) matrix. The atomistic simulations predicted a significant enhancement of all elastic constants (Young's, shear and bulk moduli and Poisson's ratio), especially for the PMMA/GO nanocomposites, which was attributed to the hydrogen bonds that develop between the oxygen atoms at the branches of PMMA chains and the OH- and -O- groups on the surface of GO particles. Wang et al. [76] have studied the effect of surface functionalization and graphene size on the interfacial load transfer in graphene–PE nanocomposites (by employing the ab initio polymer-consistent force field, PCFF) and showed that oxygen-FGS lead to larger interfacial shear force (than hydrogen-functionalized or pristine ones) during the pull-out process. Increasing the oxygen coverage and graphene size enhanced the interfacial shear force, but further increasing the oxygen coverage to about 7% led to a saturation in the interfacial shear force. Liu et al. [75] studied the effect of shape of carbon-based NPs on the toughening efficiency of a PE matrix by applying pure tensile strain experiments in MD simulations with several model systems. At the same wt. % loading in either CNTs, or fullerenes or GS, they found that the highest toughening was observed for the PE/graphene system; this should be related to the increased surface area that graphene offers for PE chain adsorption. This effect was examined in both the rubbery and glassy states (i.e., above and below the glass transition temperature of PE) and it was found that the nanocomposite in the rubbery state was tougher than the corresponding glassy material [75]. More recently, in a systematic study, Lin et al. [73] investigated how temperature and GS loading affect the mechanical performance of model PMMA/GS nanocomposites. The general trend was that Young's and shear moduli increase with increasing GS loading but decrease with increasing temperature while the temperature dependence of the two moduli is stronger in the nanocomposites with the higher concentration in GS. Comparison of their simulation results with the equations based on the simple rule of mixing for the mechanical properties of microfiber-reinforced composites revealed that the latter do not apply for graphene-reinforced nanocomposites; thus, further research is needed to address this issue [73].

Graphene dispersion

Clearly, among all NPs considered as polymer reinforcing agents, graphene is the most promising due to its outstanding features. However, because of its high specific surface area and tendency for self-adhesion (driven by the very strong van der Waals forces and π–π interactions), GS tend to agglomerate in the form of multilayer graphitic structures. Significant research has thus been undertaken in the last years to understand and improve GS dispersion, and thus realize its unique properties in practice. The degree of GS dispersion in nanocomposites affects primarily their mechanical performance. Indeed, several studies have shown that the mechanical properties of graphene-based polymer nanocomposites increase as their content in SL graphene increases [39, 77–78]. Motivated by these findings, in the next paragraphs of this chapter we provide a brief overview of the methods employed nowadays, typically based on covalent or noncovalent graphene functionalization, to achieve homogeneous GS dispersions [79–80].

Techniques based on covalent functionalization make use of small molecules or entire macromolecules that are attached on the surface of GS through covalent bonding. Then, owing to strong chemical interactions developing between the grafted groups and the GS, very homogeneous and stable dispersions can be obtained. However, covalent functionalization disrupts the perfect crystal structure of graphene, which can affect many of its outstanding properties, especially the electrical ones. An attractive alternative is noncovalent functionalization. This is based on the use of functional groups that are attached on the surface of graphene through favorable π–π interactions without disturbing its electronic network. Noncovalent methods are by default nondestructive; however, the forces that develop between the wrapping molecules and the graphene surface are much weaker compared to those in the case of covalent functionalization; this can limit the degree of dispersion eventually achieved.

Covalent functionalization of graphene has been exploited quite widely in the literature for the improvement of the mechanical properties of graphene-based nanocomposites [81–85]. Using mostly peripheral ester linkages that are extensively used in tissue engineering applications, Sayyar et al. [84] managed to successfully link polycarbonate to GO. The fabricated nanocomposite had a very homogeneous and stable GS dispersion and exhibited several good properties: improved Young's modulus, improved tensile strength and high electrical conductivity (~14 orders of magnitude compared to the conductivity of the pure bulk polymer). In an earlier study, Cheng and coworkers [81] had produced PVA nanocomposites by incorporating PVA-grafted GO fillers, which resulted in an enhancement of Young's modulus by 150% and of tensile strength by 88%, for 1 wt.% loading of the nanocomposite in GO.

The tendency of GS to agglomerate and form graphitic structures is witnessed most easily in atomistic or coarse-grained simulations. Park and Aluru [86] have reported GS self-assembly in atomistic MD simulations of GS in water solutions, but this was probably an expected result due to hydrophobic nature of graphene. In a later study, Li and coworkers [61] found that several carbon allotropes (buckyballs, graphenes, etc.) agglomerate in a PE matrix, thus lowering the degree of their interaction (contacts) with the host matrix. Ju et al. [87] studied the degree of GS miscibility in PMMA as a function of NP volume fraction with the help of dissipative particle dynamics (DPD) simulations using different values of the DPD-repulsive interaction parameter to effectively account for surface functionalization. GS agglomeration has also been reported by Karatasos et al. [88–89] in all-atom MD simulations of linear and hyperbranched polymer matrices. Guo et al. [90] have studied the effect of GS agglomeration on the mechanical properties of polymer nanocomposites by examining cases where several GS were intercalated (polymer chains were "sandwiched" between two consecutive GS) or stacked (in the form of graphite-like structure) in the sample. By investigating their mechanical behavior in a series of uniaxial tension experiments, they concluded that intercalated systems exhibit a higher storage modulus compared to stacked ones due to stronger interactions between the polymer and the GS resulting from the higher surface-to-volume ratio in the intercalated design. More recently [91], atomistic MD simulations were combined with advanced chemistry techniques to produce graphene-based PMMA solutions and nanocomposites of high degree of GS dispersion. Instead of graphene covalent functionalization or incorporation of small molecules as dispersing agents in the solutions, in the new method a fraction of PMMA chains were covalently functionalized by a procedure that allows adding pyrene molecules preferentially to the two ends of the polymer chain [91]. The accompanying MD simulations provided very useful information about the interaction of these functionalized PMMA chains with GS and how these are kept well separated in the polymer matrix, thereby achieving a stable, highly homogeneous dispersion.

5.2 Simulation methods at the atomistic level: the molecular dynamics method

In the last few decades, molecular simulations have emerged as an excellent tool for explaining many of the microscopic mechanisms behind the measured macroscopic properties of complex materials, also for connecting the predictions of theoretical models with experimentally measured properties. Being able, in particular, to predict important material properties from the chemical composition and molecular architecture of the chemical formulation is of paramount importance because it can guide design efforts both at the level of materials synthesis and at the level of materials processing. With their inherent potential to predict important physical properties directly from the underlying atomistic structure, molecular simulations are considered today as very useful, virtual experiments that can replace in many cases actual laboratory measurements.

Molecular simulations encompass two main techniques: MD and Monte Carlo (MC) [92–93]. MD is based on the solution of Newton's equations of motion in an appropriate statistical ensemble. These are numerically integrated, and the result of integration yields the positions and velocities of each atomistic unit in the system in time (this is also known as the trajectory of the system in phase space). In this way, we can monitor how the model system involves in time (under the macroscopic constraints imposed by the statistical ensemble in which the simulation is carried out) and thus extract information about its thermodynamic, structural, conformational, dynamic and rheological properties. To run an MD simulation, we need initial conditions, namely atomic positions and atomic velocities at zero time. The trajectory followed by the system depends then crucially on the potential energy function describing intra- and intermolecular interactions.

In contrast to MD which is a deterministic method, MC is based on the design of (artificial or even unphysical) moves to sample new states that are selected randomly with the help of an appropriate acceptance criterion. Thanks, in particular, to the design of some very clever moves, MC helps the system tunnel through large potential energy barriers, and this can accelerate system equilibration by orders of magnitude compared to MD. The major drawback of the method is that it does not offer any dynamic information about the system. This happens because the system evolves stochastically in configuration space based on predefined acceptance criterion for each attempted move. Indeed, in a MC simulation, only atomic positions matter. Our discussion in the remainder of this chapter is devoted solely to the MD method.

In general, the complexity of polymers prohibits the analytical solution of the corresponding statisticomechanical problem. Molecular simulations overcome this by solving the corresponding problem numerically, given a mathematical model for the molecular geometry and description of interatomic interactions. Two general types of force fields are typically employed for the molecular simulation of polymers: Explicit atom (EA) models where every atom is considered as an independent interaction site or entity, and united atom (UA) models where hydrogen atoms are neglected by considering a relatively larger, spherically interacting particle that embodies the contributions of both the hydrogen atoms and of the atom (e.g., the carbon) to which the hydrogens are bonded. EA models offer a very detailed and accurate description of the problem. Their main drawback is that, for the same number of total molecules considered in the simulation box, they involve almost twice as many interacting sites as the corresponding UA model; as a result, molecular simulations with EA models are significantly more CPU time demanding than simulations with UA models.

Given the type of molecular model adopted, an initial configuration for the system and initial velocities for all atoms (assigned according to the Maxwell–Boltzmann distribution), what remains to be defined next is the potential energy function describing interatomic interactions. In general, these are distinguished between bonded and nonbonded. Bonded interactions typically include contributions from

- bond length stretching (Figure 5.2a),
- bond angle bending (Figure 5.2b),
- proper dihedral angles (Figure 5.2c) and
- improper dihedral angles (Figure 5.2d).

Nonbonded interactions, on the other hand, include

- inter- and intramolecular van der Waals interactions (Figure 5.2e) and
- inter- and intramolecular Coulomb interactions (Figure 5.2f).

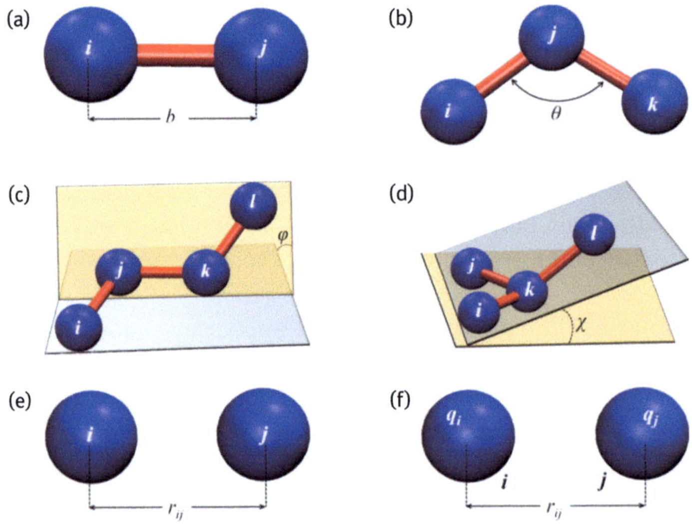

Figure 5.2: Schematic representations of (a) bond length stretching, (b) bond angle bending, (c) proper dihedral, (d) improper dihedral, (e) van der Waals and (f) Coulomb interactions to the potential energy of the system.

A typical mathematical form of the total potential energy function [94] is

$$U_{\text{pot}} = \overbrace{\sum_{\text{bonds}} k_b(b-b_0)^2 + \sum_{\text{angles}} k_\theta(\theta-\theta_0)^2 + \sum_{\text{dihedrals}} k_\phi[1+\cos(n\varphi-\delta)] + \sum_{\text{impropers}} k_\chi(\chi-\chi_0)^2}^{\text{Bonded energetic terms}} \\ + \underbrace{\sum_{\text{vdW}} 4\varepsilon_{ij}\left[\left(\frac{\sigma_{ij}}{r_{ij}}\right)^{12} - \left(\frac{\sigma_{ij}}{r_{ij}}\right)^{6}\right] + \sum_{\text{Coulomb}} \frac{q_i q_j}{4\pi\varepsilon_0 r_{ij}}}_{\text{Nonbonded energetic terms}} \tag{5.1}$$

where the first and second sums on the right-hand side correspond to a harmonic potential for the description of bond length stretching and bond angle bending contributions, the third term is a cosine function describing contributions from proper dihedral angles (formed by four consecutive atoms along the chain) and the fourth term is a harmonic function describing contributions due to improper dihedral (or out-of-plane) angles formed by four nonconsecutive atoms along the chain (they help maintain the stereoregularity of the fourth atom with respect to the plane of the other three atoms at branch points). The fifth term in eq. (5.1) corresponds to the total van der Waals energy of the system based on a typical 12–6 Lennard–Jones (LJ) potential while the last term is the total electrostatic energy due to Coulomb interactions between charged atoms.

Force fields described by eq. (5.1) are called classical force fields. Some examples are the DREIDING [94], the OPLS [95], the CHARMM [96] and the AMBER [97].

The functional form of eq. (5.1) includes the minimum number of terms (bonded and nonbonded), providing a satisfactory description of the total potential energy of the system. However, during the last 30 years, more detailed atomistic-level force fields have been developed based on ab initio calculations parameterized on the basis of a large body of experimental data for several organic compounds consisting of carbon, oxygen, hydrogen, nitrogen and sulfur atoms. These second-generation potentials are called class II potentials and differ from the classical potentials in that they include extra (coupling) terms between the various contributions in eq. (5.1), thus affording a more accurate representation of the total potential energy. A typical expression reads [98]

$$\begin{aligned}
U_{\text{pot}} = {} & \left.\begin{aligned}
& \sum_{b} k_2^b (b-b_0)^2 + k_3^b (b-b_0)^3 + k_4^b (b-b_0)^4 \\
& + \sum_{\theta} k_2^{\theta} (\theta-\theta_0)^2 + k_3^{\theta} (\theta-\theta_0)^3 + k_4^{\theta} (\theta-\theta_0)^4 \\
& + \sum_{\varphi} k_1^{\phi} [1-\cos\varphi] + k_2^{\phi} [1-\cos 2\varphi] + k_3^{\phi} [1-\cos 3\varphi] \\
& + \sum_{\chi} k_{\chi} (\chi-\chi_0)^2
\end{aligned}\right\} \text{bonded energetic terms} \\
& \left.\begin{aligned}
& + \sum_{b,b'} k_{b,b'} (b-b_0)(b'-b'_0) + \sum_{b,\theta} k_{b,\theta} (b-b_0)(\theta-\theta_0) \\
& + \sum_{b,\varphi} (b-b_0) \left(\sum_{n=1}^{3} k_n^{b,\varphi} \cos(n\varphi) \right) \\
& + \sum_{\theta,\theta'} k_{\theta,\theta'} (\theta-\theta_0)\left(\theta'-\theta'_0\right) + \sum_{\theta,\varphi} (\theta-\theta_0) \left(\sum_{n=1}^{3} k_n^{\theta,\varphi} \cos(n\varphi) \right) \\
& + \sum_{\theta,\theta',\varphi} k_{\theta,\theta',\varphi} (\theta-\theta_0)\left(\theta'-\theta'_0\right) \cos\varphi
\end{aligned}\right\} \begin{array}{l}\text{cross} - \text{coupling} \\ \text{terms between} \\ \text{bonded interactions}\end{array} \\
& \underbrace{+ \sum_{\text{vdW}} \varepsilon_{ij} \left[2\left(\frac{r_{ij}^0}{r_{ij}}\right)^9 - 3\left(\frac{r_{ij}^0}{r_{ij}}\right)^6 \right] + \sum_{\text{Coulomb}} \frac{q_i q_j}{4\pi\varepsilon_0 r_{ij}}}_{\text{Nonbonded energetic terms}}
\end{aligned} \tag{5.2}$$

In general, simulations with class II potentials yield excellent predictions for the majority of the physical properties of system. Their only drawback is that they are computationally more intensive compared to the simpler, classical force fields. Typical examples of class II potentials are the COMPASS [98], the CFF [99] and the PCFF.

Having available a mathematical expression for the calculation of the potential energy of the system under study, the next step is to integrate Newton's equations of motion in the relevant statistical ensemble to sample system configurations in phase space. From this point of view, MD simulations are in many aspects similar to real experiments. When a real experiment is performed, a sample of the material is prepared and connected to the instrument (a thermometer, a manometer, a viscometer, a spectrometer, etc.) to measure the desired property. The value of a specific property is computed as the average over many different measurements to increase statistical accuracy. In a similar way, in an MD simulation of N interacting atoms described by the potential energy function U_{pot}, the solution of Newton's equations of motion provides snapshots of the system in real time, from the analysis of which one can extract predictions of the relevant physical properties. Newton's equations of motion read

$$m_i \ddot{\boldsymbol{r}}_i = \boldsymbol{F}_i, \quad i = 1, 2, \ldots, N \tag{5.3}$$

or, equivalently,

$$m_i \frac{d^2 \boldsymbol{r}_i}{dt^2} = -\frac{\partial U_{pot}(\boldsymbol{r}_1, \boldsymbol{r}_2, \ldots, \boldsymbol{r}_N)}{\partial \boldsymbol{r}_i}, \quad i = 1, 2, \ldots, N \tag{5.4}$$

where i denotes any atom in the system, m_i its mass, $\boldsymbol{r}_i$ its position vector, $\boldsymbol{F}_i$ the force acting on it, and t the time. Solving the $3N$ (where N stands for the total number of atomistic units) second-order differential equations (eq. (5.4)), we track the time evolution of the system and obtain its trajectory, that is, the positions and velocities of all its atomistic units in time. Applying then basic principles of statistical mechanics [100], we can obtain estimates of several properties of interest (e.g., thermodynamic, structural, conformational and dynamic), which can be directly compared to experimental data. Figure 5.3 provides the basic structure of a typical MD algorithm.

To solve the equations of motion, a numerical integrator must be employed, which should guarantee efficiency, stability and accuracy. To this, the algorithm must:

- not require an expensively large number of force evaluations per time step (thus, very popular techniques for the solution of ordinary differential equations such as the fourth-order Runge–Kutta method become inappropriate);
- permit the use of a large time step;
- be fast;
- require little memory;
- satisfy the energy conservation law.

Over the years, several algorithms have been proposed that satisfy the above criteria. The most popular are the gear predictor-corrector methods [92], the Verlet algorithms [93, 101] and the multiple time step methods (such as the *r*eversible *RE*ference *S*ystem *P*ropagator *A*lgorithm or rRESPA [102]). In addition, thanks to the rapid growth of computing power, the development of robust parallelization techniques (based on domain [103–105] and force [106–107] decomposition and CUDA programming) and the availability of supercomputing infrastructures (computer machines with millions of CPU processors), highly complicated systems containing hundreds of thousands or even millions of interacting particles (corresponding to simulation boxes with edge

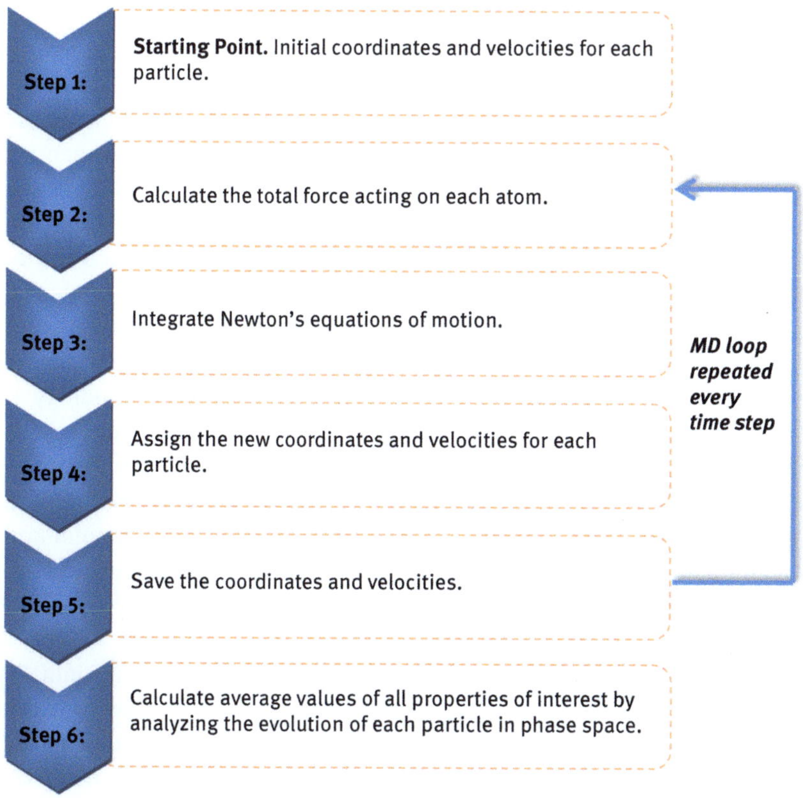

Figure 5.3: Simplified flow diagram of a typical MD algorithm.

lengths equal to hundreds of nanometers) can be simulated today in full atomistic detail for times up to several microseconds. Also, highly parallelized, user-friendly MD simulation packages are commercially available or freely distributed; we mention:

- *L*arge-scale *A*tomic/*M*olecular *M*assively *P*arallel *S*imulator, *LAMMPS* [107]
- *GRO*ningen *MA*chine for *C*hemical *S*imulations, *GROMACS* [108]
- *A*ssisted *M*odel *B*uilding with *E*nergy *R*efinement, *AMBER*
- *NA*noscale *M*olecular *D*ynamics, *NAMD* [109]
- *C*hemistry at *HA*rvard *M*acromolecular *M*echanics, *CHARMM*

5.3 Atomistic MD simulation of graphene-based PMMA nanocomposites

As a model polymer, we have chosen PMMA whose nanocomposites with GS have been studied in detail over the years with several techniques. For example,

Ramanathan et al. [39] used sonication to break rigid nanoplatelets of expanded graphite apart into thinner platelets, which were dispersed next in a PMMA solution using high-speed shearing methods. This led to an increase in the T_g by 29 °C at 0.05 wt.% loading of the matrix in FGS, and up to an 80% enhancement of the Young's modulus at 1 wt.% loading in FGS. Similar observations have been reported by Li and McKenna [45] for GO/PMMA nanocomposites. One reason for the extraordinary mechanical properties of FGS-PMMA nanocomposites is the enhanced interfacial interactions of oxygen functionalities across the surface of graphene with PMMA chains. FGS contain pendant hydroxyl groups, which may form hydrogen bonds with the carbonyl groups of PMMA. Additional enhancement comes from the nanoscale surface roughness of FGS, the defects caused during thermal exfoliation of the precursor graphite oxide and their wrinkled topology at the nanoscale due to their extremely small thickness. These can enhance mechanical interlocking with the polymer chains, which also leads to better adhesion.

Atomistic MD simulations [62–63] have shown strong adhesion of PMMA chains (especially of its side groups) on graphene and considerably slower segmental and chain mobility in the interfacial area. According to simulation data, local mass density, segmental dynamics and chain terminal relaxation differ from the bulk behavior up to several nanometers from the graphene surface.

In the remaining of this chapter, we will focus on a methodology [110], initially proposed for a simpler class of systems (glassy vinyl polymers such as polypropylene and polystyrene) [110], which allows the determination of the mechanical properties of PMMA nanocomposites filled with GS (functionalized or nonfunctionalized), based on small-strain deformation experiments on the computer of microscopically detailed model structures. The procedure involves several modeling and mathematical steps and allows computing the elastic constants (Young's modulus E, bulk modulus B, shear modulus G and Poisson's ratio v) of a polymeric glass under the assumption that vibrational contributions of the hard degrees of freedom are not significant; as a result, estimates of the elastic constants can be obtained by computing changes only in the total potential energy of static microscopic structures subjected to simple deformation modes. For glassy atactic polypropylene for which the method was first developed and implemented by Theodorou and Suter [110], elastic constants were predicted within 15% of the experimentally measured values.

All the simulations have been performed with an all-atom force field, allowing for a direct comparison of the computed with available experimental data. We chose DREIDING [95] because it combines simplicity with accuracy (for acrylic polymers). Since DREIDING [95] does not provide information about the values of partial charges of PMMA atoms, these were borrowed by the OPLS-AA [96] force field. Additional technical details (such as the parameter values of all bonded and nonbonded interactions describing intra- and interatomic contributions to potential energy) can be found in two published articles [72, 91]. Figure 5.4a–d provides typical

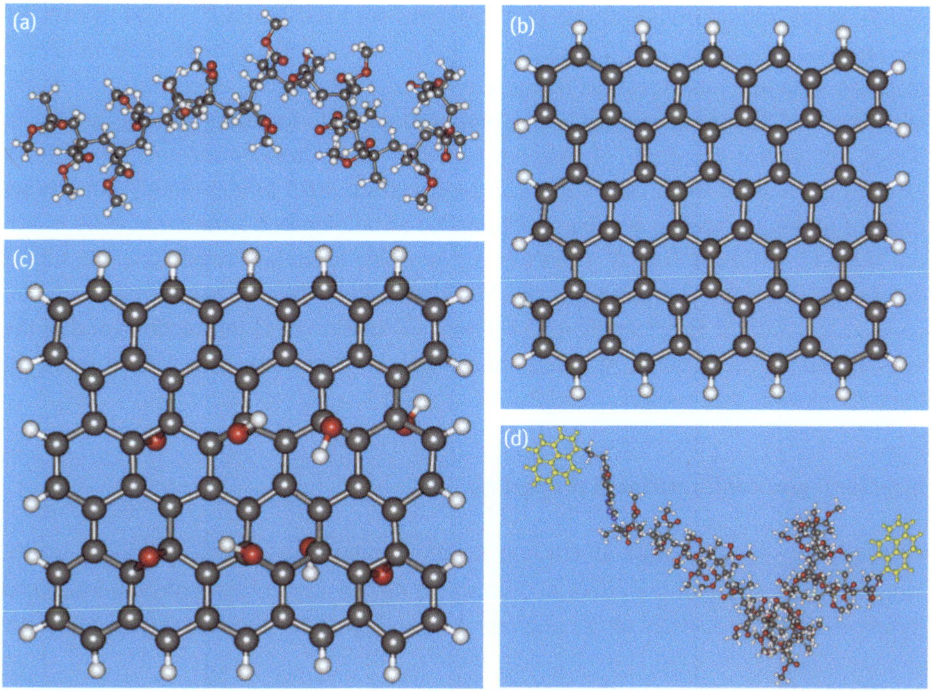

Figure 5.4: Typical atomistic structures of (a) an sPMMA chain, (b) a nonfunctionalized graphene sheet (GS), (c) a functionalized graphene sheet (FGS) and (d) a functionalized (py-sPMMA-py) chain.

atomistic structures of a PMMA chain, a GS, a GO and a functionalized PMMA chain with pyrene groups added to its two ends.

Systems simulated and simulation strategy

We focus on sPMMA, at P = 1 atm. The simulations were performed with strictly monodisperse samples with the model system consisting of 27 chains of degree of polymerization X = 15 (corresponding to a molecular weight of 1,503.75 g/mol). Unfunctionalized and functionalized GS had lateral dimensions 12 Å × 12 Å. Three model systems were studied: (a) the neat sPMMA matrix (no GS added; it will be denoted as sPMMA in the following), (b) its nanocomposite with three unfunctionalized monolayer GS (it will be denoted as GS-sPMMA in the following) corresponding to 5.67 wt.% concentration in GS and (c) its nanocomposite with three functionalized monolayer GS or GO (it will be denoted as FGS-sPMMA in the following) corresponding to 6.54 wt.% concentration in GO. The surface concentration of GO in hydroxyl (-OH) and epoxy (-O-) groups in the latter system was chosen to match as closely as possible the experimentally determined concentration reported by Ramanathan et al. [39] through elemental analysis.

To build initial configurations of all systems we used MAPS [111] and to execute the MD simulations we used LAMMPS [107]. All initial configurations were subjected to static structure optimization using a molecular mechanics algorithm to remove overlaps, and the resulting minimum potential energy structures were annealed to 500 K for several hundreds of nanoseconds to render them completely amorphous prior to quenching them down to room temperature, also to completely equilibrate them at all length scales. We used rectangular parallelepiped simulation cells of initial sides 40 Å × 40 Å × 40 Å subject to full periodic boundary conditions. Technical details regarding the execution of the MD simulations (type of thermostat–barostat used, calculation of electrostatic interactions, calculation of LJ interactions and of the tail corrections, integration of equations of motion, time step, etc.) can be found in the two relevant publications [72, 91].

Structural and conformational properties

From the equilibration runs at T = 500 K, we calculated several properties that provided a good picture of the effect of GS and GO on the structural, conformational and thermodynamic properties of the polymer matrix. Figure 5.5 shows a typical atomistic

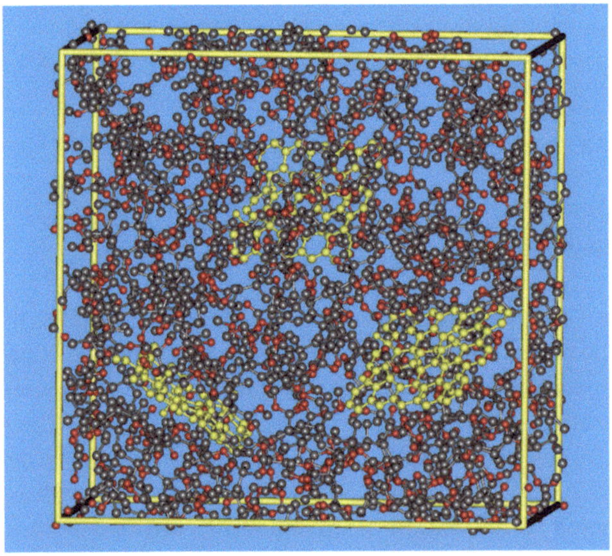

Figure 5.5: Typical atomistic snapshot from the simulation with the FGS-sPMMA nanocomposite at 6.54 wt.% loading. The simulation cell contains 27 PMMA chains and three FGS with five hydroxyl groups and three oxygen atoms on their surface. Initial cell dimensions 40 Å × 40 Å × 40 Å. Carbon (sPMMA), carbon (FGS) and oxygen atoms are represented with gray, yellow and red colors, respectively. Hydrogen atoms have been omitted for clarity.

configuration of the FGS-sPMMA nanocomposite at the end of the MD simulation with this system at T = 500 K. A first quantity that can be easily calculated from an MD simulation in the *NPT* ensemble is the density ρ. Our predictions are ρ = 1.065 g/cm^3 for the sPMMA, ρ = 1.082 g/cm^3 for the GS-sPMMA and ρ = 1.091 g/cm^3 for the FGS-sPMMA system. The experimentally determined value for infinite molecular weight PMMA at the same temperature (500 K) is 1.072 g/cm^3 [112].

We also calculated (see Figure 5.6) the variation of polymer mass density with distance from a GS.

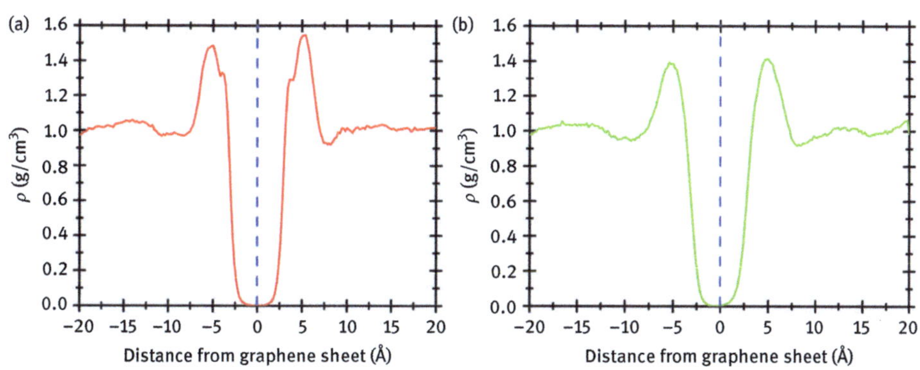

Figure 5.6: Local mass density normal to graphene sheets as obtained from the present MD simulations (T = 500 K, P = 1 atm) for the cases of pristine graphene sheets (a) and graphene oxide (b). The blue dashed line at the zero value of the horizontal axis denotes the average position of the graphene sheet midplane.

The high-density values (up to ~30% compared to the bulk density value) observed at distances up to ~5 Å from the surface of the GS indicate that sPMMA chains adsorb strongly on GS (either unfunctionalized GS or GO sheets). An example of a typical conformation of an adsorbed sPMMA chain on a GO is displayed in Figure 5.7.

An interesting point is that the local mass density of sPMMA is enhanced less in the FGS-sPMMA system than in the GS-sPMMA system. This should be attributed to the relative roughness of GO sheets (compared to the perfectly smooth surfaces of pristine GS) because of the presence of the characteristic -O- and -OH groups, which leads to the adsorption of less sPMMA molecules on GO than on GS. However, as we will see below, the -OH groups present on the surface of GO help the system develop a non-negligible number of hydrogen bonds with the oxygen atoms of sPMMA ester branches; this will be shown to have a strong impact on the overall mechanical performance of the nanocomposite. Figure 5.8 shows an example of a GO NP, which has developed two hydrogen bonds with one sPMMA chain on its one side and one hydrogen bond with another chain on its other side. The hydrogen bonds are highlighted with dashed circles in the figure. We clarify that the formation

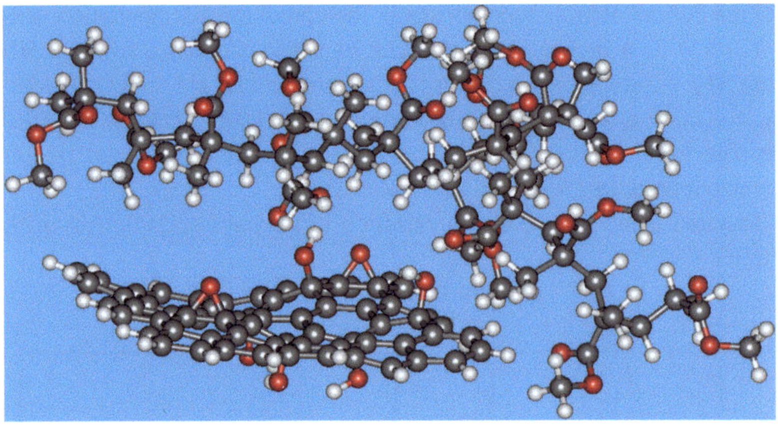

Figure 5.7: Atomistic snapshot from the MD simulation at $T = 500$ K showing a typical sPMMA chain configuration next to a GO.

of hydrogen bonds is not imposed directly in our simulations, but it is the indirect result of the employed force field (particularly of the partial charges assigned to the various atoms).

Mechanical properties

For the estimation of the mechanical properties of the simulated systems, we followed the methodology first proposed by Theodorou and Suter [110] for an amorphous glassy polymer, to which the reader is kindly referred for more details. The method involves the selection of several (about 15) completely equilibrated configurations of the system, which are then submitted to deformation experiments, from which one can calculate in a rigorous way the elastic properties of the sample. At the temperature and pressure conditions of interest here (T = 300 K and P = 1 atm), sPMMA and its GS- or GO-nanocomposites are in the glassy state, implying that one cannot directly use MD to sample well-equilibrated system configurations because the simulations will be nonergodic (the system will be trapped in configurations characterized by a local minimum in their potential energy). One way to overcome this is to equilibrate the system at a higher temperature (above the melting point), where equilibration is much easier to achieve, select a good number of relaxed configurations from this simulation and subject them to cooling runs down to the lower temperature (T = 300 K), followed by a short MD run for the density and local structure to equilibrate further. The resulting glassy structures will then be good candidates to use in the subsequent computational deformation experiments for the estimation of the elastic constants (Young's, bulk, shear moduli and Poisson's ratio). The results can be significantly improved by averaging over all configurations subjected to deformation.

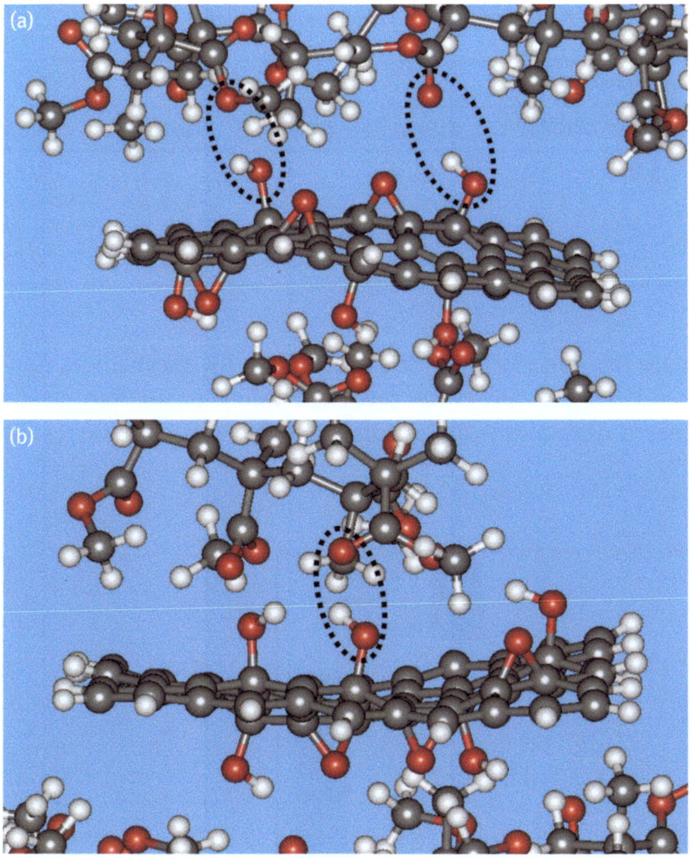

Figure 5.8: An example of a situation where three hydrogen bonds develop between a GO and the surrounding sPMMA chains. We can observe the formation of two hydrogen bonds with the same chain on the one side of the GO (a) and of one hydrogen bond with a different chain on the other side of the GO (b).

Applying the methodology [110] to the systems studied here, we obtained the results shown in Table 5.1. From the numerical data presented in Table 5.1, we can draw several conclusions:

1) The predicted values of the Young's modulus, shear modulus, bulk modulus and Poisson's ratio for the pure sPMMA system are in an excellent quantitative agreement with reported experimental values, which are summarized in ref. 72.
2) A significant enhancement of the mechanical properties of both types of nanocomposites (GS-PMMA and FGS-PMMA) is observed. This is more pronounced for the FGS-based ones, which should be attributed to the development of hydrogen bonds between the filler (GO) and the polar chains (sPMMA).

Table 5.1: Predicted elastic constants for all simulated systems from our study.

System	Lamé constants		Elastic constants			
	μ (GPa)	λ (GPa)	E (GPa)	B (GPa)	G (GPa)	v
Pure sPMMA	1.3	3.4	3.4	4.2	1.3	0.36
Experimental values	–	–	2.07–6.2	5.9–6.17	1.78–2.3	0.34–0.36
GS-sPMMA	1.4	3.7	3.8	4.7	1.4	0.36
FGS-sPMMA	2.3	3.5	6.0	5.0	2.3	0.31

To further appreciate the effect of graphene and GO on the mechanical reinforcement of PMMA, we have normalized the predicted values of the four elastic constants with the values corresponding to the pure sPMMA matrix, and the results are shown in Figure 5.9.

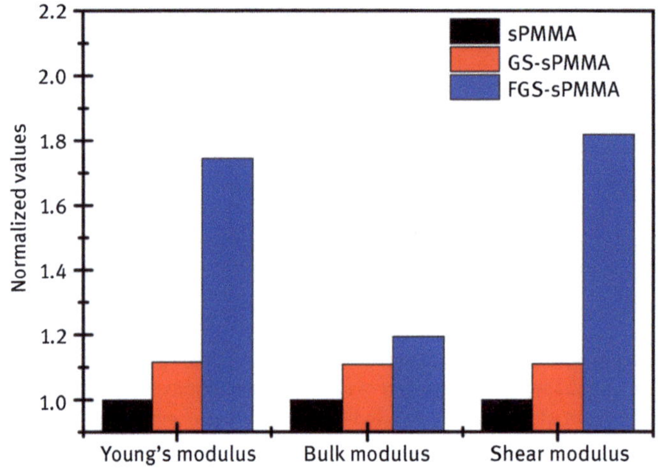

Figure 5.9: Summary of property improvement for the elastic constants of GS-sPMMA and FGS-sPMMA nanocomposites (T = 300 K, P = 1 atm). Numerical values have been normalized with the corresponding value of the neat PMMA matrix at the same conditions (E = 3.4 GPa, B = 4.2 GPa, G = 1.3 GPa and v = 0.36).

Overall, our results are in a good qualitative agreement with the experimental work of Ramanathan and coworkers [39] who reported an increase in the Young's modulus of ~80% in PMMA samples modified with FGS. However, a direct comparison of our simulation results with the experimental data is difficult to make because the molecular weight of the polymer matrix and the size of graphene flakes used in the experimental measurements are too large to address with atomistic MD simulations.

Graphene agglomeration in the PMMA matrix and how to prevent it

The homogeneous dispersion of GS in a polymer matrix is perhaps the most important and challenging issue in the fabrication of graphene-based polymer nanocomposites. In principle, GS fine dispersion offers larger areas for the effective adsorption of polymer chains, thus also for the better improvement of the properties of the nanocomposites.

All calculations reported in the previous section were carried out with model systems characterized by a uniform distribution of GS in the polymer matrix. However, several times during the MD equilibrations at T = 500 K, GS agglomeration was observed to occur at long times, as GS exhibited a strong tendency to come close to each other and form graphitic (π–π stacking) structures. To gain a better understanding of such a phenomenon, we conducted an additional simulation study with a larger system containing 100 atactic PMMA (aPMMA) chains with degree of polymerization X = 30 and six GS of size 60 Å × 60 Å (system 1), and we monitored the time evolution of the positions of the six GS inside the simulation box. The results are shown in Figure 5.10a, where aPMMA chains have been omitted for clarity. We see that already from the first two nanoseconds of the simulation, a pair of GS have come close to each other to form an agglomerate.

To overcome the problem of GS agglomeration, we proposed [91] a novel methodology, which relies on the functionalization not of GS but of a good fraction of PMMA matrix chains by adding pyrene groups to their ends. The functionalized chains are noted as py-PMMA-py (see Figure 5.4d). The key idea is that pyrene groups adsorb strongly on the surface of GS due to very favorable π–π interactions developing between their four benzene rings and the corresponding hexagonal structures of GS; then, the intervening polymer mass between the two GS prevents them from approaching one or the other, thus self-assembly is avoided. To test the idea, we repeated the simulation of system 1 at the same temperature and pressure conditions (T = 500 K and P = 1 atm) by replacing approximately 40% of the aPMMA chains in the matrix with functionalized py-PMMA-py chains (system 2). The time evolution of the positions of GS for such a system is shown in Figure 5.10b. To enable a one-to-one comparison between the two systems (system 1 and system 2), the initial positions and initial velocities of all atoms in GS in the simulation with system 2 matched exactly those in the simulation with system 1. Then, according to Figure 5.10b, for the entire duration of the simulation (~500 ns), the six GS remained homogeneously dispersed in the polymer matrix. It is also clear that many of the pyrene groups of the functionalized py-PMMA-py chains were adsorbed strongly on the surface of GS.

To shed additional light in our MD findings, we calculated the time evolution of the distances between the centers of mass of all GS pairs in the two simulated systems. The results are shown in Figure 5.11a and b for system 1 and system 2, respectively. For system 1, we see that the distances between GS in pairs 1–4, 2–3 and 5–6 suddenly dropped down to 3.4 Å at t = 3, 10 and 120 ns, respectively, which is the

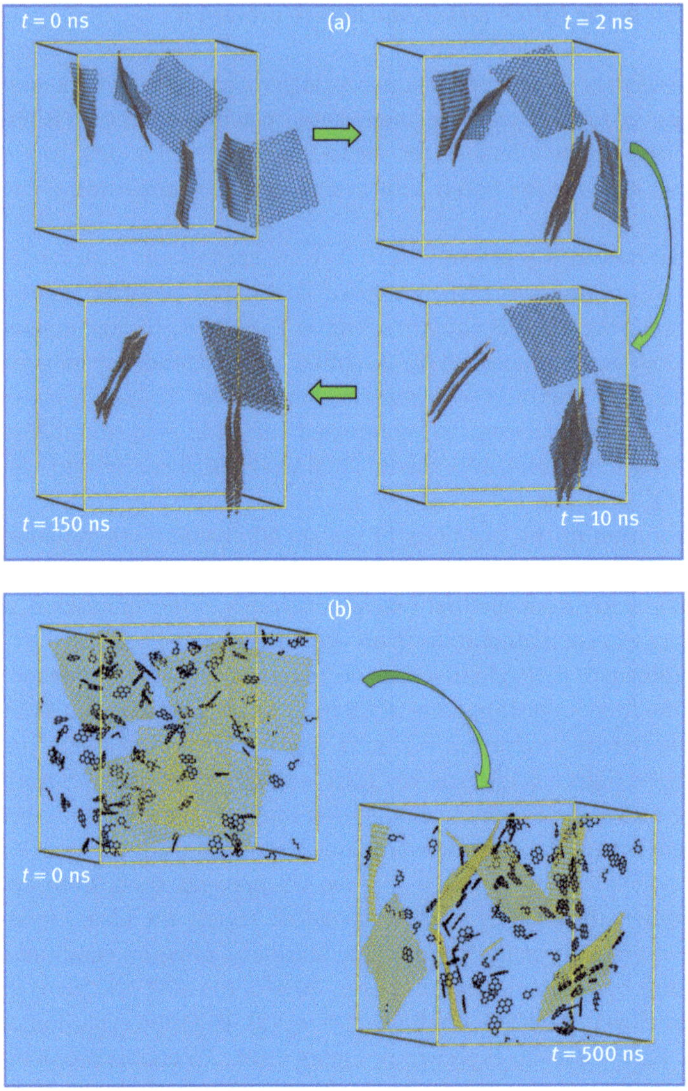

Figure 5.10: (a) Evolution of GS self-assembly during the MD simulation with the aPMMA-GS nanocomposite in the absence of py-PMMA-py chains. (b) Same as with (a) but for the case where 40% of aPMMA chains have been replaced by functionalized py-PMMA-py chains.

characteristic distance between consecutive graphene planes in a typical graphitic structure, and remained to 3.4 Å throughout the simulation. In contrast, for system 2, no such phenomenon was observed. The reason for this behavior is the strong adsorption of functionalized py-PMMA-py chains on the surface of GS by their end-pyrene groups. A characteristic example is depicted in Figure 5.12, showing a GS on the surface of which 10 pyrene groups from different py-PMMA-py chains have been adsorbed.

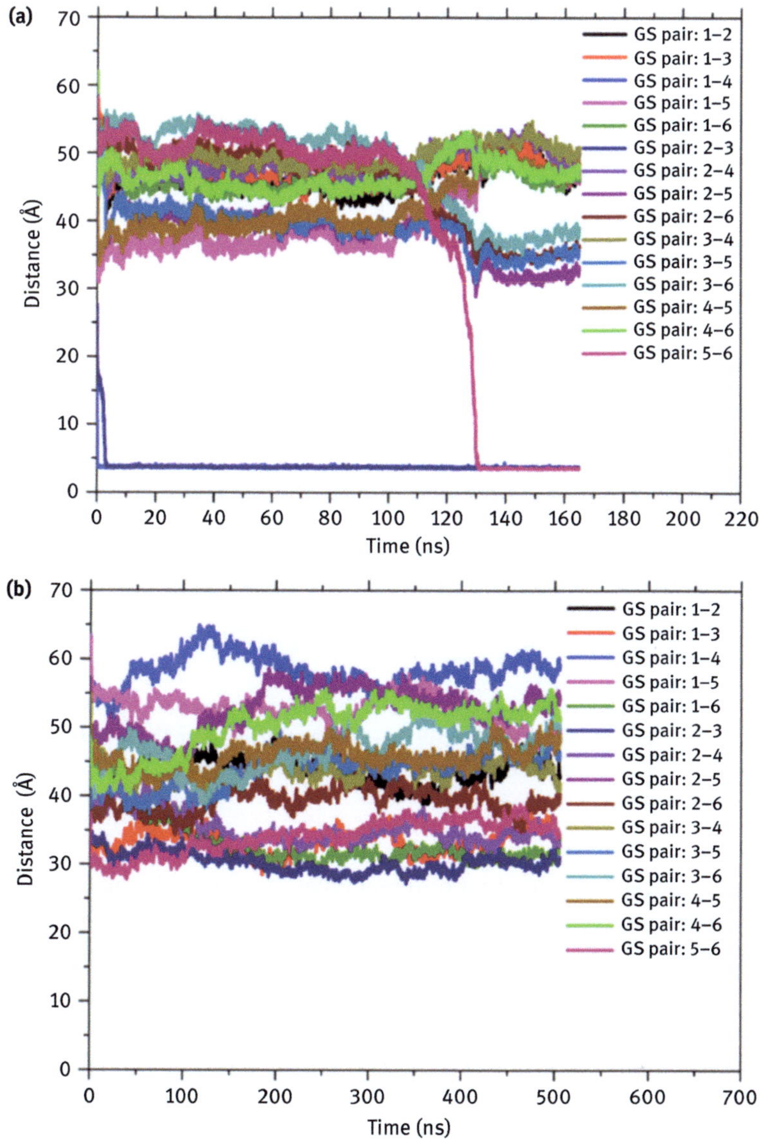

Figure 5.11: Time evolution of the distances between the centers of mass of graphene sheets in all GS pairs in the course of the MD simulation with (a) system 1 and (b) system 2.

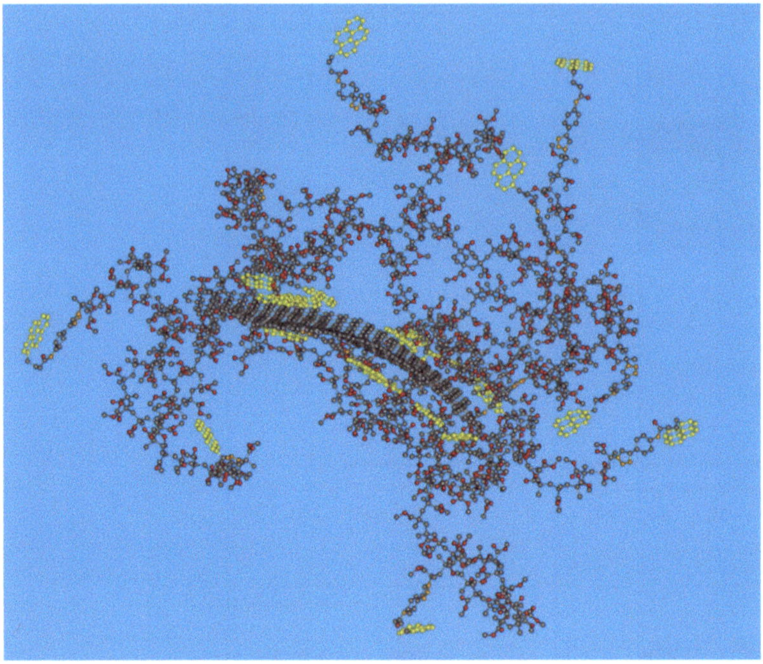

Figure 5.12: Multiple pyrene adsorption on the two faces of a GS. We can discern the parallel arrangement of pyrene groups on the surface of the GS due to π–π stacking, with the rest of the polymer chain extending away from the GS. Hydrogen atoms have been omitted for clarity and adsorbed pyrene groups are depicted in yellow.

A more detailed analysis revealed four different types of adsorbed structures:

a) py-PMMA-py chains adsorbed on a GS by only one of their two ends (the other end remains free) forming a *dangling end*. A typical example is shown in Figure 5.13a.
b) py-PMMA-py chains adsorbed on the same face of the same GS by both of their end-pyrene groups forming a *loop*. A typical example is shown in Figure 5.13b.
c) py-PMMA-py chains adsorbed on both faces of the same GS forming an *extended loop*. A typical example is shown in Figure 5.13c.
d) py-PMMA-py chains adsorbed with their end-pyrene groups on two different GS forming a *bridge*. A typical example is shown in Figure 5.13d.

The time evolution of the population of the four different types of adsorbed conformations in the course of the MD simulation is depicted in Figure 5.14. We see that, initially, all py-PMMA-py chains are free (i.e., nonadsorbed). This was done on purpose since we did not want to bias the py-PMMA-py chains to be close to GS.

Then, during the first 15 ns of the simulation with system 2, the population of *dangling ends* increases rapidly, and this continues (albeit with a smoother rate) at

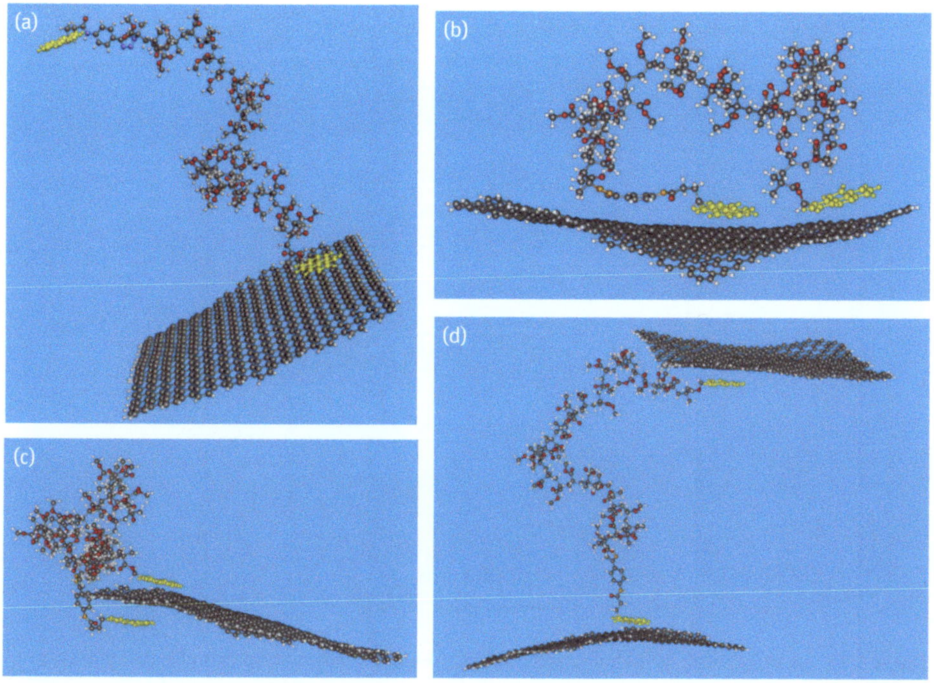

Figure 5.13: Examples of typical py-PMMA-py conformations adsorbed on GS: (a) a *dangling end*, (b) a *loop*, (c) an *extended loop* and (d) a *bridge*.

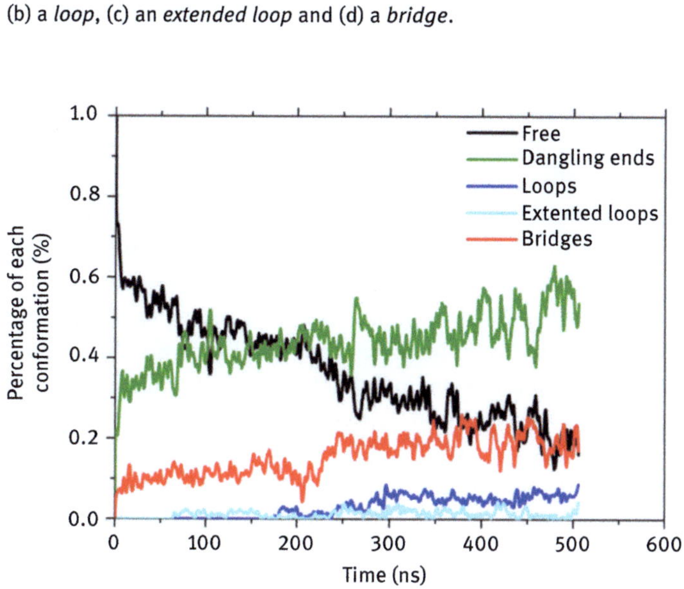

Figure 5.14: Time evolution of the relative population of adsorbed (*dangling ends*, *loops*, *extended loops* and *bridges*) and nonadsorbed (i.e., *free*) py-PMMA-py conformations in the course of the MD simulation.

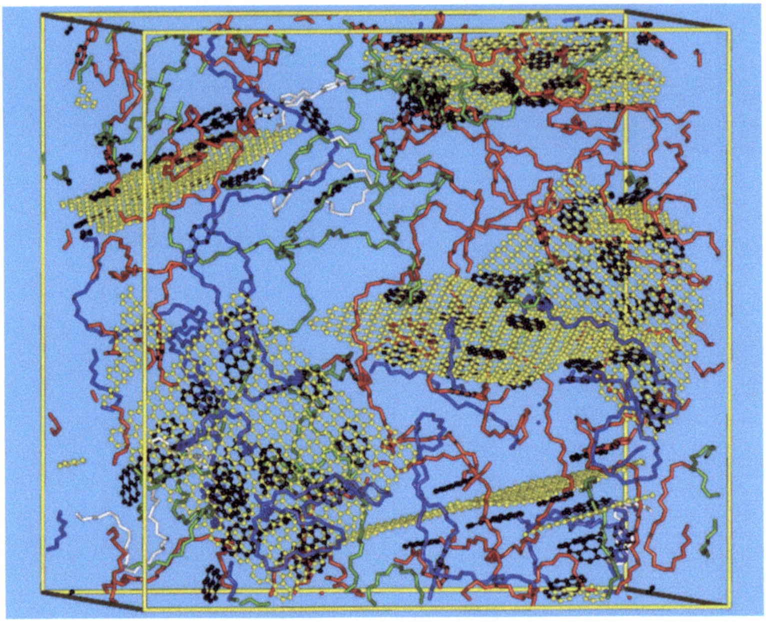

Figure 5.15: Typical conformations of adsorbed and free py-PMMA-py chains on GS (in yellow) in the simulation cell after several nanoseconds of simulation time. Different conformations are shown in different colors: *closed* and *extended closed loops* in blue, *bridges* in red, *dangling ends* in green and *free* chains in white.

all later times. Also increasing with time are the numbers of *loops*, *extended loops* and *bridges*. At the same time, the fraction of nonadsorbed chains decreases.

Figure 5.15 presents the configuration of system 2 at the end of the (approximately 500 ns long) MD simulation. Only graphene, pyrene and backbone atoms of the py-PMMA-py chains are shown for clarity. With blue, red and green colors, we depict the py-PMMA-py chains that participate in *loops*, *bridges* and *dangling ends*, with white color we depict *free* py-PMMA-py chains, and with yellow and black colors we depict carbon atoms in GS and pyrenes, respectively. At the end of the simulation, only few py-PMMA-py chains have remained *free* (i.e., nonadsorbed); the rest have formed *bridges*, *extended bridges* and *dangling ends*.

5.4 Conclusions

Within the space limitations of this chapter we hope that we have given a good account of the unique, extraordinary properties of graphene that have justified why it is widely considered as the material of the future. We have highlighted its great potential for new applications and have presented the recent progress in the design of graphene-based polymer nanocomposites both from an experimental and a computational point of view.

We have also presented the basic concepts of molecular simulations focusing particularly on the MD technique, and simulation results for the structural, conformational and mechanical properties of a test nanocomposite system based on PMMA filled with two different types of GS: simple (i.e., unfunctionalized) GS and functionalized (i.e., GO). Our study was driven by recent experimental reports according to which addition of a small fraction of GO in a polar polymer matrix such as PMMA can lead to a remarkable enhancement of its elastic constants.

We started with simulations at a high temperature (T = 500 K) well above the melting point of PMMA, where system equilibration was easier to achieve and studied several physicochemical properties of the corresponding nanocomposites (density, polymer dimensions, local structure, type and statistics of adsorbed conformations). We found that for both types of fillers, the polymer local density in the proximity of GS is significantly higher than in the pure bulk. In the case of the GO-sPMMA nanocomposite, we observed the formation of a large population of hydrogen bonds between the ester branches of sPMMA and the hydroxyl groups of GO NPs.

In a second step, a good number of selected configurations from these high-temperature simulations were cooled down to room temperature, re-equilibrated and subjected to a rigorous statisticomechanical analysis which provided predictions of their elastic constants. For the pure glassy sPMMA matrix, these were found to be in very favorable agreement with values measured in the experiment. For the nanocomposites, we found that GO sheets, even when present in very small concentrations in the sPMMA matrix, lead to a remarkable enhancement of the mechanical performance. This is due to the surface chemistry of GO which promotes strong interfacial interactions (mainly through hydrogen bonds) with the polymer matrix, thus significantly increasing the cohesive energy of the system and therefore its mechanical strength. Unfunctionalized GS also enhance mechanical behavior but to a much lesser degree than GO.

We also highlighted the tendency of GS to come close to each other due to very favorable π–π interactions and form stacked (layered) structures in the polymer matrix. To overcome this, a new methodology was discussed based on the end-functionalization of a good fraction of PMMA chains in the nanocomposite with pyrene functional groups to form py-PMMA-py chains. Our atomistic simulations showed that with this new nondestructive and thus highly attractive technique (covalent functionalization of graphene is completely avoided) GS self-assembly is avoided. This happens because pyrene groups adsorb quickly on GS and, owing to the intervening polymer mass, GS are kept separated, forming a stable and uniform dispersion. Our simulations revealed the creation of several interesting adsorbed structures in such a system: (a) *closed loops* (formed by py-PMMA-py chains adsorbed on the same face of a GS through both of their end pyrene groups), (b) *extended closed loops* (formed by py-PMMA-py chains adsorbed on the two faces of the same GS), (c) *bridges* (formed by py-PMMA-py chains adsorbed on two different GS) and (d) *dangling ends* (corresponding to the free ends of py-PMMA-py chains adsorbed on a GS with only one of their pyrene groups). All these characteristic structures, especially *loops* and *bridges*, keep GS apart and well separated in the PMMA matrix, even at very high wt.% loadings of the matrix in GS, ensuring a homogeneous GS dispersion in the nanocomposite.

Acknowledgments: The authors are grateful to the Limmat Foundation, Zürich, Switzerland, for generous financial support through the project "Multiscale Simulations of Complex Polymer Systems" (MuSiComPS). The work was supported by computational time granted from the Greek Research & Technology Network (GRNET) in the National HPC facility – ARIS under project name COACERVATE (pr004011).

References

[1] Geim, A. K., Novoselov, K. S., The rise of graphene. Nat Mater 2007, 6, 3, 183 DOI: 10.1038/Nmat1849.
[2] Novoselov, K. S., Geim, A. K., Morozov, S. V., Jiang, D., Zhang, Y., Dubonos, S. V., Grigorieva, I. V., Firsov, A. A., Electric field effect in atomically thin carbon films. Science 2004, 306, 56–96, 666 DOI: 10.1126/science.1102896.
[3] Allen, M. J., Tung, V. C., Kaner, R. B., Honeycomb Carbon: A Review of Graphene. Chem Rev 2010, 110, 1, 132 DOI: 10.1021/cr900070d.
[4] Castro Neto, A. H., Guinea, F., Peres, N. M. R., Novoselov, K. S., Geim, A. K., The electronic properties of graphene. Rev Mod Phys 2009, 81, 1, 109 DOI: 10.1103/RevModPhys.81.109.
[5] Choi, W., Lahiri, I., Seelaboyina, R., Kang, Y. S., Synthesis of graphene and its applications: A review. Crit Rev Solid State 2010, 35, 1), 52 DOI: 10.1080/10408430903505036.
[6] Geim, A. K., Graphene: Status and prospects. Science 2009, 324, 5934, 1530 DOI: 10.1126/science.1158877.
[7] Zhu, Y. W., Murali, S., Cai, W. W., Li, X. S., Suk, J. W., Potts, J. R., Ruoff, R. S., Graphene and graphene oxide: Synthesis, properties, and applications. Adv Mater 2010, 22, 46), 52–26 DOI: 10.1002/adma.201001068.
[8] Kim, H., Abdala, A. A., Macosko, C. W., Graphene/polymer nanocomposites. Macromolecules 2010, 43, 16), 6515 DOI: 10.1021/ma100572e.
[9] Kuilla, T., Bhadra, S., Yao, D. H., Kim, N. H., Bose, S., Lee, J. H., Recent advances in graphene based polymer composites. Prog Polym Sci 2010, 35, 11, 1350 DOI: 10.1016/j.progpolymsci.2010.07.005.
[10] Potts, J. R., Dreyer, D. R., Bielawski, C. W., Ruoff, R. S., Graphene-based polymer nanocomposites. Polymer 2011, 52, 1, 5 DOI: 10.1016/j.polymer.2010.11.042.
[11] Stankovich, S., Dikin, D. A., Dommett, G. H. B., Kohlhaas, K. M., Zimney, E. J., Stach, E. A., Piner, R. D., Nguyen, S. T., Ruoff, R. S., Graphene-based composite materials. Nature 2006, 442, 7100, 282 DOI: 10.1038/nature04969.
[12] Verdejo, R., Bernal, M. M., Romasanta, L. J., Lopez-Manchado, M. A., Graphene filled polymer nanocomposites. J Mater Chem 2011, 21, 10, 3301 DOI: 10.1039/c0jm02708a.

[13] Young, R. J., Kinloch, I. A., Gong, L., Novoselov, K. S., The mechanics of graphene nanocomposites: A review. Compos Sci Technol 2012, 72, 12, 1459 DOI: 10.1016/j.compscitech.2012.05.005.
[14] de Heer, W. A., Berger, C., Wu, X. S., First, P. N., Conrad, E. H., Li, X. B., Li, T. B., Sprinkle, M., Hass, J., Sadowski, M. L., Potemski, M., Martinez, G., Epitaxial graphene. Solid State Commun 2007, 143, 1–2, 92 DOI: 10.1016/j.ssc.2007.04.023.
[15] Ni, Z. H., Chen, W., Fan, X. F., Kuo, J. L., Yu, T., Wee, A. T. S., Shen, Z. X., Raman spectroscopy of epitaxial graphene on a SiC substrate. Phys Rev B 2008, 77, 11 DOI: 10.1103/Physrevb.77.115416.
[16] Sutter, P. W., Flege, J. I., Sutter, E. A., Epitaxial graphene on ruthenium. Nat Mater 2008, 7, 5, 406 DOI: 10.1038/nmat2166.
[17] Li, X. S., Cai, W. W., An, J. H., Kim, S., Nah, J., Yang, D. X., Piner, R., Velamakanni, A., Jung, I., Tutuc, E., Banerjee, S. K., Colombo, L., Ruoff, R. S., Large-area synthesis of high-quality and uniform graphene films on copper foils. Science 2009, 324, 5932, 1312 DOI: 10.1126/science.1171245.
[18] Wang, Y., Chen, X. H., Zhong, Y. L., Zhu, F. R., Loh, K. P., Large area, continuous, few-layered graphene as anodes in organic photovoltaic devices. Appl Phys Lett 2009, 95, 6 DOI: 10.1063/1.3204698.
[19] Jiao, L. Y., Zhang, L., Wang, X. R., Diankov, G., Dai, H. J., Narrow graphene nanoribbons from carbon nanotubes. Nature 2009, 458, 7240, 877 DOI: 10.1038/nature07919.
[20] Kosynkin, D. V., Higginbotham, A. L., Sinitskii, A., Lomeda, J. R., Dimiev, A., Price, B. K., Tour, J. M., Longitudinal unzipping of carbon nanotubes to form graphene nanoribbons. Nature 2009, 458,7240, 872-U5 DOI: 10.1038/nature07872.
[21] Bourlinos, A. B., Georgakilas, V., Zboril, R., Steriotis, T. A., Stubos, A. K., Liquid-phase exfoliation of graphite towards solubilized graphenes. Small 2009, 5, 16, 1841 DOI: 10.1002/smll.200900242.
[22] Hernandez, Y., Nicolosi, V., Lotya, M., Blighe, F. M., Sun, Z. Y., De, S., McGovern, I. T., Holland, B., Byrne, M., Gun'ko, Y. K., Boland, J. J., Niraj, P., Duesberg, G., Krishnamurthy, S., Goodhue, R., Hutchison, J., Scardaci, V., Ferrari, A. C., Coleman, J. N., High-yield production of graphene by liquid-phase exfoliation of graphite. Nat Nanotechnol 2008, 3, 9, 563 DOI: 10.1038/nnano.2008.215.
[23] Stankovich, S., Dikin, D. A., Piner, R. D., Kohlhaas, K. A., Kleinhammes, A., Jia, Y., Wu, Y., Nguyen, S. T., Ruoff, R. S., Synthesis of graphene-based nanosheets via chemical reduction of exfoliated graphite oxide. Carbon 2007, 45, 7, 1558 DOI: 10.1016/j.carbon.2007.02.034.
[24] Schniepp, H. C., Li, J. L., McAllister, M. J., Sai, H., Herrera-Alonso, M., Adamson, D. H., Prud'homme, R. K., Car, R., Saville, D. A., Aksay, I. A., Functionalized single graphene sheets derived from splitting graphite oxide. J Phys Chem B 2006, 110, 17, 8535 DOI: 10.1021/jp060936f.
[25] Lee, C., Wei, X. D., Kysar, J. W., Hone, J., Measurement of the elastic properties and intrinsic strength of monolayer graphene. Science 2008, 321, 5887, 385 DOI: 10.1126/science.1157996.
[26] Gomez-Navarro, C., Burghard, M., Kern, K., Elastic properties of chemically derived single graphene sheets. Nano Lett 2008, 8, 7, 20–45 DOI: 10.1021/nl801384y.
[27] Ansari, S., Giannelis, E. P., Functionalized graphene sheet-polyvinylidene fluoride) conductive nanocomposites. J Polym Sci Pol Phys 2009, 47, 9, 888 DOI: 10.1002/polb.21695.
[28] Cai, D. Y., Yusoh, K., Song, M., The mechanical properties and morphology of a graphite oxide nanoplatelet/polyurethane composite. Nanotechnology 2009, 20, 8 DOI: 10.1088/0957-4484/20/8/085712.

[29] Morozov, S. V., Novoselov, K. S., Geim, A. K., Electron transport in graphene. Phys-Usp+ 2008, 51, 7, 744 DOI: 10.1070/PU2008v051n07ABEH006575.

[30] Morozov, S. V., Novoselov, K. S., Katsnelson, M. I., Schedin, F., Elias, D. C., Jaszczak, J. A., Geim, A. K., Giant intrinsic carrier mobilities in graphene and its bilayer. Phys Rev Lett 2008, 100, 1, DOI: 10.1103/Physrevlett.100.016602.

[31] Durkop, T., Getty, S. A., Cobas, E., Fuhrer, M. S., Extraordinary mobility in semiconducting carbon nanotubes. Nano Lett 2004, 4, 1, 35 DOI: 10.1021/nl034841q.

[32] Chen, J. H., Jang, C., Adam, S., Fuhrer, M. S., Williams, E. D., Ishigami, M., Charged-impurity scattering in graphene. Nat Phys 2008, 4, 5, 377 DOI: 10.1038/nphys935.

[33] Chen, J. H., Jang, C., Xiao, S. D., Ishigami, M., Fuhrer, M. S., Intrinsic and extrinsic performance limits of graphene devices on SiO2. Nat Nanotechnol 2008, 3, 4, 206 DOI: 10.1038/nnano.2008.58.

[34] Balandin, A. A., Ghosh, S., Bao, W. Z., Calizo, I., Teweldebrhan, D., Miao, F., Lau, C. N., Superior thermal conductivity of single-layer graphene. Nano Lett 2008, 8, 3, 902 DOI: 10.1021/nl0731872.

[35] Kim, P., Shi, L., Majumdar, A., McEuen, P. L., Thermal transport measurements of individual multiwalled nanotubes. Phys Rev Lett 2001, 87, 21 DOI: 10.1103/Physrevlett.87.215502.

[36] Pop, E., Mann, D., Wang, Q., Goodson, K. E., Dai, H. J., Thermal conductance of an individual single-wall carbon nanotube above room temperature. Nano Lett 2006, 6, 1, 96 DOI: 10.1021/nl052145f.

[37] Seol, J. H., Jo, I., Moore, A. L., Lindsay, L., Aitken, Z. H., Pettes, M. T., Li, X. S., Yao, Z., Huang, R., Broido, D., Mingo, N., Ruoff, R. S., Shi, L., Two-dimensional phonon transport in supported graphene. Science 2010, 328, 5975), 213 DOI: 10.1126/science.1184014.

[38] Moniruzzaman, M., Winey, K. I., Polymer nanocomposites containing carbon nanotubes. Macromolecules 2006, 39, 16, 5194 DOI: 10.1021/ma060733p.

[39] Ramanathan, T., Abdala, A. A., Stankovich, S., Dikin, D. A., Herrera-Alonso, M., Piner, R. D., Adamson, D. H., Schniepp, H. C., Chen, X., Ruoff, R. S., Nguyen, S. T., Aksay, I. A., Prud'homme, R. K., Brinson, L. C., Functionalized graphene sheets for polymer nanocomposites. Nat Nanotechnol 2008, 3, 6, 327 DOI: 10.1038/nnano.2008.96.

[40] Fang, M., Wang, K. G., Lu, H. B., Yang, Y. L., Nutt, S., Covalent polymer functionalization of graphene nanosheets and mechanical properties of composites. J Mater Chem 2009, 19, 38, 7098 DOI: 10.1039/b908220d.

[41] Lee, S. H., Dreyer, D. R., An, J. H., Velamakanni, A., Piner, R. D., Park, S., Zhu, Y. W., Kim, S. O., Bielawski, C. W., Ruoff, R. S., Polymer brushes via controlled, surface-initiated atom transfer radical polymerization (atrp) from graphene oxide. Macromol Rapid Comm 2010, 31, 3, 281 DOI: 10.1002/marc.200900641.

[42] Kim, H., Macosko, C. W., Processing-property relationships of polycarbonate/graphene composites. Polymer 2009, 50, 15, 3797 DOI: 10.1016/j.polymer.2009.05.038.

[43] Steurer, P., Wissert, R., Thomann, R., Mulhaupt, R., Functionalized Graphenes and Thermoplastic nanocomposites based upon expanded graphite oxide. Macromol Rapid Comm 2009, 30, 4–5, 316 DOI: 10.1002/marc.200800754.

[44] Kumar, A., Sharma K., Dixit A. R., A review of the mechanical and thermal properties of graphene and its hybrid polymer nanocomposites for structural applications. J Mater Sci 2019, 54, 8, 5992 DOI: 10.1007/s10853-018-03244-3.

[45] Li, X. G., McKenna, G. B., Considering viscoelastic micromechanics for the reinforcement of graphene polymer nanocomposites. ACS Macro Lett 2012, 1, 3, 388 DOI: 10.1021/mz200253x.

[46] Liao, K. H., Aoyama, S., Abdala, A. A., Macosko, C., Does graphene change Tg of nanocomposites? Macromolecules 2014, 47, 23, 8311 DOI: 10.1021/ma501799z.

[47] Liao, K. H., Kobayashi, S., Kim, H., Abdala, A. A., Macosko, C. W., influence of functionalized graphene sheets on modulus and glass transition of PMMA. Macromolecules 2014, 47, 21, 7674 DOI: 10.1021/ma501709g.

[48] Cano, M., Khan, U., Sainsbury, T., O'Neill, A., Wang, Z. M., McGovern, I. T., Maser, W. K., Benito, A. M., Coleman, J. N., Improving the mechanical properties of graphene oxide based materials by covalent attachment of polymer chains. Carbon 2013, 52, 363 DOI: 10.1016/j.carbon.2012.09.046.

[49] Zhao, X., Zhang, Q. H., Chen, D. J., Lu, P., Enhanced mechanical properties of graphene-based poly(vinyl alcohol) composites. Macromolecules 2010, 43, 5, 2357 DOI: 10.1021/ma902862u.

[50] Kim, H., Macosko, C. W., Morphology and properties of polyester/exfoliated graphite nanocomposites. Macromolecules 2008, 41, 9, 3317 DOI: 10.1021/ma702385h.

[51] Liang, J. J., Wang, Y., Huang, Y., Ma, Y. F., Liu, Z. F., Cai, J. M., Zhang, C. D., Gao, H. J., Chen, Y. S., Electromagnetic interference shielding of graphene/epoxy composites. Carbon 2009, 47, 3, 922 DOI: 10.1016/j.carbon.2008.12.038.

[52] Stankovich, S., Piner, R. D., Nguyen, S. T., Ruoff, R. S., Synthesis and exfoliation of isocyanate-treated graphene oxide nanoplatelets. Carbon 2006, 44, 15, 3342 DOI: 10.1016/j.carbon.2006.06.004.

[53] Eda, G., Chhowalla, M., Graphene-based composite thin films for electronics. Nano Lett 2009, 9, 2, 814 DOI: 10.1021/nl8035367.

[54] Van Lier, G., Van Alsenoy, C., Van Doren, V., Geerlings, P., Ab initio study of the elastic properties of single-walled carbon nanotubes and graphene. Chem Phys Lett 2000, 326, 1–2, 181 DOI: 10.1016/S0009-2614(00)00764-8.

[55] Liu, F., Ming, P. M., Li, J., Ab initio calculation of ideal strength and phonon instability of graphene under tension. Phys Rev B 2007, 76, 6 DOI: 10.1103/Physrevb.76.064120.

[56] Kalosakas, G., Lathiotakis, N. N., Galiotis, C., Papagelis, K., In-plane force fields and elastic properties of graphene. J Appl Phys 2013, 113, 13, DOI: 10.1063/1.4798384.

[57] Jiang, J. W., Wang, J. S., Li, B. W., Young's modulus of graphene: a molecular dynamics study. Phys Rev B 2009, 80, 11, DOI: 10.1103/Physrevb.80.113405.

[58] Mattausch, A., Pankratov, O., Ab initio study of graphene on SiC. Phys Rev Lett 2007, 99, 7, DOI: 10.1103/Physrevlett.99.076802.

[59] Ma, Y. D., Dai, Y., Guo, M., Niu, C. W., Huang, B. B., Graphene adhesion on MoS2 monolayer: An ab initio study. Nanoscale 2011, 3, 9, 3883 DOI: 10.1039/c1nr10577a.

[60] Awasthi, A. P., Lagoudas, D. C., Hammerand, D. C., Modeling of graphene-polymer interfacial mechanical behavior using molecular dynamics. Model Simul Mater Sci 2009, 17, 1, DOI: 10.1088/0965-0393/17/1/015002.

[61] Li, Y., Kroger, M., Liu, W. K., Nanoparticle geometrical effect on structure, dynamics and anisotropic viscosity of polyethylene nanocomposites. Macromolecules 2012, 45, 4, 20–99 DOI: 10.1021/ma202289a.

[62] Rissanou, A. N., Harmandaris, V., Structure and dynamics of poly(methylmethacrylate)/graphene systems through atomistic molecular dynamics simulations. J Nanopart Res 2013, 15, 5, DOI: 10.1007/S11051-013-1589-2.

[63] Rissanou, A. N., Harmandaris, V., Dynamics of various polymer-graphene interfacial systems through atomistic molecular dynamics simulations. Soft Matter 2014, 10, 16, 2876 DOI: 10.1039/c3sm52688g.

[64] Rissanou, A. N., Power, A. J., Harmandaris, V., Structural and dynamical properties of polyethylene/graphene nanocomposites through molecular dynamics simulations. Polymers-Basel 2015, 7, 3, 390, DOI: 10.3390/polym7030390.

[65] Bacova, P., Rissanou, A. N., Harmandaris, V., Edge-functionalized graphene as a nanofiller: Molecular dynamics simulation study. Macromolecules 2015, 48, 24, 9024 DOI: 10.1021/acs.macromol.5b01782.
[66] Guryel, S., Walker, M., Geerlings, P., De Proft, F., Wilson, M. R., Molecular dynamics simulations of the structure and the morphology of graphene/polymer nanocomposites. Phys Chem Chem Phys 2017, 19, 20, 12959, DOI: 10.1039/c7cp01552f.
[67] Gulde, M., Rissanou, A. N., Harmandaris, V., Muller, M., Schafer, S., Ropers, C., Dynamics and structure of monolayer polymer crystallites on graphene. Nano Lett 2016, 16, 11, 6994 DOI: 10.1021/acs.nanolett.6b03079.
[68] Lv, C., Xue, Q. Z., Xia, D., Ma, M., Xie, J., Chen, H. J., Effect of chemisorption on the interfacial bonding characteristics of graphene-polymer composites. J Phys Chem C 2010, 114, 14, 6588 DOI: 10.1021/jp100110n.
[69] Karatasos, K., Kritikos, G., Characterization of a graphene oxide/poly(acrylic acid) nanocomposite by means of molecular dynamics simulations. RSC Adv 2016, 6, 111, 109–267 DOI: 10.1039/c6ra22951d.
[70] Xue, Q. Z., Lv, C., Shan, M. X., Zhang, H. X., Ling, C. C., Zhou, X. Y., Jiao, Z. Y., Glass transition temperature of functionalized graphene-polymer composites. Comp Mater Sci 2013, 71, 66 DOI: 10.1016/j.commatsci.2013.01.009.
[71] Azimi, M., Mirjavadi, S. S., Hamouda, A. M. S., Makki, H., Heterogeneities in polymer structural and dynamic properties in graphene and graphene oxide nanocomposites: Molecular dynamics simulations. Macromol Theor Simul 2017, 26, 2, DOI: 10.1002/Mats.201600086.
[72] Skountzos, E. N., Anastassiou, A., Mavrantzas, V. G., Theodorou, D. N., Determination of the mechanical properties of a poly(methyl methacrylate) nanocomposite with functionalized graphene sheets through detailed atomistic simulations. Macromolecules 2014, 47, 22, 8072 DOI: 10.1021/ma5017093.
[73] Lin, F., Xiang, Y., Shen, H. S., Temperature dependent mechanical properties of graphene reinforced polymer nanocomposites – A molecular dynamics simulation. Compos Part B-Eng 2017, 111, 261 DOI: 10.1016/j.compositesb.2016.12.0041359-8368.
[74] Liu, F., Hu, N., Ning, H. M., Liu, Y. L., Li, Y., Wu, L. K., Molecular dynamics simulation on interfacial mechanical properties of polymer nanocomposites with wrinkled graphene. Comp Mater Sci 2015, 108, 160 DOI: 10.1016/j.commatsci.2015.06.023.
[75] Liu, J., Shen, J. X., Zheng, Z. J., Wu, Y. P., Zhang, L. Q., Revealing the toughening mechanism of graphene-polymer nanocomposite through molecular dynamics simulation. Nanotechnology 2015, 26, 29, DOI: 10.1088/0957-4484/26/29/291003.
[76] Wang, M. C., Lai, Z. B., Galpaya, D., Yan, C., Hu, N., Zhou, L. M., Atomistic simulation of surface functionalization on the interfacial properties of graphene-polymer nanocomposites. J Appl Phys 2014, 115, 12 DOI: 10.1063/1.4870170.
[77] Schopp, S., Thomann, R., Ratzsch, K. F., Kerling, S., Altstadt, V., Mulhaupt, R., Functionalized Graphene and Carbon materials as components of styrene- butadiene rubber nanocomposites prepared by aqueous dispersion blending. Macromol Mater Eng 2014, 299, 3, 319 DOI: 10.1002/mame.201300127.
[78] Yin, B., Zhang, X. M., Zhang, X., Wang, J. Y., Wen, Y. W., Jia, H. B., Ji, Q. M., Ding, L. F., Ionic liquid functionalized graphene oxide for enhancement of styrene-butadiene rubber nanocomposites. Polym Advan Technol 2017, 28, 3, 293 DOI: 10.1002/pat.3886.
[79] Georgakilas, V., Otyepka, M., Bourlinos, A. B., Chandra, V., Kim, N., Kemp, K. C., Hobza, P., Zboril, R., Kim, K. S., Functionalization of graphene: Covalent and non-covalent approaches, derivatives and applications. Chem Rev 2012, 112, 11, 6156 DOI: 10.1021/cr3000412.

[80] Lonkar, S. P., Deshmukh, Y. S., Abdala, A. A., Recent advances in chemical modifications of graphene. Nano Res 2015, 8, 4, 1039 DOI: 10.1007/s12274-014-0622-9.
[81] Cheng, H. K. F., Sahoo, N. G., Tan, Y. P., Pan, Y. Z., Bao, H. Q., Li, L., Chan, S. H., Zhao, J. H., Poly(vinyl alcohol) nanocomposites filled with poly(vinyl alcohol)-grafted graphene oxide. ACS Appl Mater Inter 2012, 4, 5, 2387 DOI: 10.1021/am300550n.
[82] Goncalves, G., Marques, P. A. A. P., Barros-Timmons, A., Bdkin, I., Singh, M. K., Emami, N., Gracio, J., Graphene oxide modified with PMMA via ATRP as a reinforcement filler. J Mater Chem 2010, 20, 44, 9927 DOI: 10.1039/c0jm01674h.
[83] Mallakpour, S., Abdolmaleki, A., Borandeh, S., Covalently functionalized graphene sheets with biocompatible natural amino acids. Appl Surf Sci 2014, 307, 533 DOI: 10.1016/j.apsusc.2014.04.070.
[84] Sayyar, S., Murray, E., Thompson, B. C., Gambhir, S., Officer, D. L., Wallace, G. G., Covalently linked biocompatible graphene/polycaprolactone composites for tissue engineering. Carbon 2013, 52, 296 DOI: 10.1016/j.carbon.2012.09.031.
[85] Xue, Y. H., Liu, Y., Lu, F., Qu, J., Chen, H., Dai, L. M., Functionalization of Graphene Oxide with Polyhedral Oligomeric Silsesquioxane (POSS) for Multifunctional Applications. J Phys Chem Lett 2012, 3, 12, 1607 DOI: 10.1021/jz3005877.
[86] Park, J. H., Aluru, N. R., Self-assembly of graphenes. Surf Sci 2011, 605, 17–18, 16–16 DOI: 10.1016/j.susc.2011.02.011.
[87] Ju, S. P., Wang, Y. C., Huang, G. J., Chang, J. W., Miscibility of graphene and poly(methyl methacrylate) (PMMA): Molecular dynamics and dissipative particle dynamics simulations. Rsc Adv 2013, 3, 22, 8298 DOI: 10.1039/c3ra22879g.
[88] Karatasos, K., Graphene/hyperbranched polymer nanocomposites: Insight from molecular dynamics simulations. Macromolecules 2014, 47, 24, 8833 DOI: 10.1021/ma502123a.
[89] Roussou, R. E., Karatasos, K., Graphene/poly(ethylene glycol) nanocomposites as studied by molecular dynamics simulations. Mater Design 2016, 97, 163, DOI: 10.1016/j.matdes.2016.02.078.
[90] Guo, Y. S., Liu, J., Wu, Y. P., Zhang, L. Q., Wang, Z., Li, Y., Molecular insights into the effect of graphene packing on mechanical behaviors of graphene reinforced cis-1,4-polybutadiene polymer nanocomposites. Phys Chem Chem Phys 2017, 19, 33, 22417 DOI: 10.1039/c7cp02945d.
[91] Papadimitriou, K. D., Skountzos, E. N., Gkermpoura, S. S., Polyzos, I., Mavrantzas, V. G., Galiotis, C., Tsitsilianis, C., Molecular modeling combined with advanced chemistry for the rational design of efficient graphene dispersing agents. ACS Macro Lett 2016, 5, 1, 24 DOI: 10.1021/acsmacrolett.5b00755.
[92] Allen, M. P., Tildesley, D. J., Computer simulation of liquids. Clarendon Press, Oxford University Press: Oxford/Oxfordshire/New York, 1987, xix, 385 p.
[93] Frenkel, D., Smit, B., Understanding molecular simulation: from algorithms to applications. Academic Press: San Diego, 1996, xviii, 443 p.
[94] Mayo, S. L., Olafson, B. D., Goddard, W. A., Dreiding – a generic force-field for molecular simulations. J Phys Chem-Us 1990, 94, 26, 8897 DOI: 10.1021/J100389a010.
[95] Jorgensen, W. L., Maxwell, D. S., TiradoRives, J., Development and testing of the OPLS all-atom force field on conformational energetics and properties of organic liquids. J Am Chem Soc 1996, 118, 45, 11225 DOI: 10.1021/Ja9621760.
[96] MacKerell, A. D., Bashford, D., Bellott, M., Dunbrack, R. L., Evanseck, J. D., Field, M. J., Fischer, S., Gao, J., Guo, H., Ha, S., Joseph-McCarthy, D., Kuchnir, L., Kuczera, K., Lau, F. T. K., Mattos, C., Michnick, S., Ngo, T., Nguyen, D. T., Prodhom, B., Reiher, W. E., Roux, B., Schlenkrich, M., Smith, J. C., Stote, R., Straub, J., Watanabe, M., Wiorkiewicz-Kuczera, J., Yin,

D., Karplus, M., All-atom empirical potential for molecular modeling and dynamics studies of proteins. J Phys Chem B 1998, 102, 18, 35–86 DOI: 10.1021/Jp973084f.
[97] Cornell, W. D., Cieplak, P., Bayly, C. I., Gould, I. R., Merz, K. M., Ferguson, D. M., Spellmeyer, D. C., Fox, T., Caldwell, J. W., Kollman, P. A., A second generation force field for the simulation of proteins, nucleic acids, and organic molecules. J Am Chem Soc 1996, 118, 9, 2309, DOI: 10.1021/Ja955032e.
[98] Sun, H., COMPASS: An ab initio force-field optimized for condensed-phase applications – Overview with details on alkane and benzene compounds. J Phys Chem B 1998, 102, 38, 7338, DOI: 10.1021/Jp980939v.
[99] Maple, J. R., Hwang, M. J., Stockfisch, T. P., Dinur, U., Waldman, M., Ewig, C. S., Hagler, A. T., Derivation of class-ii force-fields .1. methodology and quantum force-field for the alkyl functional-group and alkane molecules. J Comput Chem 1994, 15, 2, 162 DOI: 10.1002/jcc.540150207.
[100] McQuarrie, D. A., Statistical mechanics. Harper and Row: New York, London, 1976, xvii, 641 p.
[101] Swope, W. C., Andersen, H. C., Berens, P. H., Wilson, K. R., A computer-simulation method for the calculation of equilibrium-constants for the formation of physical clusters of molecules – application to small water clusters. J Chem Phys 1982, 76, 1, 637 DOI: 10.1063/1.442716.
[102] Tuckerman, M., Berne, B. J., Martyna, G. J., Reversible multiple time scale molecular-dynamics. J Chem Phys 1992, 97, 3, 1990 DOI: 10.1063/1.463137.
[103] Brown, D., Clarke, J. H. R., Okuda, M., Yamazaki, T., A domain decomposition parallel-processing algorithm for molecular-dynamics simulations of polymers. Comput Phys Commun 1994, 83, 1, 1 DOI: 10.1016/0010-4655(95)00007-3.
[104] Putz, M., Kolb, A., Optimization techniques for parallel molecular dynamics using domain decomposition. Comput Phys Commun 1998, 113, 2–3, 145 DOI: 0.1016/S0010-4655(98)00074-5.
[105] Wilson, M. R., Allen, M. P., Warren, M. A., Sauron, A., Smith, W., Replicated data and domain decomposition molecular dynamics techniques for simulation of anisotropic potentials. J Comput Chem 1997, 18, 4, 478.
[106] Plimpton, S., Hendrickson, B., A new parallel method for molecular dynamics simulation of macromolecular systems. J Comput Chem 1996, 17, 3, 326.
[107] Plimpton, S., Fast parallel algorithms for short-range molecular-dynamics. J Comput Phys 1995, 117, 1, 1 DOI: 10.1006/jcph.1995.1039.
[108] Lindahl, E., Hess, B., van der Spoel, D., GROMACS 3.0: A package for molecular simulation and trajectory analysis. J Mol Model 2001, 7, 8, 306 DOI: 10.1007/s008940100045.
[109] Phillips, J. C., Braun, R., Wang, W., Gumbart, J., Tajkhorshid, E., Villa, E., Chipot, C., Skeel, R. D., Kale, L., Schulten, K., Scalable molecular dynamics with NAMD. J Comput Chem 2005, 26, 16, 1781 DOI: 10.1002/jcc.20289.
[110] Theodorou, D. N., Suter, U. W., Atomistic modeling of mechanical-properties of polymeric glasses. Macromolecules 1986, 19, 1, 139 DOI: 10.1021/Ma00155a022.
[111] Scienomics, MAPS platform, version 3.4.2, France 2015. See also: http://scienomics.com/.
[112] Brandrup, J., Immergut, E. H., Grulke, E. A., Polymer handbook. 4th ed., Wiley: New York, 1999.

Pravin Jagdale, Gemma Rius, Krishna Rajan, Jijeesh Ravi Nair, Massimo Rovere, Alberto Tagliaferro, Claudio Gerbaldi

6 Carbon from waste source for Li-ion battery

6.1 Introduction

Energy generation and use depends heavily on fossil fuels, and it severely impacts the global economics and living sustainability [1]. Completely shifting to clean energy production and consumption is vital for the present and future generations. Electrochemical energy storage and conversion can be implemented by various devices such as (rechargeable) batteries, fuel cells and electrochemical capacitors (ECs). They are considered fundamental elements for developing a sustainable energy society. In particular, a battery [2] is an electrochemical power source comprising several cells, which are connected in series or parallel or in combination in order to deliver the required output voltage and energy.

In ancient Greece, Thales observed that rubbing amber could generate an electric charge [3]. In between 250 BC and AD 640, there are evidences for the use of a storage system known as the "Baghdad batteries" [4]. In 1748, Benjamin Franklin first coined the term "battery" to describe an array of charged glass plates. In the 1780s, Luigi Galvani demonstrated the electrical basis of nerve impulses, which is widely called as the frog leg experiment. The Italian physicist Alessandro Volta in 1800 invented the chemical battery, Voltaic pile. Volta discovered that an electrical current is generated when metals come in contact with chemicals. John Frederic Daniell in 1820 improved the voltaic cell using copper (Cu) and zinc (Zn) plates as electrodes and sulfates of respective metals as electrolytes, and termed as the Daniell cell, which was later used for operating telegraphs and doorbells. In the 1830s, Michael Faraday conducted experiments with a ferrite ring and device called a battery. Carbon–zinc battery was developed by Georges Leclanché in the 1860s. The wet cell consisted of electrodes that were plunged into a body of electrolyte fluid, which was manufactured easily and exhibited acceptable shelf life [5]. An improved version known as a dry cell was later developed by sealing the cell and changing the fluid electrolyte to a wet paste. Raymond Gaston Planté in the 1860s invented the first practical storage lead acid battery. He immersed two thin solid lead plates separated by rubber sheets in a dilute sulfuric acid solution to make a

Pravin Jagdale, Krishna Rajan, Jijeesh Ravi Nair, Massimo Rovere, Alberto Tagliaferro, Claudio Gerbaldi, Department of Applied Science and technology (DISAT), Politecnico di Torino, Torino, Italy
Pravin Jagdale, Italian Institute of technology (IIT), Torino, Italy
Gemma Rius, Institute of Microelectronics of Barcelona, Bellaterra, Spain

https://doi.org/10.1515/9783110479133-006

secondary battery. Later in 1900, Thomas Edison developed the nickel (Ni) storage battery and later in 1905 he developed the Ni-iron (Fe) battery [6].

A cell is made of three major components: a positive electrode (cathode), a negative electrode (anode) and a separator. A separator is soaked in organic carbonate-based liquid electrolyte. The electrolyte contains dissociated salts, which facilitate ion transfer between the two electrodes. Once these electrodes are externally connected to a device, the chemical reactions occur at the electrodes. The process generates electrons and enables the current to flow. The terms anode and cathode are applicable generally for primary batteries, but are not suitable for secondary (rechargeable) batteries. During charging and discharging the direction of electron flow changes, however, the electrodes retain their polarity. Thus, the terms negative and positive electrodes are most suited to describe the terminals of a rechargeable battery system.

The energy storage system can be primary or secondary. Primary cells are designed to discharge once and can be discarded after use. If one attempts to recharge a primary cell, then it can present serious safety hazards. The examples of primary cells include carbon–zinc and alkaline cells. Primary cells having lithium (Li) as an anode are called Li cells irrespective of the nature of the cathode. Such cells are separated into two types: cells with solid cathodes (e.g., Li–MnO_2) and cells with liquid cathodes (e.g., Li-SO_2) [7]. Reserve batteries are a special-purpose primary cell, which are used for applications that require a long inactive shelf period followed by intense discharge and unusual operational conditions. The oldest form of a rechargeable battery system in use is the wet cell lead acid battery. It represents one of the worst energy-to-weight ratios. However, it delivers good power-to-weight ratio along with good energy-to-volume ratio compared to other type of batteries. It is cheap and can supply high surge currents needed in starter motors along with low manufacturing cost.

Li metal [8] has been selected as a good choice due to its low mass density, smaller size compared to other known battery materials as charge carrier, and a high mobility and fast transport speed. These features give Li-based secondary batteries a distinct advantage over other battery systems such as Ni-cadmium (Cd) and lead acid batteries. Cd is highly toxic and heavier, and the same problems are applied to lead as well. The discovery (1970s) of several intercalating inorganic compounds was monumental in the development and widespread use of high-energy rechargeable Li-based batteries. Whittingham et al. introduced the concept of electrochemical intercalation and its potential use [9]. The discovery and the understanding of the intercalating compounds led to the development of rechargeable Li batteries using Li insertion compounds as positive electrodes and Li metal as the negative electrode (Li metal battery). The first Li metal battery cell was introduced by Exxon et al. [10] in which they used titanium disulfide (TiS_2) as positive electrode, Li metal as negative electrode and Li perchlorate in dioxolane as the electrolyte. The safety concerns related to the use of Li metal as anode was addressed via alternative approaches. It led to the modification of either the electrode or the electrolyte. Murphy et al. [11] and Scrosati et al. [12] demonstrated that Li metal can be replaced by an insertion host having low standard reduction potential. In the early 1990s, the concept of using electrodes having Li^+-ion insertion capabilities led to the development of today's Li^+-ion batteries. The presence of Li in the anode in ionic form rather than in metallic form of a Li^+-ion cell resolved the dendrite problem, which was inherited by Li metal cells.

In the 1980s, Goodenough et al. [13, 14] proposed the high-voltage metal oxides (Li_xMO_2, where M = Ni, cobalt (Co) or manganese (Mn)) as cathodes for Li^+-ion cells. The discovery of highly reversible, low-voltage Li^+-ion intercalation/deintercalation process in graphite led to the creation of the first Li^+-ion cell based on $C/LiCoO_2$, which was commercialized by Sony Corporation. Such cells exhibited an operational voltage of 3.6 V (3 times vs. alkaline systems), gravimetric energy densities of 150 Wh/kg (2–3 times vs. Ni–Cd (cadmium) batteries) and a cycle life of at least 500 discharges.

Presently, Li^+-ion batteries are the fastest growing storage devices. The salient features of today's commercial Li^+-ion batteries are
- high operating voltage between 3.6 and 4.2 V,
- compact,
- lightweight,
- high energy density,
- fast charging,
- low self-discharge,
- not suffer from memory effect,
- use between −20 and +60 °C and
- recharge for more than 1,000 cycles.

The Li^+-ion rechargeable cell working mechanism depends on an "intercalation/de-intercalation" process [15]. The storage processes, which do not involve intercalation are called "insertion/de-insertion" process. Indeed, the term "intercalation" is used for layered-type host substances (e.g., graphite, TiS_2, molybdenum disulfide (MoS_2) or lithium cobalt oxide ($LiCoO_2$)), while "insertion" is used for 3-D materials that have narrow channel cavities. The charge/discharge process is reversible where the Li^+ ions move back and forth between the anode and cathode. The process presented the Li^+-ion battery names such as *rocking chair* [16], *swing* [17, 18] or *shuttlecock* [19].

In general, for a Li^+-ion cell:
- The electrode materials must possess high specific capacity and durability.
- The electrolyte must exhibit high ionic conductivity at ambient and subambient temperatures.
- The electrolyte must have good interfacial properties.
- The materials should be economical, eco-friendly, easily available and safe.

Thus, the selection of right materials is important to produce a sustainable and an efficient battery pack.

A successful electrode should meet a number of criteria:

- The intercalation/insertion compound should have large number of available Li sites.
- The Li^+ ions insertion/de-insertion process should be reversible.
- The host material should not undergo structural changes.
- The host material should be accessing the multiple oxidation states of the transition metals.
- The compounds should possess good electronic and ionic conduction.
- The electrodes must be stable over the working voltage range.
- The electrodes should not undergo dissolution into the electrolyte.
- It should be environmentally friendly and easy to produce.

Li metal is an ideal anode for rechargeable batteries due to its high potential and specific capacity. It is widely used as an anode in primary Li cells. However, dendritic growth, cost, safety hazards and poor recharge ability prevents its practical use in secondary batteries. The issues are addressed by the use of Li alloys (with Al or Si). Unfortunately, the alloy anodes undergo severe volume changes during the cycling process. Nevertheless, the urge for the development of solid-state (ceramic/polymer) electrolytes are a promising approach to realize the use of safe Li metal rechargeable batteries. The standard and commercialized anode for Li^+-ion batteries is carbon in its different layered structures.

6.2 Role of carbon material in batteries

Li metal anodes with an organic electrolyte result in nonuniform formation of a passive film on the anode surface [20–22], which causes dendrite growth of Li metal. Longer charging–discharging cycles are one of the important requirements for making a commercially viable Li-ion battery. This problem is tackled by utilizing the electrochemical intercalation of Li in carbon. It produces a significant negative potential close to that of Li, but this is less reactive and easily reversible [23].

Various kinds of carbon materials have been employed as anodes in various electrolytes for Li rechargeable batteries. Carbon offers advantages over other types of materials used in rechargeable batteries such as high operating voltage, longer life cycle and lack of "memory" effects [24–26]. The structural characteristics of carbon are believed to be the major factor that controls the performance of Li^+-ion batteries. The intercalation of Li^+ ion with carbon depends upon many factors such as the preparative history, carbon precursor and structure. The variations in the average layer spacing and the voltage profile for the carbon with intercalating depend on the degree of graphitization [27]. Literature survey on carbon in battery development revealed that graphite is the most commonly used anode for Li^+-ion secondary battery. Many studies employed graphite as an anode, which has theoretical capacity of 372 mAh/g based on the stoichiometry of Li carbon [28–34].

6.3 Graphene materials for highly performing electrodes

The widespread interest in graphene includes its potential as a suitable material for energy storage and particularly for Li^+-ion batteries. Intrinsically, graphene is aimed at providing excellent performance based on a combination of physicochemical properties, its extremely high surface area of 2,630 m^2/g [35], high electrical conductivity and chemical inertness [36].

Crystalline carbon nanomaterial, carbon nanotube [37, 38] and graphene had their expected applications in electrochemical energy storage. This was precisely envisioned for both Li^+-ion and Li-sulfur batteries and supercapacitor devices [39]. Advantageously with respect to carbon nanotube, graphene is the basic unit of sp^2 carbons (graphitic), which are the most conventionally/commonly used carbons for batteries (or other high surface area applications).

Production wise, the cost of using a certain graphene material could be compared with conventional and cost-efficient traditional macroscopic methods, based on top-down approach or on raw materials, typical chemical or metallurgical approaches to obtain graphite (and graphene-like materials) synthesis and cell-making methods. Examples of current trends and recent achievements include pyrolytic graphite from graphene oxide [40] and kish graphite [41], which are obtained from waste graphite flakes from steel production. Scalable recycling and reusing approach applied to these abundant raw materials will be cited in the following sections.

In the following section, a nonexhaustive collection of bottom-up methods, selected techniques for production of high-quality graphene materials as well as examples of graphene-decorated electrodes are described. For a comprehensive topic review on graphene synthesis, there exist a number of extended resources such as how to achieve specialized morphologies, particularly, those specifically useful/appropriate for EC energy storage [42]. Interestingly, some general vision and hints on using waste and recycled or recyclable carbon sources are included for the described methods. Besides, the quest for valorizing waste materials for energy storage is not new or exclusive for novel nanomaterials [43]. The use of hierarchical structures such as combining graphene with other materials, for example, oxides [44], is beyond the scope of this text.

In summary, graphene is still considered to ultimately outperform existing solutions as it would provide improvements in terms of increased surface area (i.e., enabling higher capacity and also lighter devices), conductivity (faster charge–discharge and efficiency) and increased chemical and mechanical resistance. Relevant evaluation and benchmarking of graphene electrodes as electrochemical energy storage elements in affordable graphite materials, to critically assess their real potential and, eventually, quantify the advantages of graphene in operation that is needed in both energy and power density are carried out. The expected exponential increase in demand and performance of batteries, resulting from their use in electrified vehicles, requires advanced and sustainable solutions. Fundamental investigations on graphitic structure from waste material for electrochemical energy storage are focused on elucidating whether they provide unique features and advantages to affordable carbons, rather than being competitive to existing commercial solutions. Therefore, they are generally based on the "ideal" concept of graphene. High-quality graphene material, in contrast to common graphite/graphitic materials and low-quality graphene materials, can be defined as pure carbon, large crystal domain size and isolated films, that is, typically consisting of monolayer graphene or, eventually, bilayer or few layers' graphene. In fact, classification of graphene based on these three parameters has been established for proper reference in scientific papers and enabling some standards for industrial purposes, such as production [45]. Advantageously, laser power seems to provide a way to transform a variety of heterogeneous materials into two graphitic forms [46]. Similarly, one could envision that the resources potentially include waste materials.

Recently, attention has been drawn to use disordered carbon materials. It may store Li via a mechanism that is completely different from that associated with graphite. Additional studies indicate that Li insertion into an unorganized carbon loses the layer structure [47]. Variation in the polarization resistance with intercalation revealed the processes in various disordered carbons. It demonstrated the single-phase reactions with different stoichiometry of Li intercalation. Electrochemical intercalation of Li into carbons has been studied using mesophase pitch-based carbon fibers with different heat-treatment temperatures, coke [48] and graphite as anodes [27, 47, 49]. It has been found that the activation energy for the diffusion process decreases by increasing the degree of graphitization [27]. The studies also demonstrated that intercalation of Li with camphor-based carbon nanobeads stabilize the cell for 10–20 days [50, 51]. The reversible Li^+-ion intercalation capacity was found to be 45–61% of that obtained with graphite [51]. Researchers also reported the result of Li intercalation by carbon nanomaterial obtained from the pyrolysis of biochar such as tealeaves [52].

Research on Li diffusion in carbon is of both theoretical and practical interest. The process was largely studied in connection with the development of Li^+-ion batteries, whose negative electrode is based on graphite or some other carbon material [27, 53–61]. In some works, Li diffusion was investigated during the reverse process, that is, the anodic deintercalation [47, 62–66]. A complex impedance model for spherical particles was used to determine the Li^+-ion diffusion coefficient in graphite as a function of the state of charge and temperature [67]. Practically, in all cases, studies were concentrated on the diffusion of Li, which was electrochemically incorporated during cathodic polarization of a carbon material in an aprotic electrolyte [68].

However, carbon pyrolysis at low temperature shows attractive behavior as their theoretical capacity exceeds that of graphite as anode material [52, 69–75]. This observation indicates that there may be some other different mechanisms operating in these carbons in terms of Li intercalation. However, the density of low-temperature carbon is lower than that of graphite [76]. Therefore, the capacity of low-temperature carbon should be large enough to compensate for its smaller density. The capacity of an anode for Li^+-ion battery depends on the amount of carbon that can be placed within the limited volume of a typical battery container. Carbon obtained from the same precursor at different temperatures has different structural orientation. Therefore, trial and error is often needed to establish a suitable preparation method for optimizing the properties of carbon [77, 78]. Many researchers studied the relation between the properties of carbon and its electrochemical characteristics relevant to Li^+-ion rechargeable [79–85].

As for the source of atomic carbon, methane is the commonly used gas as its decomposition products are the simplest. The reuse of greenhouse gases such as the very methane for graphene production would follow automatically. Similarly, green approach is also aimed at reusing carbon dioxide [86], yet it is more complex in terms of chemistry and efficiency of the reaction. Likewise, a number of additional works have actually been reported for the (re)use of greener carbon sources, which include molecules or polymers [87], as well as various recycled waste materials, such as plastics but also organic materials or industrial coal (Figure 6.1) [88–90].

In this study, we examined the effects of carbonization atmosphere on the charge–discharge characteristics of carbon from waste source as an anode for a Li^+-ion rechargeable battery. This work targeted the “green” approach for energy storage. The carbonaceous materials such as carbon nanowalls (CNWs) and carbon fiber sheet used in this study were made from different precursors such as organic waste (methane) and polymer waste [nanoscale microfibrillated cellulose (NMFC)].

6.4 Vertically oriented micro-mesoporous CNWs

Graphene materials obtained by chemical vapor deposition (CVD) present several limitations for EC energy storage. First, morphologically the effective graphene surface could be reduced up to half of its theoretical value as CVD-obtained graphene is bonded/stuck onto the metal substrate. Besides, the high temperature used in CVD is not desirable to obtain wrinkled or crumbled carbon, especially if power

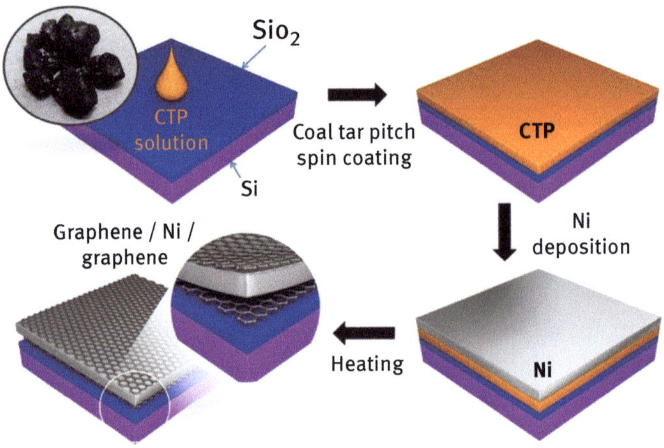

Figure 6.1: Graphene films are converted from coal tar pitch on the top and bottom surface of Ni layer [90]. (Image courtesy: https://www.nature.com/articles/srep16710/figures/1)

consumption reduction is aimed, but also for process compatibility, with any substrate such as plastics. Apart from providing better volume–area ratio, an increase in charge storage capacity is expected from the large amount of graphene flake edges to the electrolyte. Consequently, alternatives of producing porous or microstructured carbon materials, ideally based on scalable techniques, have been studied [91].

As an analogy with CNTs, this is the case of using physically assisted CVD systems, such as fitted with radiofrequency and microwave (MW) sources. In particular, for certain solid sources, the reactivity of physically assisted CVD such as based on the use of a powerful MW field may restrict the growth process to be operated in conditions of relatively low pressure. For instrument and operator's safety in the presence of chemical elements (carbon source), many systems and synthesis principles do make use of hydrogen as a carries gas to avoid oxygen contamination. With the plasma systems, the reuse of methane and other recycled carbon sources may be possible to be applied. Although the compatibility of recycled carbon sources could be more difficult in general, it has actually been achieved [92]. Yet, more examples can be found in the literature [93].

Figure 6.2 provides the image of CNWs deposited on Cu substrate, which has the possibility of being used as anode for Li^+-ion batteries. Owing to the orientation and morphologies, obtained materials are typically referred in the literature with a variety of terms, including carbon nanosheets [94] or CNWs [95]. These graphene flakes or CNWs are randomly distributed, perpendicularly oriented to the substrate and/or curvy. Their dimensions are maximum of approximately 1–1.5 µm in height and 10–20 nm in thickness, which accounts for multilayered

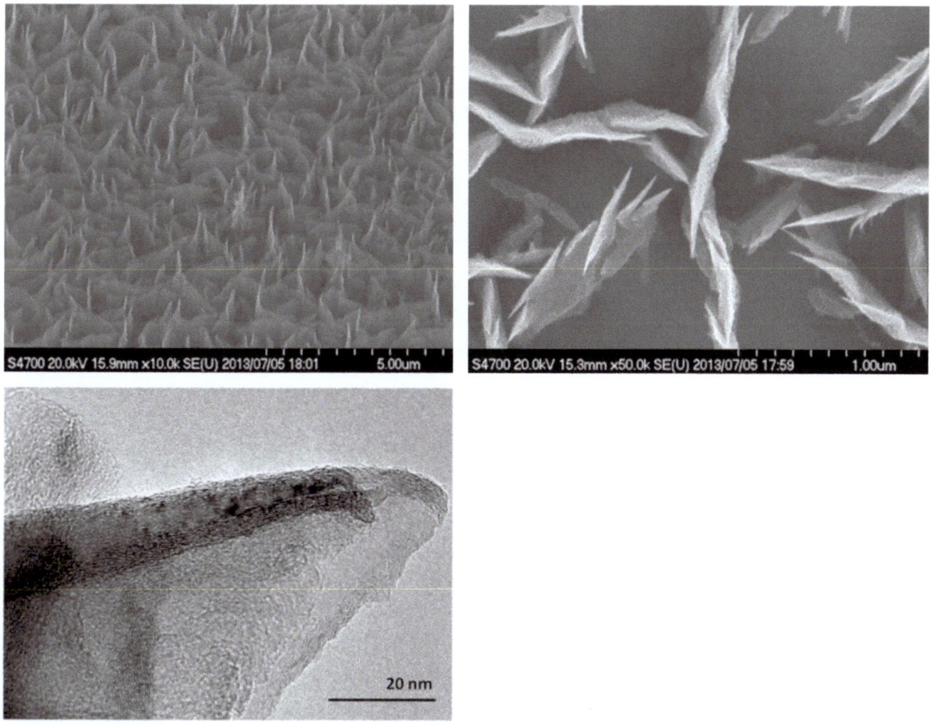

Figure 6.2: Microscopic images of CNWs deposited on Cu: (left and center) FESEM and (right) TEM image.

graphene, and are characterized by very small crystal size, in the order of nanometers. This is further confirmed with micro-Raman analysis.

The observed Raman spectra (Figure 6.3) is typical of a good quality graphitic material. Both first- and second-order Raman effects can be seen. First-order Raman peak is presented as a so-called G-peak around 1,580 cm^{-1}. This peak is commonly observed in all graphitic materials being often the sharpest and intense one. It arises from the first Brillouin zone center of the graphene unit cell and it is due to the double degeneration of in-plane transverse optic [96, 97] and longitudinal optic modes [98, 99]. All other peaks are due to second-order Raman effects. The main peak belonging to this category is D-peak around 1,350 cm^{-1}. This peak is normally associated with the disorder of the graphitic structure. The principal reasons for the presence of this peak are the combination of grain boundaries, vacancies and other structural defects. The D-peak is particularly intense and very sharp: this can be explained not mainly by the density defects but related to the particular orientation of CNWs, which are vertically aligned or oriented (mainly edges exposed to laser beam). This statement is further confirmed by looking at the D + G peak. It does not show any abnormally great number of defects. In conclusion, D-peak intensity is

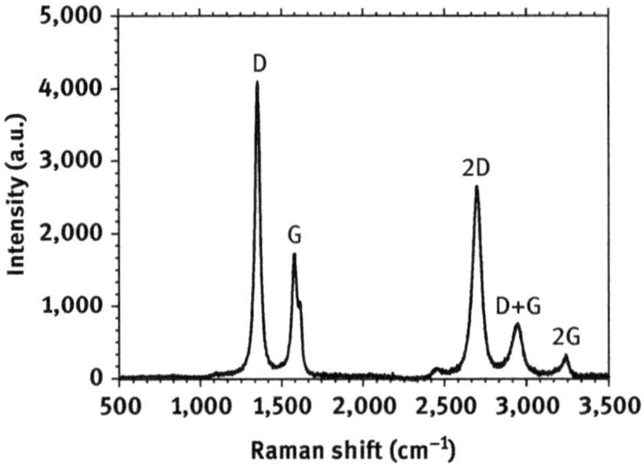

Figure 6.3: Raman analysis of CNWs. (Image courtesy by Electrochimica Acta (Elsevier) to reuse in a book under the copyright license number 4454750182186)

correlated with the number of defects in the structure, while the full-width half-maxima of D-peak is correlated with the grain size. The presence of a right-shouldered G-peak around 1,620 cm^{-1} is called second-order D'-peak which appears as a shoulder peak of G-peak. This peak is observed in some disordered graphitic carbon structures, including graphene, but cannot be seen in defect-free structure. Its intensity is usually lower as compared to the G-peak but becomes more important especially in CNW structure. Other second-order features are indicated by the remaining peaks named 2D, D + G, 2G. The peak around 2,700 cm^{-1} is often called 2D. Generally, this peak fall around a zone of the Raman graph where the 2D Raman shift doubles the Raman shift of the D-peak [100, 101]. The origin of peaks is not related to the presence of defects or grain boundary. It arises from the interaction of two phonons. The peak is very intense, especially in planar structure like graphene, CNWs. The peak shape of Raman can give us information about the number of layers present in the structure [102]. The peak around 2,700 cm^{-1} observed here is D + G peak, which is generally associated with the damaged graphene. It is less evident or almost absent in high-quality and defect-free carbon structure [103]. The peak around 3,240 cm^{-1} is sometimes called 2G and is present in both planar (graphene, graphite, CNWs) and nonplanar (CNT) structures [104].

6.5 Growth of CNWs

Vertically arranged graphene materials, arrays of CNWs, can be grown onto as-purchased flexible metal polycrystalline Cu foil. It is directly tested as anode in Li^+-ion batteries without any prior treatment [70, 105]. A commercial MW plasma-enhanced

CVD technique (*CVD-CN-100 from Ulvac, Japan Ltd.*) [106–108] is used for the growth of CNWs. The whole process and growth conditions typically consist of, first, exposure of Cu foil to a hydrogen (H_2) flux; then, methane (CH_4) as a reaction gas is introduced in combination with H_2 (H_2:CH_4) with the ratio of 20:20 sccm. At growth stage, bias voltage of 200 V is applied. The whole process is done at a constant pressure of 1.5 mTorr pressure. In the process, hydrocarbon gas decomposes into graphitic carbon. The process needs to be catalyzed and is therefore applied on affordable Cu-metal surface, which provides a better control on the number of layers with electrochemically active surface. Figure 6.4 shows the charge–discharge cycling of CNWs on Cu at different current rates. The behavior of current is slightly different from the standard behavior of carbon. This may be due to the presence of disordered carbon in the CNW deposits (*as learnt by Raman spectroscopy*). However, one of the highlights of the material is that even if the current rate is increased to higher orders, the capacity is not dropping down.

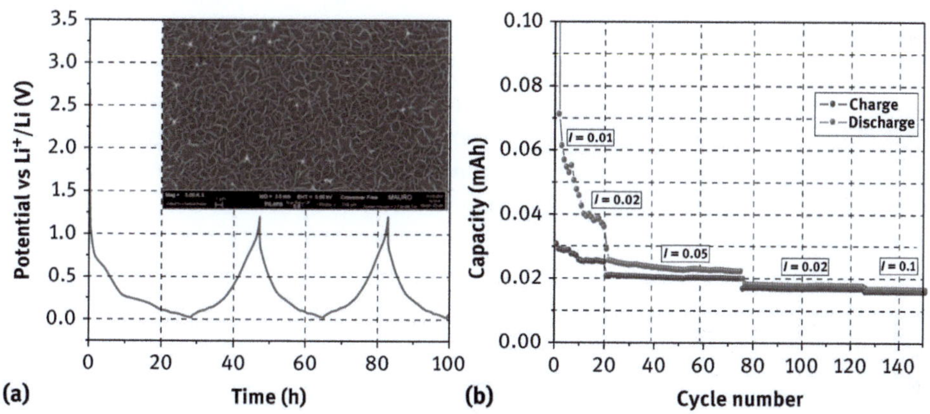

Figure 6.4: (a) Graph showing the galvanostatic charge/discharge curves of CNWs deposited on Cu: inset – morphological characterization of the nanostructured carbons deposited on Cu by FESEM (*ZEISS Supra 40*); (b) capacity versus number of cycles in a cell Li/electrolyte/CNW-Cu at room temperature. (Image courtesy by Electrochimica Acta(Elsevier) to reuse in a book under the copyright license number 4454750182186)

6.6 Carbonized cellulose paper

Use of paper in our everyday life has turned out to be an original answer to the demand of eco-friendly, flexible, lightweight and cost-effective materials for electronics [108, 109] and energy-storage devices [110, 111]. There are promising reports on paper/cellulose-based composite electrodes for supercapacitors and Li batteries

[112, 113]; the rough and porous surface of paper has turned out to be a great advantage, where both high surface area and electrolyte absorption are in demand. Recyclability and biodegradability are significant benefits, considering that cost of waste management and impact on the environment could be considerably low. It has been proven by Liu et al. [109] that paper printed circuit boards have about two orders of magnitude less impact on the environment than the ordinary printed ones.

The work on this material has demonstrated the feasibility of manufacturing graphite-based negative electrodes on a pilot-line exploiting microfibrillated cellulose as binder, through the spray-coating method, and the use of water-based papermaking technologies for large-scale production of Li batteries [114]. Beneventi et al. have successfully assembled a flexible, truly solid Li^+-ion cell based on self-standing electrodes manufactured using conducting paper as substrate. This approach makes use of a carbonization step of the nanocellulose-binded electrodes at high temperature to obtain conducting carbon nanofiber binder concurrently removing water and impurities, so that no successive drying step was necessary, thus making it more cost-effective in terms of energy and time. An NMFC-reinforced polymeric membrane which was activated by incorporating a freshly prepared room temperature ionic liquid-based green and safe electrolyte solution was used, and the resulting truly solid system assembled in a "pouch cell" envelop demonstrated a remarkably stable cycling behavior upon prolonged cycling at ambient temperature even at elevated current rates as well as in bent configuration. It was also noted that all cell components could be easily recovered using common water-based recycling procedures used in paper mill. The waste was minimal; the process was of low cost and low energy demanding, which led to the worldwide sustainable Li-based battery use and recovery.

6.7 Preparation of self-standing electrodes

Figure 6.5 demonstrates the different processes involved in obtaining pyrolyzed nanocellulose-binded electrodes, which are then used to assemble the solid polymeric Li^+-ion cell. The anode slurry was made of graphite powder (GP), carbon black (CB), NMFC, carboxy methyl cellulose (CMC) and alum. The cathode slurry was composed of lithium iron phosphate ($LiFePO_4$)/carbon, CB, NMFC, CMC and alum. Both the slurries, having a dry solid content ranging between 25 and 15 wt.%, were prepared by dissolving CMC in the aqueous NMFC slurry, followed by the addition of CB.

A commercial airless spray system (*Wagner, Project Pro 119*) was used to spray the anode slurry on both sheets and on other substrates (i.e., smooth polyester films and copy paper). An aqueous suspension of carbonized NMFC (2 wt.%) was

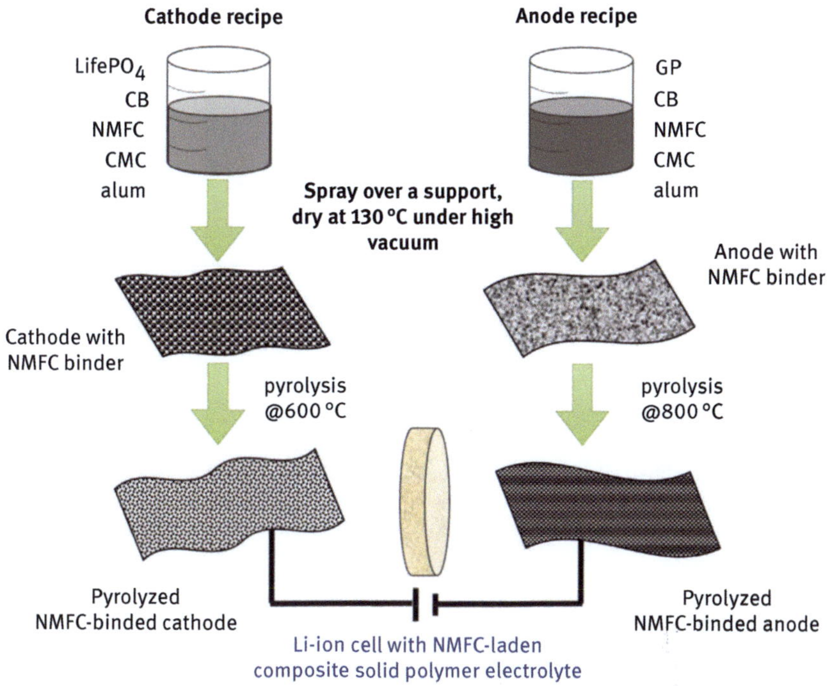

Figure 6.5: Different phases of the process involved in obtaining pyrolyzed nanocellulose-binded electrodes, which are then used to assemble the solid polymeric Li^{+}-ion cell. (Image courtesy by Carbon (Elsevier) to reuse in a book under the copyright license number 4454160899455)

prepared for reinforcing the polymer electrolyte and also the same was used as a binder for retaining the active materials within the electrode films (Figure 6.6).

6.8 Pyrolysis of cellulose paper

Pyrolysis method [115] is adopted for carbonization of waste NMFC. This involves the pyrolysis of self-standing anode and cathode samples at 800 and 600 °C, respectively. To avoid thermal degradation of the $LiFePO_4$ active material, lower pyrolysis temperature was applied to the cathode material. The pyrolysis chamber is normally flushed with argon inert gas to avoid oxygen traces. The furnace is heated with 10 °C/min. Pyrolysis is carried out for 30 min after achieving the required temperature. These steps ensured the completion of the carbonization procedure. Figure 6.7 shows the different forms of an NMFC sheet that has undergone pyrolysis. The pyrolysis method is normally followed by the cooling of the furnace to room temperature to recover the carbonized materials [69].

Figure 6.6: Different steps illustrating the facile recovery of pyrolyzed nanocellulose-binded electrode materials by exploiting the standard water-based recycling procedure used in paper industries. (Image courtesy by Carbon (Elsevier) to reuse in a book under the copyright license number 4454160899455)

Figure 6.7: Self-standing ability of (a) pristine NMFC paper, (b) pyrolyzed NMFC paper, (c) pyrolyzed NMFC paper in its bend configuration, (d) representative nanocellulose-binded $LiFePO_4$-based cathode in its top and (e) bottom side views, (f) pyrolyzed $LiFePO_4$-based cathode in its bottom and (g) top side views, (h) pyrolyzed GP-based anode. (Image courtesy by Carbon (Elsevier) to reuse in a book under the copyright license number 4454160899455)

6.9 Analysis of pyrolyzed cellulose paper

Figures 6.8 and 6.9 shows the FESEM images of GP-based anode pyrolyzed at 800 °C and $LiFePO_4$-based cathode pyrolyzed at 600 °C. The structure of the carbonized electrode (carbonized anode) is more accessible with respect to the pristine spray-deposited paper electrodes, and the presence of both $LiFePO_4$ and CB particles is well evidenced. The carbonized cellulose fiber network successfully binds the electrode components effectively, maintaining the preburning structure, despite slight shrinkage. It was observed that $LiFePO_4$ particles perfectly maintained their characteristics after heat treatment without suffering any kind of modification (Figure 6.9d). The fundamental triphasic system composed of the active material grain ($LiFePO_4$ particle), the electronic conductivity enhancer (CB spherical particle, locally enhancing the electronic conductivity of $LiFePO_4$) and the binder (electronic conductive carbonized cellulose fiber, which in turn also ameliorates the electronic conductivity of the system) was still present as highlighted by the yellow circle.

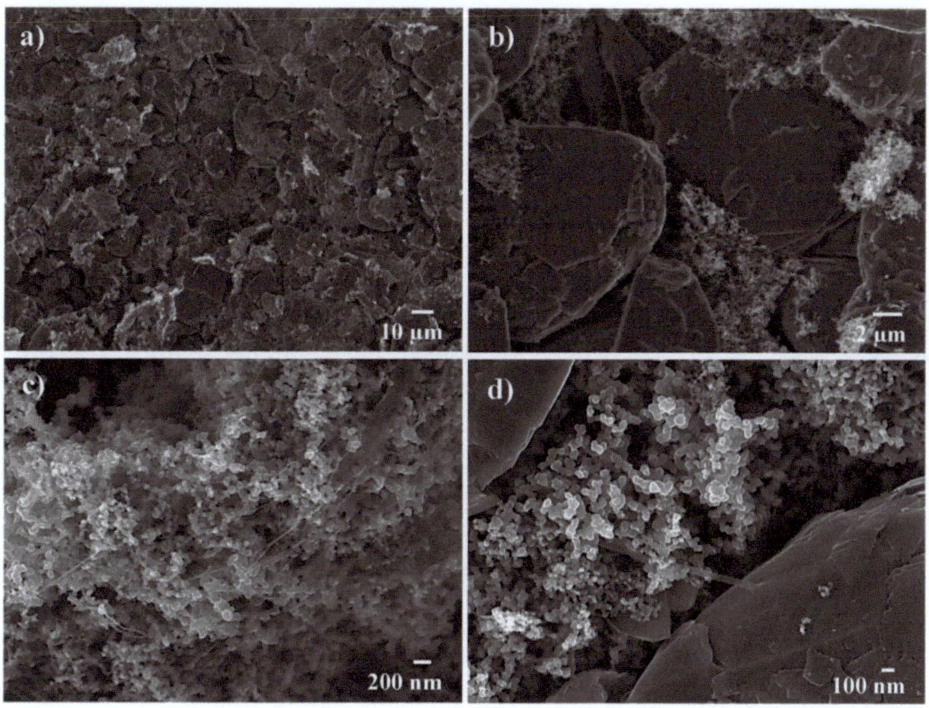

Figure 6.8: FESEM at different magnifications of GP anode pyrolyzed at 800 °C. (Image courtesy by Carbon (Elsevier) to reuse in a book under the copyright license number 4454160899455)

Figure 6.9: FESEM at different magnifications of $LiFePO_4$ cathode pyrolyzed at 600 °C. (Image courtesy by Carbon (Elsevier) to reuse in a book under the copyright license number 4454160899455)

Figure 6.10 shows the Raman spectrum of GP-based anode, $LiFePO_4$-based cathode and NMFC paper. A-800-TS shows a good separation and sharp D, G and G_O peaks when compared with A-800-BS. It evidences that the carbon material in sample A-800-TS is better graphitized than that of A-800-BS. Raman analysis of top rough (C-600-TS) and bottom smooth (C-600-BS) surfaces of the $LiFePO_4$-based cathode shows different behavior (Figure 6.10c and d). Intensity ratios shown in Figure 6.10f indicate that C-600-TS has lower graphitized matrix as compared to C-600-BS. The Raman spectrum of pyrolyzed NMFC paper (Figure 6.10e) shows no peak intensity and positioning variation on either side of the paper due to the uniform characteristics of both the surfaces. In general, the Raman analysis of pyrolyzed NMFC paper shows similar graphitized structure as that of C-600-BS. It can be stated that A-800-TS and C-800-TS have lower degrees of graphitization after analyzing the Raman spectra. Raman analysis also confirmed that the temperature variation did not reflect on the variation of graphitization of the sample. Raman spectroscopy by itself does not provide a complete picture of carbon quality.

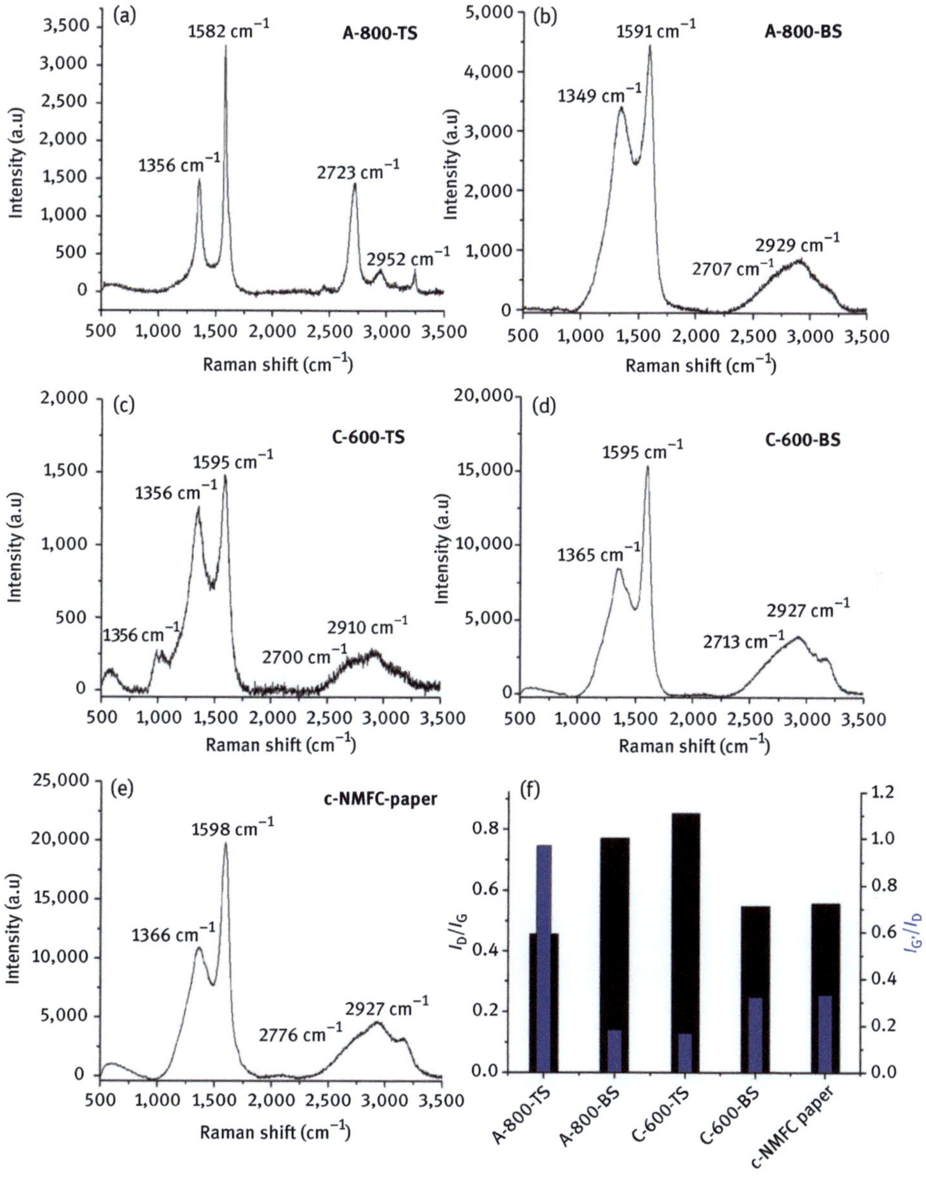

Figure 6.10: Raman spectra of (a) GP-based anode pyrolyzed at 800 °C analyzed on top and (b) bottom surfaces, (c) $LiFePO_4$-based cathode pyrolyzed at 600 °C analyzed on top and (d) bottom surfaces, (e) NMFC paper carbonized at 600 °C. (f) Samples versus I_D/I_G and $I_{G'}/I_D$ ratios. (Image courtesy by Carbon (Elsevier) to reuse in a book under the copyright license number 4454160899455)

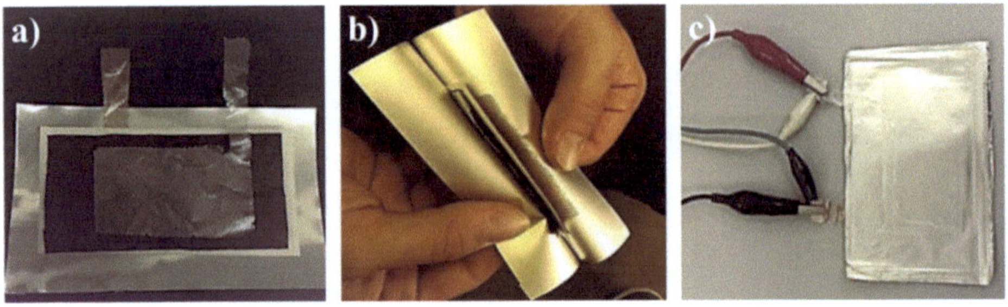

Figure 6.11: Different assembly stages of an all-paper-based solid polymer cell in pouch configuration: (a) assembly of the electrodes on the current collector grids, (b) packing followed by sealing, (c) final Li^+-ion cell connected to the instrument for testing. (Image courtesy by Carbon (Elsevier) to reuse in a book under the copyright license number 4454160899455)

Figure 6.11 provides the different assembly stages of an all-paper-based solid polymer cell in pouch configuration. The composite polymer electrolyte is sandwiched between the carbonized nanocellulose-binded $LiFePO_4$-based cathode and graphite-based anode paper electrodes and housed in a "pouch cell" envelop with Al and Cu grid current collectors (for cathode and anode, respectively). Different phases of the assembly are shown in Figure 6.11a–c. The cell is kept for ambient temperature galvanostatic charge/discharge cycling at various current rates.

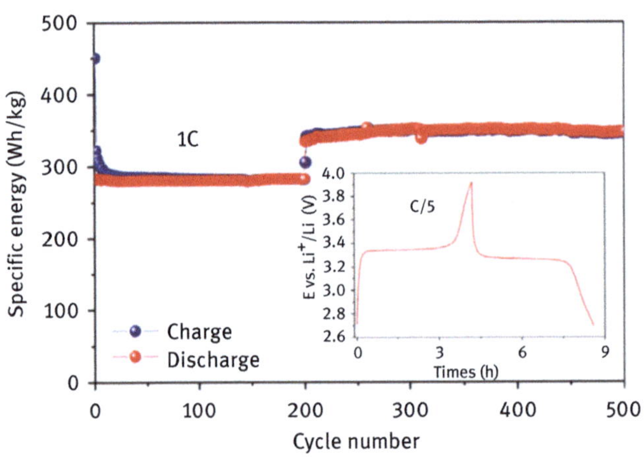

Figure 6.12: Current charge/discharge cycling behavior in terms of specific energy versus cycle number at different current regimes and charge and discharge potential versus time profiles extracted from the 400th cycle at C/5 rate. (Image courtesy by Carbon (Elsevier) to reuse in a book under the copyright license number 4454160899455)

Figure 6.12 provides the electrochemical response of the Li^+-ion polymer cell in terms of specific energy (Wh/kg calculated on the weight of the active cell components excluding current collectors and packaging) versus cycle number at different rates. The potential versus time profiles (400th constant current charge and discharge profiles at C/5 current rate) are very well consistent with the potential trend of the corresponding cell with natural cellulose bonded electrodes and liquid electrolyte [116]. The potential drop between charge and discharge plateaus was found to be limited, which accounts for low internal resistance, thus good electrode/electrolyte interface. The cell showed good rate capability and high coulombic efficiency approaching 100%. Specific energy values were found to be definitely interesting for a solid polymeric device, approaching 300 Wh/kg at 1 C and even exceeding 350 Wh/kg when reducing the current rate to C/5. Even more important, the cycling response is maintained steady for at least 500 cycles. This observation is definitely remarkable for a solid polymeric device. The exceptional capacity retention indicates that no contact loss occurred between the active material particles. The interface between active material particles and pyrolyzed nanocellulose binder remains intact even after over 500 charge/discharge cycles.

Figure 6.13 provides the electrochemical stability window of the composite polymer electrolyte evaluated by linear sweep voltammetry. It was found to exceed 4.6 V versus Li^+/Li in the anodic region. The Li plating/stripping indicates an overall good cathodic electrochemical stability, which is evident at around 0 V versus Li^+/Li.

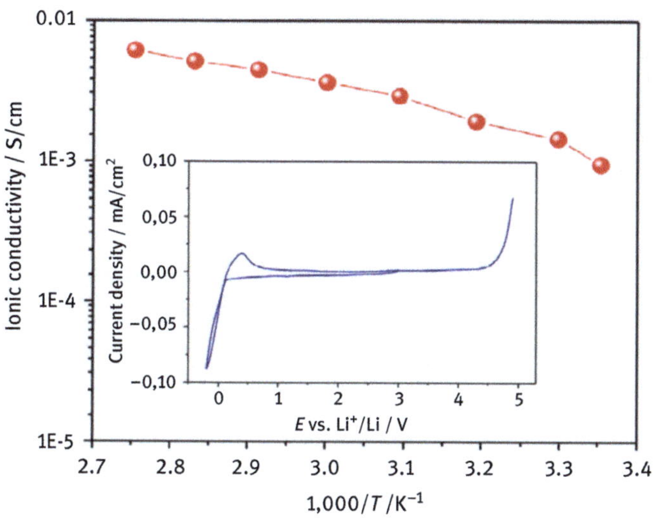

Figure 6.13: Arrhenius plot showing the ionic conductivity versus temperature; inset: electrochemical stability window of the polymer electrolyte. (Image courtesy by Carbon (Elsevier) to reuse in a book under the copyright license number 4454160899455)

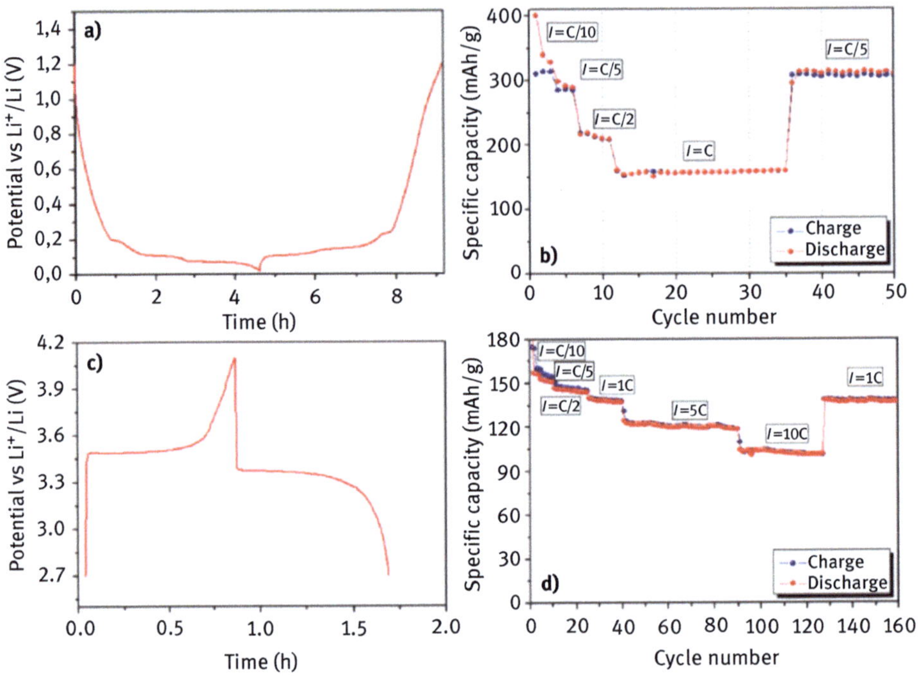

Figure 6.14: Ambient temperature electrochemical characterization of the carbonized electrodes in Li metal cell with standard liquid electrolyte: (a) potential versus time plot of a GP-based anode pyrolyzed at 800 °C showing typical charge and discharge profiles at C/5; (b) specific capacity versus cycle number plot of the same anode at different discharge/charge current regimes; (c) potential versus time plot of a $LiFePO_4$-based cathode pyrolyzed at 600 °C showing charge and discharge profiles at 1C; (d) specific capacity versus cycle number plot of the same cathode at different charge/discharge current regimes. (Image courtesy by Carbon (Elsevier) to reuse in a book under the copyright license number 4454160899455)

Figure 6.14 provides the potential versus time plot of the carbonized GP-based anode. Figure 6.14a clearly reflects the characteristics of the electrode active material. It shows the typical discharge (intercalation of Li^+ ions) and charge (deintercalation of Li^+ ions) profiles expected for a highly crystalline graphite material, where the different insertion steps toward the so-called stage-1 Li-graphite intercalation compound, that is LiC_6, are clearly visible [117]. From the analysis shown in Figure 6.14b, the irreversible capacity loss during the first cycle can be attributed to the expected side reactions with the components of the electrolyte (solid electrolyte interface layer formation). Typically, organic solvent molecules and salt anions are reduced on the active material's surface. Thus, forming insoluble Li salt that precipitates to form a passivating film. This is helpful in preventing further reaction between components of the electrolyte and graphite active material [118].

The reversible specific capacity obtained at low C/10 rate is close to the theoretical value for graphite, reflecting the practical specific capacity of the commercial material used. This can be explained by the fact that the carbon obtained upon cellulose pyrolysis also takes part in the electrochemical process to a certain extent. After the initial cycles, the Coulombic efficiency rapidly increases to above 99%. Subsequently, it remains highly stable throughout the cycles. This indicates that the formed surface film lasted intact and showed excellent reversible cycling after surface reactions were completed. At a relatively high current of 1C, the anode can deliver a capacity of 150 mAh/g, which is definitely higher if compared to the results obtained for the same pristine electrode reported by Beneventi et al. [114]. Thus, it can be summarized that the carbonization procedure at high temperature successfully enhances the electrochemical behavior of paper electrodes. This can be due to the improved electronic conductivity resulting from the presence of the carbonized network of fibers, which contributes to the overall electrode specific capacity. The performance of carbonized $LiFePO_4$-based cathode in lab-scale Li cell was outstanding in terms of both overall specific capacity and rate capability (Figure 6.14d). The capacity obtained at low C/10 rate was around 160 mAh/g, which is the specific capacity of the commercial material used. It is also interesting to note that after the specific carbonization procedure adopted, this kind of paper-like cathode is able to operate at very high 10C current regime, still providing specific capacity values exceeding 100 mAh/g, thus allowing a higher power output of the final device. These results are difficult to be achieved by the corresponding "as-prepared" spray deposited $LiFePO_4$-based electrodes as described by Zolin et al. [116]. This signifies the role of pyrolyzed nanocellulose fibers, which combine the functions of strong binder and conducting additive without negatively affecting cycling stability and rate performance [119]. It is worthy to note that this system does not show any performance decay; indeed, reducing the C-rate after more than 120 charge/discharge cycles completely restores the specific capacity, thus achieving excellent stability.

The overall observation concludes that the accomplishment of the "12 principles of Green Chemistry" is a fundamental goal to be pursued by means of designing/optimizing environmentally "conscious" processes, materials and devices to reduce their environmental impact throughout all the phases of their life cycle, thus benefiting the economy, protecting people and achieving the real goal of a truly sustainable world. This goal can be achieved by exploiting carbonized nanocellulosic binders and water-based papermaking coupled with high-performing green composite polymer electrolyte.

Acknowledgments: This work was partially supported by grant 685844 (project MODCOMP) in the framework of European H2020 program. The authors thank Mr. Mauro Raimondo, Politecnico di Torino, for FESEM analysis of the samples.

References

[1] Winter, M. and Brodd, R. J., What are batteries, fuel cells, and supercapacitors?. Chem. Rev 2004, 104, 10, 4245–4270, Oct.

[2] Linden, David, *Handbook of Batteries*, 3rd ed. McGraw-Hill, New York City, 1994.

[3] Lacks, D. J. and Mohan Sankaran, R., Contact electrification of insulating materials. J. Phys. D. Appl. Phys 2011, 44, 45, 453001, Nov.

[4] Khundkar, R., Malic, C., and Estela, C., Burned by a battery–coin short circuit: Old concept for a new burn, Burns 2010, 36, 1, e4–e5, Feb.

[5] Seroglou, F., Koumaras, P., and Tselfes, V., History of science and instructional design: The case of electromagnetism, Sci. Educ 1998. 7, 3, 261–280

[6] Whittingham, M. S., History, evolution, and future status of energy storage, Proc. IEEE 2012, 100, Special Centennial Issue, 1518–1534.

[7] George Blomgren, J. H., "Batteries, primary cells," in *Kirk-Othmer Encyclopedia of Chemical Technology*, Kirk-Othmer, Ed. Wiley, 2004, 880.

[8] Tarascon, J.-M. and Armand, M., Issues and challenges facing rechargeable lithium batteries, Nature 2001, 414, 359, Nov.

[9] M. S. W.-R. H. and Armand, M., *Fast Ion Transport in Solids*, W. Van Gool, Ed. New York, NY, United States, 1973, 645–665.

[10] Whittingham, M. S., Electrical energy storage and intercalation chemistry. Science (80-.) 1976 192, 42–44, 1126–1127.

[11] Murphy, D. W., Di Salvo, F. J., Carides, J. N., and Waszczak, J. V., Topochemical reactions of rutile related structures with lithium. Mater. Res. Bull 1978, 13, 12,1395–1402, Dec.

[12] Lazzari, M. and Scrosati, B., A cyclable lithium organic electrolyte cell based on two intercalation electrodes, J. Electrochem. Soc 1980, 127, 3, 773, Mar.

[13] Mizushima, K., Jones, P. C., Wiseman, P. J., and Goodenough, J. B., LixCoO2 (0. Mater. Res. Bull 1980, 15, 6, 783–789, Jun.

[14] Thackeray, M. M., David, W. I. F., Bruce, P. G., and Goodenough, J. B., Lithium insertion into manganese spinels. Mater. Res. Bull 1983, 18, 4, 461–472, Apr.

[15] Boehm, H. P., Setton, R., and Stumpp, E., Nomenclature and terminology of graphite intercalation compounds. Carbon N. Y 1986, 24, 2, 241–245, Jan.

[16] Armand, M., "Intercalation Electrodes," in *Materials for Advanced Batteries*, 1st ed., D. Murphy, Ed. New York: Springer US, 1980, 373.

[17] Brandt, K., Historical development of secondary lithium batteries. Solid State Ionics 1994, 69, 3–4, 173–183, Aug.

[18] Brandt, K., Herr, K., Hoge, D., DECHEMA Monogr 1993, 128, 279.

[19] Ohzuku, T., Iwakoshi, Y., and Sawai, K., Formation of lithium-graphite intercalation compounds in nonaqueous electrolytes and their application as a negative electrode for a lithium ion (Shuttlecock) cell. J. Electrochem. Soc 1993, 140, 9, 24–90, Sep.

[20] Koch, V. R., Specular lithium deposits from lithium hexafluoroarsenate/diethyl ether electrolytes, J. Electrochem. Soc 1982, 129, 1, 1.

[21] Daniel, C. and Besenhard, J. O., *Handbook of Battery Materials, 2.Auflage*. Wiley, 2011.

[22] Hirai, T., Effect of additives on lithium cycling efficiency, J. Electrochem. Soc, 1994, 141, 9, 2300.

[23] Flandrois, S. and Simon, B., Carbon materials for lithium-ion rechargeable batteries. Carbon N. Y 1999, 37, 2, 165–180.

[24] Z. Q. von S. U. Dahn JR, Sleigh AK, Shi H, Way BM, Weydanz WJ, Reimers JN, "Carbons and graphites as substitutes for the lithium anode in lithium batteries: New materials,

developments and perspectivesitle," in Industrial chemistry library, 5th ed., P. G, Ed. London: Elsevier, 1994, 1–47.
[25] Agubra, Victor and Fergus, Jeffrey, Review lithium ion battery anode aging mechanisms. Materials (Basel) 2013, 6, 1310–1325.
[26] Kodama, T. and Sakaebe, H., Present status and future prospect for national project on lithium batteries. J. Power Sources 1999, 81–82, 144–149, Sep.
[27] Takami, N., Structural and kinetic characterization of lithium intercalation into carbon anodes for secondary lithium batteries. J. Electrochem. Soc 1995, 142, 2, 371.
[28] Wang, H., Ikeda, T., Fukuda, K., Yoshio, M., Effect of milling on the electrochemical performance of natural graphite as an anode material for lithium-ion battery," J. Power Sources 1999, 83, 1–2, 141–147, Oct.
[29] Liu, Z., Yu, A., and Lee, J. Y., Modifications of synthetic graphite for secondary lithium-ion battery applications. J. Power Sources 1999, 81–82, 187–191, Sep.
[30] McMillan, R., Slegr, H., Shu, Z.,Wang, W., Fluoroethylene carbonate electrolyte and its use in lithium ion batteries with graphite anodes. J. Power Sources 1999, 81–82, 20–26, Sep.
[31] Andersson, A. M., Edström, K., Thomas, J. O., Characterisation of the ambient and elevated temperature performance of a graphite electrode. J. Power Sources 1999, 81–82, 8–12, Sep.
[32] Yazami,R., Surface chemistry and lithium storage capability of the graphite–lithium electrode. Electrochim. Acta 1999, 45, 1–2, 87–97, Sep.
[33] W. X. and Dahn, J. R., Study of irreversible capacities for Li Insertion in hard and graphitic carbons. J. Electrochem. Soc 1997, 144, 4, 1195–1201.
[34] Y. I. and K. S. Ohzuku, Tsutomu, Formation of lithium-graphite intercalation compounds in nonaqueous electrolytes and their application as a negative electrode for a lithium ion (Shuttlecock) cell J. Electrochem. Soc 1993, 140, 9, 2490–2498.
[35] Geim, A. K. and Novoselov, K. S., The rise of graphene," Nat. Mater 2007, 6, 3, 183–191, Mar.
[36] Han, J. et al., Engineering graphenes from the nano- to the macroscale for electrochemical energy storage. Electrochem. Energy Rev 2018, 1, 2, 139–168.
[37] Niu, C., Sichel, E. K., Hoch, R., Moy, D., Tennent, H., High power electrochemical capacitors based on carbon nanotube electrodes. Appl. Phys. Lett 1997, 70, 11, 1480–1482, Mar.
[38] Lee, S. W. et al., High-power lithium batteries from functionalized carbon-nanotube electrodes. Nat. Nanotechnol 2010, 5, 531, Jun.
[39] Kim, H., Park, K.-Y., Hong, J., Kang, K., All-graphene-battery: bridging the gap between supercapacitors and lithium ion batteries. Sci. Rep 2014, 4, 5278, Jun.
[40] Cheng,Q. , Okamoto, Y., Tamura, N., Tsuji, M., Maruyama, S., and Matsuo, Y., Graphene-like-graphite as fast-chargeable and high-capacity anode materials for lithium ion batteries. Sci. Rep 2017, 7, 1, 14782.
[41] Wang, S., Kravchyk, K. V, Krumeich, F., and Kovalenko, M. V, Kish graphite flakes as a cathode material for an aluminum chloride–graphite battery. ACS Appl. Mater. Interfaces 2017, 9, 34, 28478–28485, Aug.
[42] Yanwu, Zhu, Shanthi, Murali, Weiwei, Cai, Xuesong, Li, Ji Won, Suk, Jeffrey R, Potts, Rodney S., Ruoff, Graphene and graphene oxide: Synthesis, properties, and applications. Adv. Mater 2010, 22, 35, 3906–3924.
[43] Olivares-Marín, M. et al., Cherry stones as precursor of activated carbons for supercapacitors. Mater. Chem. Phys 2009, 114, 1, 323–327, Mar.
[44] Vivekchand, S. R. C., Rout, C. S., Subrahmanyam, K. S., Govindaraj, A., and Rao, C. N. R., Graphene-based electrochemical supercapacitors. J. Chem. Sci 2008, 120, 1, 9–13.
[45] Wick,P. et al., Classification framework for graphene-based materials. Angew. Chem. Int. Ed. Engl 2014.

[46] Chyan, Y., Ye, R., Li, Y., Singh, S. P., Arnusch, C. J., and Tour, J. M., Laser-Induced graphene by multiple lasing: Toward electronics on cloth, paper, and food. ACS Nano 2018, 12, 3, 2176–2183, Mar.
[47] T. O. and Norio, M. K. Takami, Asako Satoh, Large hysteresis during lithium insertion into and extraction from high-capacity disordered carbons. J. Electrochem. Soc 1998, 145, 2, 478–482.
[48] Pan, Q., Deng, Z., Zhang, X., and Wan, G., Electrochemical characteristics of lithium intercalation into natural gas coke serving as the negative electrode of a lithium battery. J. Power Sources 1999, 79, 1, 25–29, May.
[49] Sato, Y. et al., Characteristics of coke carbon modified with mesophase-pitch as a negative electrode for lithium ion batteries. J. Power Sources 1999, 81–82, 182–186, Sep.
[50] Sharon, M. et al., Camphor-based carbon nanotubes as an anode in lithium secondary batteries. J. Power Sources 2002, 104, 1, 148–153, Jan.
[51] Kumar, M., Kichambare, P., Sharon, M., Avery, N. R., and Black, K. J., Study of camphor-pyrolysed carbon electrode in a lithium rechargeable cell. Mater. Chem. Phys 2000, 66, 1, 83–89, Sep.
[52] Sunil Bhardwaj, M. S., Jaybhaye, Sandesh V., Madhuri Sharon, D. Sathiyamoorthy, K. Dasgupta, Jagadale, Pravin, Gupta, Arvind, Patil, Bhushan, Ozha, Goldie, Sunil Pandey, T. Soga, Rakesh Afre, Kalita, Golap, Carbon nanomaterial from tea leaves as an anode in lithium secondary batteries. Asian J. Exp. Sci 2008, 22, 2, 89–93.
[53] Yazami, R. and Touzain, P., A reversible graphite-lithium negative electrode for electrochemical generators. J. Power Sources 1983, 9, 3, 365–371, Jan.
[54] Skundin, A. M., Grigor'eva, O. Y., Kulova, T. L., and Pouchko, S. V, The lithium intercalation into graphite from electrolyte and from solid lithium. J. Solid State Electrochem 2003, 8, 1, 11–14.
[55] A. M. S. O. Yu. Grigor'eva, T. L. Kulova, S. V. Pushko, Lithium Intercalation into Graphite during Direct Contact and Anodic Polarization. Russ. J. Electrochem 2002, 38, 12, 1327–1333.
[56] Kulova, T. L., Skundin, A. M., Nizhnikovskii, E. A., and Fesenko, A. V, Temperature effect on the lithium diffusion rate in graphite. Russ. J. Electrochem 2006, 42, 3, 259–262.
[57] Roh, Y. B. et al., Unique charge/discharge properties of carbon materials with different structures. J. Power Sources 1997, 68, 2, 271–276, Oct.
[58] Morita, M., Nishimura, N., and Matsuda, Y., Charge/discharge cycling behavior of pitch-based carbon fiber in organic electrolyte solutions. Electrochim. Acta 1993, 38, 13, 1721–1726, Sep.
[59] Jean, R. M. M., Desnoyer, C., Tranchant, A., Electrochemical and Structural Studies of Petroleum Coke in Carbonate-Based Electrolytes. J. Electrochem. Soc 1995, 142, 7, 2122–2125.
[60] Takashi Uchida, K. S., Morikawa, Yasuyuki, Ikuta, Hiromasa, Wakihara, Masataka, Chemical Diffusion Coefficient of Lithium in Carbon Fiber. J. Electrochem. Soc 1996, 143, 8, 2606–2610.
[61] Aurbach, D., Levi, M. D., Levi, E., and Schechter, A., Failure and Stabilization Mechanisms of Graphite Electrodes. J. Phys. Chem. B 1997, 101, 12, 2195–2206, Mar.
[62] Levi, M. D. E. Levi, A., and Aurbach, D., The mechanism of lithium intercalation in graphite film electrodes in aprotic media. Part 2. Potentiostatic intermittent titration and in situ XRD studies of the solid-state ionic diffusion. J. Electroanal. Chem 1997, 421, 1–2, 89–97, Jan.
[63] Markovsky, B., Levi, M. D., and Aurbach, D., The basic electroanalytical behavior of practical graphite–lithium intercalation electrodes. Electrochim. Acta 1998, 43, 16–17, 2287–2304, May.

[64] Barsoukov, E., Kim, J. H., Kim, J. H., Yoon, C. O., and Lee, H., Kinetics of lithium intercalation into carbon anodes: In situ impedance investigation of thickness and potential dependence. Solid State Ionics 1999, 116, 3–4, 249–261, Jan.
[65] Churikov, A. V, Chronoamperometric determination of the rate of lithium transfer in carbon electrodes. Ehlektrokhimiya 2002, 38, 1, 120–125.
[66] C. D. and S. M. Piao, Tiehua, Parka, Su-Moon, Intercalation of Lithium Ions into Graphite Electrodes Studied by AC Impedance Measurements. J. Electrochem. Soc 1999, 146, 8, 2794–2798.
[67] J. A. R. and R. E. W. Ping Yu, B. N. Popov, Determination of the lithium ion diffusion coefficient in graphite. J. Electrochem. Soc 1999, 146, 1, 8–14.
[68] Lau, K. C., Assary, R. S., and Curtiss, L. A. , "Aprotic Electrolytes in Li–Air Batteries BT – Electrolytes for Lithium and Lithium-Ion Batteries," Jow, T. R., Xu, K., Borodin, O., and Ue, M., Eds. New York, NY: Springer New York, 2014, 445–466.
[69] Zolin, L. et al., A simple route toward next-gen green energy storage concept by nanofibres-based self-supporting electrodes and a solid polymeric design. Carbon N. Y 2016, 107, 811–822.
[70] Nair, J. R. et al., Remarkably stable high power Li-ion battery anodes based on vertically arranged multilayered-graphene. Electrochim. Acta 2015, 182.
[71] Wang, Y.-G., Chang, Y.-C., Ishida, S., Korai, Y., and Mochida, I., Stabilization and carbonization properties of mesocarbon microbeads (MCMB) prepared from a synthetic naphthalene isotropic pitch. Carbon N. Y 1999, 37, 6, 969–976, Jan.
[72] Guerin, K., Fevrier-Bouvier, A., Flandrois, S., Simon, B., and Biensan, P., On the irreversible capacities of disordered carbons in lithium-ion rechargeable batteries, Electrochim. Acta 2000, 45, 10, 1607–1615, Jan.
[73] Buiel, E. and Dahn, J., Li-insertion in hard carbon anode materials for Li-ion batteries. Electrochim. Acta 1999, 45, 1–2, 121–130, Sep.
[74] Appetecchi, G., Croce, F., Marassi, R., Persi, L., Romagnoli, P., Scrosati, B., Lithium insertion into carbonaceous materials and transition metal oxides from high performance polymer electrolytes. Electrochim. Acta 1999, 45, 1–2, 23–30, Sep.
[75] Sato, E. M. Noguchi, K, Demachi, M, A, Oki N, A mechanism of lithium storage in disordered carbons. Science (80-.) 1994, 22, 264 (5158), 556–568.
[76] Walker, S., Holliday AK, Hughes, G., "Comprehensive inorganic chemistry," in Carbon, T.-D. AF, Ed. Oxford: Pergamon Pres, 1973, 1173–1294.
[77] Liu, Y., Xue, J. S., Zheng, T., and Dahn, J. R., Mechanism of lithium insertion in hard carbons prepared by pyrolysis of epoxy resins. Carbon N. Y 1996, 34, 2, 193–200, Jan.
[78] Larcher, D., Mudalige, C., Gharghouri, M., and Dahn, J. R., Electrochemical insertion of Li and irreversibility in disordered carbons prepared from oxygen and sulfur-containing pitches. Electrochim. Acta 1999, 44, 23, 4069–4072, Jul.
[79] Gong, J., Wu, H., and Yang, Q., "Structural and electrochemical properties of disordered carbon prepared by the pyrolysis of poly(p-phenylene) below 1000 C for the anode of a lithium-ion battery. Carbon N. Y 1999, 37, 9, 1409–1416, Jan.
[80] Buiel, E. R., George, A. E., and Dahn, J. R., Model of micropore closure in hard carbon prepared from sucrose. Carbon N. Y 1999, 37, 9, 1399–1407, Jan.
[81] Dahn, J. R., Xing, W., and Gao, Y., The 'falling cards model' for the structure of microporous carbons. Carbon N. Y 1997, 35, 6, 825–830, Jan.
[82] Zheng, T., Xing, W., and Dahn, J. R., Carbons prepared from coals for anodes of lithium-ion cells. Carbon N. Y 1996, 34, 12, 1501–1507, Jan.
[83] Gibaud, A., Xue, J. S., and Dahn, J. R., A small angle X-ray scattering study of carbons made from pyrolyzed sugar. Carbon N. Y 1996, 34, 4, 499–503, Jan.

[84] Zheng, T. and Dahn, J. R., Hysteresis observed in quasi open-circuit voltage measurements of lithium insertion in hydrogen-containing carbons. J. Power Sources 1997, 68, 2, 201–203, Oct.
[85] Zhou, P., Papanek, P., Bindra, C., Lee, R., and Fischer, J. E., High capacity carbon anode materials: Structure, hydrogen effect, and stability. J. Power Sources 1997, 68, 2, 296–300, Oct.
[86] Chakrabarti, A. et al., Conversion of carbon dioxide to few-layer graphene. J. Mater. Chem 2011, 21, 26, 9491–9493.
[87] Seo, Hong-Kyu and Lee, Tae-Woo, Graphene growth from polymers. Carbon Lett 2013, 14, 3, 145–151.
[88] Sun, Z., Yan, Z., Yao, J., Beitler, E., Zhu, Y., and Tour, J. M., Growth of graphene from solid carbon sources. Nature 2010, 468,549, Nov.
[89] Ruan, G., Sun, Z., Peng, Z., and Tour, J. M., Growth of Graphene from Food, Insects, and Waste. ACS Nano 2011, 5, 9, 7601–7607, Sep.
[90] Seo, H.-K. et al., Value-added Synthesis of Graphene: Recycling Industrial Carbon Waste into Electrodes for High-Performance Electronic Devices. Sci. Rep 2015, 5, 16710, Nov.
[91] Guo, H.-L., Wang, X.-F., Qian, Q.-Y., Wang, F.-B., and Xia, X.-H., A Green Approach to the Synthesis of Graphene Nanosheets. ACS Nano 2009 3, 9, 2653–2659, Sep.
[92] Kim, Y. S. et al., Methane as an effective hydrogen source for single-layer graphene synthesis on Cu foil by plasma enhanced chemical vapor deposition. Nanoscale 2013, 5, 3, 1221–1226.
[93] Shah, J., Lopez-Mercado, J., Carreon, M. G., Lopez-Miranda, A., and Carreon, M. L., Plasma synthesis of graphene from mango peel. ACS Omega 2018, 3, 1, 455–463, Jan.
[94] Wang, J., Zhu, M., Outlaw, R. A., Zhao, X., Manos, D. M., and Holloway, B. C., Synthesis of carbon nanosheets by inductively coupled radio-frequency plasma enhanced chemical vapor deposition. Carbon N. Y 2004, 42, 14, 2867–2872, Jan.
[95] Wu, Y., Qiao, P., Chong, T., and Shen, Z., Carbon nanowalls grown by microwave plasma enhanced chemical vapor deposition. Adv. Mater 2002, 14, 1, 64–67.
[96] Brar, V. W. et al., Second-order harmonic and combination modes in graphite, single-wall carbon nanotube bundles, and isolated single-wall carbon nanotubes. Phys. Rev. B 2002, 66, 15, 155–418, Oct.
[97] Dresselhaus, M. S., Dresselhaus, G., Saito, R., and Jorio, A., Raman spectroscopy of carbon nanotubes. Phys. Rep 2005, 409, 2, 47–99, Mar.
[98] Damen, T. C., Porto, S. P. S., and Tell, B., Raman effect in zinc oxide. Phys. Rev 142, 2, 570–574, Feb.
[99] Galeener, F. L. and Lucovsky, G., Longitudinal optical vibrations in glasses: GeO2 and SiO2. Phys. Rev. Lett 1976, 37, 22, 1474–1478, Nov.
[100] Cançado, L. G. et al., Quantifying defects in graphene via raman spectroscopy at different excitation energies. Nano Lett 2011, 11, 8, 3190–3196, Aug.
[101] Graf, D. et al., Spatially resolved raman spectroscopy of single- and few-layer graphene. Nano Lett 2007, 7, 2, 238–242, Feb.
[102] Dresselhaus, M. S., Dresselhaus, G., Saito, R., and Jorio, A., Raman spectroscopy of carbon nanotubes. Phys. Rep 2005, 409, 2, 47–99, Mar.
[103] Ferrari, A. C. and Robertson, J., Interpretation of Raman spectra of disordered and amorphous carbon, Phys. Rev. B 2000, 61, 20, 14095–14107, May.
[104] Tan, P., Dimovski, S., and Gogotsi, Y., Raman scattering of non–planar graphite: arched edges, polyhedral crystals, whiskers and cones. Philos. Trans. R. Soc. London. Ser. A Math. Phys. Eng. Sci 2004, 362, 18–24, 2289 LP-2310, Nov.

[105] Hiralal, P., Rius, G., Andrew, P., Yoshimura, M., and Amaratunga, G. A. J., Tailoring carbon nanostructure for high frequency supercapacitor operation. J. Nanomater 2014, 8, 1–7, Jan.
[106] Rius, G. and Yoshimura, M., Synthesis control for carbon nanowalls on copper supports pro development of green energy applications. e-Journal Surf. Sci. Nanotechnol 2012, 10, 305–309.
[107] Rius, G. and Yoshimura, M., Structured nanocarbon on various metal foils by microwave plasma enhanced chemical vapor deposition. J. Phys. Conf. Ser 2013, 417, 012–010, Mar.
[108] Tobjörk, D. and Österbacka, R., Paper electronics. Adv. Mater 23, 17, 1935–1961, May 2011.
[109] Liu, J. et al., Future paper based printed circuit boards for green electronics: fabrication and life cycle assessment, Energy Environ. Sci 2014, 7, 11, 3674–3682, Oct.
[110] Jabbour, L., Bongiovanni, R., Chaussy, D., Gerbaldi, C., and Beneventi, D., Cellulose-based Li-ion batteries: A review. Cellulose 2013, 20, 4, 1523–1545, Aug.
[111] Berecibar, M., Gandiaga, I., Villarreal, I., Omar, N., Van Mierlo, J., and Van den Bossche, P., Critical review of state of health estimation methods of Li-ion batteries for real applications. Renew. Sustain. Energy Rev 2016, 56, 572–587, Apr.
[112] Sharifi, F., Ghobadian, S., Cavalcanti, F. R., and Hashemi, N., Paper-based devices for energy applications. Renew. Sustain. Energy Rev 2015, 52, 1453–1472, Dec.
[113] Hu, L. and Cui, Y., Energy and environmental nanotechnology in conductive paper and textiles. Energy Environ. Sci 2012, 5, 4, 6423–6435, Mar.
[114] Beneventi, D. et al., Pilot-scale elaboration of graphite/microfibrillated cellulose anodes for Li-ion batteries by spray deposition on a forming paper sheet. Chem. Eng. J 2014, 243, 372–379, May.
[115] Blazsó, M., Pyrolysis for recycling waste composites. Manag. Recycl. Reuse Waste Compos 2010, 102–121, Jan.
[116] Zolin, L. et al., Flexible cellulose-based electrodes: Towards eco-friendly all-paper batteries. Chem. Eng. trans. 2014, 41, 361–366.
[117] Gotoh, K. et al., Properties of a novel hard-carbon optimized to large size Li ion secondary battery studied by 7Li NMR. J. Power Sources 2006, 162, 2, 1322–1328, Nov.
[118] Peled, E., The electrochemical behavior of alkali and alkaline earth metals in nonaqueous battery systems – the solid electrolyte interphase model. J. Electrochem. Soc 1979, 126, 12, 2047–2051, Dec.
[119] Kirshenbaum, K., et al. Batteries. In situ visualization of $Li/Ag_2VP_2O_8$ batteries revealing rate-dependent discharge mechanism. Science 2015, 347, 6218, 149–54, Jan.

Index

Activate 17
Activation treatment 25
Active screen plasma 18
Active species 19
Agglomeration 125, 139
Anode 164
Artificial intelligence 68

Battery 153
Biochar 158
Buckypapers (BP) 81

Carbon 156
Carbon-based materials 30
Carbon-based polymer nanocomposites 36
Carbon fibers 26–27, 52, 159
Carbon nanofibers 22
Carbon nanomaterial 157
Carbon nanotube (CNT) 34, 81, 115
Carbon nanowalls 159
Carbon paper 23
Catalysts 29
CFRP mirror 81
Chemical vapor deposition 159
Circular economy 66
Class II 129
CNTs 28
Coarse-grained 56
Composite material 27
Coulombic efficiency 173

Deformation experiments 121, 132, 136
Dispersion in nanocomposites 124
Dissipative particle dynamics 55

Effective medium approximation 61
Elastic moduli 48
Electrical conductivity 52, 54, 157
Electrical resistance 20
Electrocatalyst 24
Electrochemical surface area 25
Electrode 155
Electromagnetic compatibility (EMC) 81
Energy storage 154
Energy storage devices 22
Engineered material 66
Epoxy resins 40

Fibre Volume Fraction (FVF) 95
Finite element method 61
Flexible substrates 13
Fuel cells 23
Functional groups 27
Functionalization 30
Functionalize 17

Gas diffusion layer 23
Generalized Hooke's law 62
Glass transition 124
Glass transition temperature 120
Graphene 1, 20, 34, 157
Graphene-based devices 1
Graphene-based nanocomposites 116
Graphene-based polymer nanocomposites 123
Graphene oxide 1

Heterogeneous materials 158
Hydrogen bonds 120, 123, 132, 135–137, 145

In situ polymerization 119
Interfacial characteristics 67
Ion 155
Ion bombardment 19

Kapitza resistance 44

Lattice Boltzmann method 55
Lithium 154

Machine learning 68
Materials Modelling 38
MD algorithm 130
MD simulation packages 131
Melt processing 119
Mesoscale 57
Microscopic structure 67
Model reduction 67
Molecular dynamics 39
Molecular simulations 53, 126–127
Monte Carlo 39
Multiphysics modeling 39

https://doi.org/10.1515/9783110479133-007

Nanoscale microfibrillated cellulose 159
Nonequilibrium 40
Nonvolatile memories 1
Nonvolatile resistive memories 6

Optical transparency 20

Passive Intermodulation (PIM) 81
Percolation 54
Phonon 45
Photoresponsive materials 35
Piezoelectric materials 34
Plasma-enhanced CVD (PE-CVD) 29
Plasma–surface interaction 19
Plasma technology 18, 30
Platinum nanowires 25
Poly(acrylic acid) 123
Poly(methyl methacrylate) 120
Poly(vinyl alcohol) 120
Polyethylene 122
Polymer electrolyte 165
Polymer matrix composite 26
Polystyrene 121, 132
Polyvinylidene fluoride 122
Potential energy function 126–128, 130
Pouch cell 170
Pyrolysis 159

Raman 161
Recyclable carbon 157

Scotch tape method 116–117
Shape memory materials 33
Smart materials 33
Smoothed particle hydrodynamics 55
Solvent processing 119
Space 82
Space application 13
Specific energy 171
Spray-coating 164
Stress–strain response 48
Surface conductivity 22
Surface engineering 17–18
Surface functionalizations 47

Technology readiness level (TRL) 112
Thermal conductivity 40
Toxicity 68
Tunneling 53

VACNT 29

Waste 157
Wettability 24, 28

Bei Fragen zur Produktsicherheit wenden Sie sich bitte an:
If you have any questions regarding product safety,
please contact:

Walter de Gruyter GmbH
Genthiner Straße 13
10785 Berlin
productsafety@degruyterbrill.com

De Gruyter Series in Discrete Mathematics and Applications 1

Petrică C. Pop

Generalized Network Design Problems

Modeling and Optimization

De Gruyter

Mathematics Subject Classification 2010: Primary: 90B10, 90C27, 90C10; Secondary: 68M10, 68T20, 90C11, 78M32, 90C10.

Acknowledgement: This work was supported by a grant of the Romanian National Authority for Scientific Research, CNCS - UEFISCDI, project number PN-II-RU-TE-2011-3-0113.

ISBN 978-3-11-026758-7
e-ISBN 978-3-11-026768-6

Library of Congress Cataloging-in-Publication Data

A CIP catalog record for this book has been applied for at the Library of Congress.

Bibliographic information published by the Deutsche Nationalbibliothek

The Deutsche Nationalbibliothek lists this publication in the Deutsche Nationalbibliografie; detailed bibliographic data are available in the internet at http://dnb.dnb.de.

Translation: Peter V. Malyshev, Kiev
Typesetting: Da-TeX Gerd Blumenstein, Leipzig, www.da-tex.de
Printing and binding: Hubert & Co. GmbH & Co. KG, Göttingen
♾ Printed on acid-free paper

Printed in Germany

www.degruyter.com

Preface

Combinatorial optimization is a fascinating topic. Combinatorial optimization problems arise in a wide variety of important fields such as transportation, telecommunications, computer networking, location, planning, distribution problems, etc. Over the last few decades, important and significant results have been obtained on theory, algorithms and applications.

In combinatorial optimization, many network design problems can be generalized in a natural way by considering a related problem on a clustered graph, where the original problem's feasibility constraints are expressed in terms of the clusters, i.e. node sets instead of individual nodes. This class of problems is usually referred to as generalized network design problems (GNDPs) or generalized combinatorial optimization problems.

The express purpose of this book is to describe, in a unified manner, a series of mathematical models, methods, propositions and algorithms arising from generalized network design problems, developed in the last years. The book consists of seven chapters. In addition to an introductory chapter, the following generalized network design problems are formulated and examined:

1. generalized minimum spanning tree problem;
2. generalized traveling salesman problem;
3. railway traveling salesman problem;
4. generalized vehicle routing problem;
5. generalized fixed-charge network design problem;
6. generalized minimum vertex-biconnected network problem.

While these topics will be described in the framework of generalized combinatorial optimization problems, my intention was to make each chapter relatively self-contained, so that each can be read separately.

This book will be useful for researchers, practitioners, and graduate students in the fields of operations research, optimization, applied mathematics and computer science. Due to the substantial practical importance of some presented problems, researchers in other areas will also find this book useful.

I am indebted to Georg Still, Gunther Raidl, Walter Kern, Andrei Horvat Marc and to many other colleagues, including my co-authors of the articles cited herein.

Finally, I am especially grateful to my wife Corina and our daughter Ana Maria, for their support throughout the writing of this monograph.

Baia Mare, September 2011 Petrică C. Pop

Contents

Preface v

1 Introduction 1

1.1 Combinatorial optimization and integer programming 1

1.2 Complexity theory 3

1.3 Heuristic and relaxation methods 5

1.4 Generalized network design problems 7

2 The Generalized Minimum Spanning Tree Problem (GMSTP) 9

2.1 Definition and complexity of the GMSTP 10

2.2 An exact algorithm for the GMSTP 12

2.3 Mathematical models of the GMSTP 14

2.3.1 Formulations based on tree properties 14

2.3.2 Formulations based on arborescence properties 17

2.3.3 Flow based formulations 19

2.3.4 A model based on Steiner tree properties 23

2.3.5 Local-global formulation of the GMSTP 24

2.4 Approximation results for the GMSTP 28

2.4.1 Introduction 29

2.4.2 Positive results: the design of the approximation algorithms . . 30

2.4.3 A negative result for the GMSTP 32

2.4.4 An approximation algorithm for the GMSTP with bounded cluster size 34

2.5 Solving the GMSTP 41

2.5.1 A branch-and-cut algorithm for solving the GMSTP 42

2.5.2 A heuristic algorithm for solving the GMSTP 44

2.5.3 Rooting procedure for solving the GMSTP 46

2.5.4 Solving the GMSTP with Simulated Annealing 48

2.6 Notes 59

3 The Generalized Traveling Salesman Problem (GTSP) 60

3.1 Definition and complexity of the GTSP 61

3.2 An efficient transformation of the GTSP into the TSP 61

3.3 An exact algorithm for the Generalized Traveling Salesman Problem . 63
3.4 Integer programming formulations of the GTSP 65
3.4.1 Formulations based on the properties of Hamiltonian tours . . . 65
3.4.2 Flow based formulations 66
3.4.3 A local-global formulation 69
3.5 Solving the Generalized Traveling Salesman Problem 71
3.5.1 Reinforcing ant colony system for solving the GTSP 72
3.5.2 Computational results 75
3.5.3 A hybrid heuristic approach for solving the GTSP 78
3.5.4 Computational results 88
3.6 The drilling problem 92
3.6.1 Stigmergy and autonomous robots 93
3.6.2 Sensitive robots 94
3.6.3 Sensitive robot metaheuristic for solving the drilling problem . 94
3.6.4 Numerical experiments 97
3.7 Notes 99
4 The Railway Traveling Salesman Problem (RTSP) 100
4.1 Definition of the RTSP 100
4.2 Preliminaries 101
4.3 Methods for solving the RTSP 103
4.3.1 The size reduction method through shortest paths 103
4.3.2 A cutting plane approach for the RTSP 111
4.3.3 Solving the RTSP via a transformation into the classical TSP . 114
4.3.4 An ant-based heuristic for solving the RTSP 119
4.4 Dynamic Railway Traveling Salesman Problem 121
4.4.1 Ant colony approach to the Dynamic RTSP 122
4.4.2 Implementation details and computational results 124
4.5 Notes 126
5 The Generalized Vehicle Routing Problem (GVRP) 128
5.1 Definition and complexity of the GVRP 129
5.2 An efficient transformation of the GVRP into a capacitated arc routing problem 130
5.3 Integer linear programming formulations of the GVRP 132
5.3.1 A general formulation 132
5.3.2 A node based formulation 135
5.3.3 Flow based formulations 137
5.4 A numerical example 139

5.5 Special cases of the proposed formulations . . . 141
5.5.1 The Generalized multiple Traveling Salesman Problem . . . 141
5.5.2 The Generalized Traveling Salesman Problem . . . 143
5.5.3 The Clustered Generalized Vehicle Routing Problem . . . 144
5.6 Solving the Generalized Vehicle Routing Problem . . . 147
5.6.1 An improved hybrid algorithm for Solving the GVRP . . . 147
5.6.2 An efficient memetic algorithm for solving the GVRP . . . 154
5.6.3 Computational experiments . . . 159
5.7 Notes . . . 165

6 The Generalized Fixed-Charge Network Design Problem (GFCNDP) 167
6.1 Definition of the GFCNDP . . . 168
6.2 Integer programming formulations of the GFCNDP . . . 169
6.3 Solving the GFCNDP . . . 173
6.4 Computational results . . . 175
6.5 Notes . . . 177

7 The Generalized Minimum Edge-Biconnected Network Problem (GMEBCNP) 178
7.1 Definition and complexity of the GMEBCNP . . . 178
7.2 A mixed integer programming model of the GMEBCNP . . . 179
7.3 Solving the GMEBCNP . . . 181
7.3.1 Simple Node Optimization Neighborhood (SNON) . . . 181
7.3.2 Node Re-Arrangement Neighborhood (NRAN) . . . 182
7.3.3 Edge Augmentation Neighborhood . . . 182
7.3.4 Node Exchange Neighborhood . . . 183
7.3.5 Variable Neighborhood Descent . . . 183
7.4 Computational results . . . 185
7.5 Notes . . . 185

Bibliography 188

Index 202

Chapter 1

Introduction

1.1 Combinatorial optimization and integer programming

Combinatorial optimization is a lively field of applied mathematics, combining techniques from combinatorics, linear programming, and the theory of computation, to solve optimization problems over discrete structures. One of the main challenges of combinatorial optimization is to develop efficient algorithms, with their corresponding running times bounded by a polynomial of the same the size as their representation.

The study of combinatorial optimization owes its existence partially to the advent of modern digital computers. Most currently accepted methods for solving combinatorial optimization problems would hardly have been taken seriously 30 years ago, for the simple reason that no one could have carried out the computations involved. Moreover, the existence of digital computers has also led to a multitude of technical problems of a combinatorial character. Our ability to solve large, important combinatorial optimization problems has improved dramatically in the past decades. The availability of reliable software, extremely fast and inexpensive hardware, and high-level languages that make the modeling of complex problems much faster, have led to a much greater demand for optimization tools.

The versatility of the combinatorial optimization model stems from the fact that in many practical problems, activities and resources such as machines, airplanes and people are indivisible. Also, many problems (e.g. scheduling) have rules that define a finite number of allowable choices and consequently can be formulated appropriately using procedures that transform the logical alternatives descriptions to linear constraint descriptions where some subset of the variables is required to take on values.

Combinatorial optimization models are often referred to as integer programming models, where some or all of the variables are required to take on discrete values. In most cases, these values are integers giving rise to the name of this class of models.

In this book we consider combinatorial optimization problems for which the objective function and the constraints are linear and the variables are integers. These problems are called integer programming problems:

$$\begin{aligned} & \min c^T x \\ \text{(IP)} \quad & \text{s. t.} \quad Ax \le b, \\ & \qquad\quad x \in \mathbb{Z}^n, \end{aligned}$$

where $\mathbb{Z}^n$ is the set of integral n-dimensional vectors, $x = (x_1, \ldots, x_n)$ is an integer n-vector and c is an n-vector. Furthermore, we let m denote the number of inequality constraints, A an $m \times n$ matrix and b an m-vector. If we allow some variables x_i to be real numbers instead of integers, i.e., $x_i \in \mathbb{R}$ instead of $x_i \in \mathbb{Z}$, we obtain a *mixed integer programming problem*, denoted by (MIP). For convenience, we discuss integer linear programs that are *minimization* problems with *binary* variables, i.e., the integer variables are restricted to the values 0 or 1.

Solving integer programming problems can be a difficult task. The difficulty arises from the fact that, unlike linear programming, where the feasible region is a convex set, in integer problems one must search a lattice of feasible points or, in the mixed integer case, a set of disjoint half lines or line segments, to find an optimal solution. Therefore, unlike linear programming where, due to the convexity of the problem, we can exploit the fact that any local solution is a global optimum, integer programming problems may have many "local optima", and finding a global optimum to the problem requires one to prove that a particular solution dominates all feasible points, by arguments other than the calculus-based derivative approaches of convex programming.

When optimizing combinatorial problems, there is always a trade-off between the computational effort (and hence the running time of the algorithm) and the quality of the obtained solution. We may either try to solve the optimization problem with an *exact algorithm*, or choose an *approximation* or *heuristic algorithm*, which uses less running time but does not guarantee optimality of the solution.

Therefore, there are three main categories of algorithms for solving integer programming problems:

- **Exact algorithms**, are guaranteed to find the optimal solution, but may take an exponential number of iterations. In practice, they are typically applicable to small instances only, due to long running times caused by the high complexity. They include: branch-and-bound, branch-and-cut, cutting plane, and dynamic programming algorithms.

- **Approximation algorithms** provide a sub-optimal solution in polynomial time, together with a bounding constraint on the degree of suboptimality.

- **Heuristic algorithms** provide a suboptimal solution, but makes no guarantees about its quality. Although the running time is not guaranteed to be polynomial, empirical evidence suggests that some of these algorithms rapidly find a good solution. These algorithms are especially suitable for large instances of the optimization problems.

Because, as we have seen, there are general techniques for dealing with integer programming, it is usually better to take advantage of the structure of the specific integer programming problems and to develop special-purpose approaches to take advantage of this particular structure.

As integer programming is $\mathcal{NP}$-hard, every $\mathcal{NP}$ problem can, in principle, be formulated as an integer linear program. In fact, such problems usually admit different integer programming formulations. Finding a particularly suitable one is often a decisive step towards providing a solution of a combinatorial optimization problem.

Therefore, it is important to become familiar with a wide variety of different classes of integer programming formulations. Since there are often different ways of mathematically representing the same problem and since obtaining an optimal solution to a large integer programming problem, within a reasonable amount of computing time, may well depend on the way it is formulated, much recent research has been directed toward reformulation of combinatorial optimization problems. In this regard, it is sometimes advantageous to increase (rather than decrease) the number of integer variables, the number of constraints, or both.

When we discuss the notion of a "good" formulation, we normally think about creating an easier problem, that provides a good approximation of the objective function value of the original problem. Since the integrality restrictions on the decision variables destroy the convexity of the feasible region, the most widely used approximation removes this restriction. Such an approximation is known as the Linear Programming (LP) relaxation. However, merely removing these integrality restrictions can alter the structure so significantly, that the optimal LP solution is generally far from the optimal integer solution.

1.2 Complexity theory

The first step in studying a combinatorial problem is to find out whether the problem is "easy" or "hard". This classification is a task of complexity theory. In this section, we summarize the most important concepts in complexity theory. Most approaches are taken from the book of Grötschel, Lovász and Schrijver [60].

An *algorithm* is a list of instructions that solves every *instance* of a problem in a finite number of steps. (This also means that the algorithm detects if a particular instance of a problem has no solution).

The *size* of a problem is the amount of information needed to represent the instance. The instance is assumed to be described (encoded) by a string of symbols. Therefore, the size of an instance equals the number of symbols in the string.

The *running time* of a combinatorial optimization algorithm is measured by an upper bound on the number of elementary arithmetic operations (addition, subtraction, multiplication, division and comparison) necessary for any valid input, expressed as a function of the input size. The *input* is the data used to represent an instance of a problem. If the input size is measured by s, then the running time of the algorithm is expressed as $O(f(s))$, if there are constants b and s_0 such that the number of steps, for any instance with $s \geq s_0$, is bounded from above by $bf(s)$. We say that the running time of such an algorithm is of order $f(s)$.

An algorithm is said to be a *polynomial time algorithm* when its running time is bounded by a polynomial function, $f(s)$. An algorithm is said to be an *exponential time algorithm* when its running time is bounded by an exponential function (e.g., $O(2^{p(s)})$).

Complexity theory concerns, in the first place, decision problems. A *decision problem* is a question that can be answered only with "yes" or "no". For example, in the case of the integer programming problem (IP), the decision problem is:

> Given an instance of (IP) and an integer L, does a feasible solution x exist, such that $c^T x \leq L$?

For a combinatorial optimization problem we have the following: if one can solve the decision problem efficiently, then one can also solve the corresponding optimization problem efficiently.

Decision problems solvable in polynomial time are considered to be "easy" and the class of these problems is denoted by $\mathcal{P}$. $\mathcal{P}$ includes, for example, linear programming and the minimum spanning tree problem.

The class of decision problems, solvable in exponential, time is denoted by EXP. Most combinatorial optimization problems belong to this class. If a problem is in EXP $\setminus \mathcal{P}$, solving large size instances of this problem will be difficult. To distinguish between "easy" and "hard" problems we first describe a class that contains $\mathcal{P}$.

The complexity class $\mathcal{NP}$ is defined as the class of decision problems that are solvable by a so-called *non-deterministic algorithm*. A decision problem is said to be in $\mathcal{NP}$ if, for any input that results in a positive answer, there is a certificate from which the correctness of this answer can be derived in polynomial time.

Obviously, $\mathcal{P} \subseteq \mathcal{NP}$ holds. It is widely assumed that $\mathcal{P} = \mathcal{NP}$ is very unlikely. The class $\mathcal{NP}$ contains a subclass of problems that are considered to be the hardest problems in $\mathcal{NP}$. These problems are called $\mathcal{NP}$*-complete* problems. Before giving the definition of an $\mathcal{NP}$*-complete* decision problem, we explain the technique of polynomially transforming one problem into another. Let Π_1 and Π_2 be two decision problems.

Definition 1.1. A *polynomial transformation* is an algorithm $\mathcal{A}$ that, for every instance σ_1 of Π_1, produces an instance σ_2 of Π_2 in polynomial time, such that the following holds: for every instance σ_1 of Π_1, the answer to σ_1 is "yes" if and only if the answer to instance σ_2 of Π_2 is "yes".

Definition 1.2. A decision problem Π is called $\mathcal{NP}$*-complete* if Π is in $\mathcal{NP}$ and every other $\mathcal{NP}$ decision problem can be polynomially transformed into Π.

Clearly, if an $\mathcal{NP}$*-complete* problem can be solved in polynomial time, then all problems in $\mathcal{NP}$ can be solved in polynomial time, hence $\mathcal{P} = \mathcal{NP}$. This explains why the $\mathcal{NP}$*-complete* problems are considered the hardest problems in $\mathcal{NP}$.

Note that polynomial transformability is a transitive relation, i.e if Π_1 is polynomially transformable to Π_2 and Π_2 is polynomially transformable to Π_3, then Π_1 is polynomially transformable to Π_3. Therefore, if we want to prove that a decision problem Π is $\mathcal{NP}$*-complete*, we only have to show that

(i) Π is in $\mathcal{NP}$.

(ii) A decision problem already known to be $\mathcal{NP}$*-complete* can be polynomially transformed to Π.

We now want to focus on combinatorial optimization problems. We therefore extend the concept of polynomial transformability. Let Π_1 and Π_2 be two problems (not necessarily decision problems).

Definition 1.3. A **polynomial reduction** from Π_1 to Π_2 is an algorithm $\mathcal{A}_1$ that solves Π_1 by using an algorithm $\mathcal{A}_2$ for Π_2 as a subroutine such that, if $\mathcal{A}_2$ were a polynomial time algorithm for Π_2, then $\mathcal{A}_1$ would be a polynomial time algorithm for Π_1.

Definition 1.4. An optimization problem Π is called $\mathcal{NP}$*-hard* if there exists an $\mathcal{NP}$*-complete* decision problem that can be polynomially reduced to Π.

Clearly, an optimization problem is $\mathcal{NP}$*-hard*, if the corresponding decision problem is $\mathcal{NP}$*-complete*.

1.3 Heuristic and relaxation methods

As we have seen, once it is established that a combinatorial problem is $\mathcal{NP}$*-hard*, it is unlikely that it can be solved by a polynomial algorithm. Different approaches have been suggested in the past, to solve this kind of problems. In principle, they can be classified as exact algorithms, approximation algorithms, relaxation methods, or heuristic approaches.

When trying to find good solutions to hard minimization problems in combinatorial optimization, two issues must be taken into consideration:

- calculating an upper bound that is as close as possible to the optimum;
- calculating a lower bound that is as close as possible to the optimum.

It is a fundamental goal of computer science to find algorithms which have both provably good running times and provably good or optimal quality of the solution.

A heuristic is an algorithm that abandons one aspect, e.g. it usually finds pretty good solutions, but there is no proof that the solutions may not become arbitrarily bad. On the other hand, the algorithm may run reasonably quickly, but there is no argument that this will always be the case. Heuristics are typically used when there is no known method to find an optimal solution, under the given constraints (of time, space etc.), or at all.

Several families of heuristic algorithms have been proposed for solving combinatorial optimization problems, which can be divided into two main classes: classical heuristics and metaheuristics. The first class contains most standard construction and improvement procedures. Additionally, for any particular problem, we may well have techniques specific to the problem being solved.

The term "metaheuristic" was first introduced by Glover [51] and refers to a number of high-level strategies or concepts that can be used to define heuristic methods, which can be applied to a wide set of different optimization problems. In other words, a metaheuristic may be seen as a general algorithmic framework, which can be applied to different optimization problems with relatively few modifications, in order adept them to a specific problem. Examples of metaheuristics include simulated annealing (SA), tabu search (TS), iterated local search (ILS), genetic algorithms (EC), variable neighborhood search (VNS), ant colony optimization (ACO), memetic algorithms (MA), genetic programming (GP), etc. We also remark that several of the leading methods are hybrid approaches, which combine different solution techniques [174]. Some research on them has been done in the last years, and the resulting publications document the success of hybrid algorithms in various application domains. However, a lot of work still has to be done, in order to find skilled combinations of metaheuristics with other optimization techniques, that will lead to more efficient behavior and higher flexibility.

Throughout this book, special attention will be given to hybrid algorithms for selected GNDPs, that combine exact methods based on integer linear programming with local search based metaheuristics. Both techniques have their particular advantages and disadvantages, which can be seen as complementary to a large degree. After analyzing them, it appears natural to combine them, to obtain more effective algorithms.

On the question of lower bounds, the most well known techniques are:

- **Linear Programming (LP) relaxation.**
  In LP relaxation we take an integer (or mixed-integer) programming formulation of the problem and relax the integrality requirement on the variables. This gives a linear program which can either be solved exactly, using a standard algorithm (simplex or interior point), or heuristically (dual ascent). The solution value obtained for this linear program gives a lower bound on the optimal solution to the original minimization problem.

- **Lagrangian relaxation.**
  The general idea of Lagrangian relaxation is to "relax" (dualize) some (or all) constraints, by adding them to the objective function with the use of Lagrangian multipliers. In practice, the Lagrangian relaxed problem can be solved more easily than the original problem. The problem of maximizing the Lagrangian function of

the dual variables (the Lagrangian multipliers) is called the Lagrangian dual problem. Choosing "good" values for the Lagrangian multipliers is of key importance in terms of quality of the generated lower bound. We refer to [110] for more details on this technique.

- **Semidefinite programming. (SDP) relaxation.**
  Combinatorial optimization problems often involve binary (0, 1 or $+1, -1$) decision variables, which can be modeled with use of quadratic constraints $x^2 - x = 0$, and $x^2 = 1$, respectively. Using the positive semidefinite matrix $X = xx^T$, we can lift the problem into a higher dimensional matrix space and obtain a Semidefinite Programming Relaxation problem, by ignoring the rank-one restriction on X, see e.g. [115]. These semidefinite relaxations provide tight bounds for many classes of hard problems. In addition, they can be solved efficiently by interior-point methods.

1.4 Generalized network design problems

In combinatorial optimization, many network design problems can be generalized in a natural way, by considering a related problem on a clustered graph, where the original problem's feasibility constraints are expressed in terms of the clusters, i.e., node sets, instead of individual nodes. This class of problems is usually referred to as generalized network design problems (GNDPs) or generalized combinatorial optimization problems.

In the literature, several GNDPs have already been considered, such as the generalized minimum spanning tree problem, the generalized traveling salesman problem, the generalized vehicle routing problem, the generalized (subset) assignment problem, the generalized fixed-charge network design problem, etc. All such problems belong to the class of $\mathcal{NP}$-complete problems, and in practice are typically harder to solve than their original counterparts. Nowadays, they are intensively studied due to their interesting properties and important real-world applications in telecommunication, network design, resource allocation, transportation problems, etc. Nevertheless, many practitioners are still reluctant to use them for modeling and solving practical problems, because of the complexity of finding optimal or near-optimal solutions.

For all the generalized network design problems described in this book, we are given an n-node undirected weighted graph $G = (V, E)$, with node set V and edge set E. The nodes are partitioned into a given number of node sets called *clusters* (i.e., $V = V_1 \cup V_2 \cup \ldots \cup V_m$ and $V_l \cap V_k = \emptyset$ for all $l, k \in \{1, \ldots, m\}$ with $l \neq k$) and with each edge $e \in E$, we associate a nonnegative cost c_e. Let $e = (i, j)$ be an edge with $i \in V_l$ and $j \in V_k$. If $l \neq k$, then e is called an *inter-cluster* edge; otherwise, e is called an *intra-cluster* edge.

The goal in solving these problems is to find a subgraph $F = (S, T)$ of G, where the subset of nodes $S = \{v_1, \ldots, v_m\} \subset V$ contains exactly one node from each cluster, with different requirements to be fulfilled by the subset of edges $T \subset E$, depending on the actual optimization problem. The cost of the subgraph S is given by its total edge costs, $c(S) = \sum_{e \in T} c_e$ and the objective is to identify a solution with a minimum cost.

A different class of GNDPs aims to find a subgraph F, containing at least one node from each cluster. For further details, we refer to [39]. In the present book, with the exception of the railway traveling salesman problem, we confine ourselves to the GNDPs of choosing exactly one node from each cluster and different requirements, depending on the actual optimization problem.

In this book, we will study the following generalized network design problems, based on recent developments and the increasing attention paid to them by researchers:

- the Generalized Minimum Spanning Tree Problem (GMSTP);
- the Generalized Traveling Salesman Problem (GTSP);
- the Generalized Vehicle Routing Problem (GVRP);
- the Railway Traveling Salesman Problem (RTSP);
- the Generalized Fixed-Charge Network Design Problem (GFCNDP);
- the Generalized Minimum Edge-Biconnected Network Problem (GMEBCNP).

Aside from some other GNDPs, the generalized minimum perfect matching problem [39], the generalized shortest path problem [111], the generalized minimum clique problem [99], the generalized Steiner tree problem [176], the generalized Chinese postman problem [29], etc., have been considered in the literature.

Some other classic combinatorial optimization problems have been recast and formulated by Dror et al. [29] in a broader framework, such as the generalized (subset) assignment problem, the generalized machine scheduling problems, the generalized knapsack and bin-packing problems, etc. A generalization of the classical graph coloring problem was introduced by Demange et al. [21] and was called *selective graph coloring problem*.

Chapter 2

The Generalized Minimum Spanning Tree Problem (GMSTP)

The minimum spanning tree (MST) problem has a venerable history in the field of combinatorial optimization. The problem was first formulated in 1926 by Boruvka, who is said to have learned about it during the rural electrification of the Southern Moraria, where he provided a solution to find the most economical layout of a power-line network. The importance and popularity of the MST problem stem from several facts: it allows an efficient solution, which makes the problem practical to solve for large graphs. Some polynomial-time algorithms for solving the MST problem were developed by Kruskal [100], Prim [171], Dijkstra [25] and Sollin [189] and have been applied to many combinatorial optimization problems such as transportation problems, telecommunication network design, distribution systems, and so on.

However, most attractive to many researchers has been the extension of the MST problem, in the last decade, to e.g., the degree-constrained minimum spanning tree problem (dc-MST) by Narula and Ho [130], the probabilistic minimum spanning tree problem (p-MST) by Bertsimas [11], the stochastic minimum spanning tree problem (s-MST) by Ishii et al. [89], the quadratic minimum spanning tree problem (q-MST) by Xu [204], the minimum Steiner tree (MStT) problem by Maculan [117], the generalized minimum spanning tree problem (GMSTP) by Myung, Lee and Tcha [128], etc. The extensions of the MST problem are generally $\mathcal{NP}$*-hard* problems and therefore polynomial-time solutions do not exist.

In this section, we are concerned with the generalized version of the minimum spanning tree problem (MST), called the generalized minimum spanning tree problem (GMSTP). Given an undirected graph, whose nodes are partitioned into a number of subsets (clusters), the GMSTP aims to find a minimum-cost tree, spanning a subset of nodes, which includes *exactly* one node from each cluster. Therefore, the MST problem is a special case of the GMSTP, with each cluster consisting of exactly one node.

The GMSTP was introduced by Myung et al. [128] and has several real world applications in the design of metropolitan area networks [47] and regional area networks [173] and in determining the locations of regional service centers [142], energy transportation [135], agricultural irrigation [30], etc.

Two variants of the generalized minimum spanning tree problem were considered in the literature: in the first one, in addition to the cost attached to the edges, we also have costs attached to the nodes. This is called the prize collecting generalized minimum spanning tree problem, see [54, 151]. The second one consists in finding a minimum cost tree, spanning at least one node from each cluster, denoted by L-GMSTP, which was introduced by Dror et al. [30]. The same authors have proved that the L-GMSTP is NP-hard. Integer programming formulations for the L-GMSTP have been presented by Pop et al. [149].

2.1 Definition and complexity of the GMSTP

The *generalized minimum spanning tree problem* (GMSTP) requires finding a minimum-cost tree T, spanning a subset of nodes, which includes *exactly* one node from each cluster V_i, $i \in \{1, \ldots, m\}$. We will call such a tree a *generalized spanning tree.* In the next figure, we present an example of a generalized spanning tree.

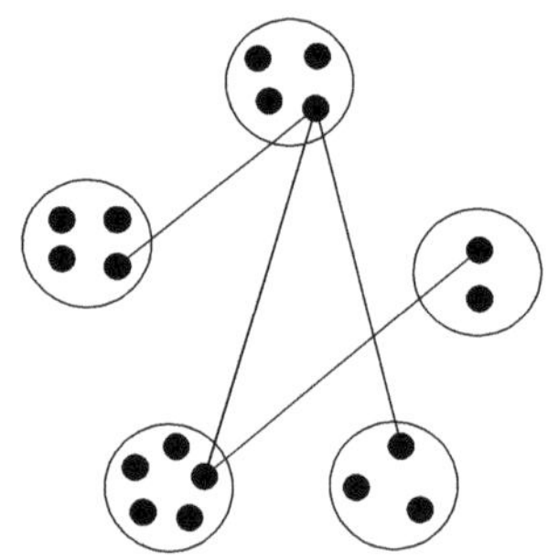

Figure 2.1. An example of a generalized spanning tree.

Myung et al. [128] proved that the GMSTP is an $\mathcal{NP}$*-hard* problem by reducing the vertex cover problem.

Garey and Johnson [45] have shown that, for certain combinatorial optimization problems, the simple structure of trees can offer algorithmic advantages for efficient solution. Indeed, a number of problems that are $\mathcal{NP}$*-complete*, when formulated on a general graph, become polynomially solvable when the graph is a tree. Unfortunately, this is not the case for the GMSTP. We will prove a stronger result concerning the complexity of the problem, namely that the GMSTP even defined on trees is $\mathcal{NP}$*-hard.*

To show that the GMSTP on trees is $\mathcal{NP}$*-hard* we introduce the so-called *set cover problem* which is known to be $\mathcal{NP}$*-complete* (see [45]).

Given a finite set $X = \{x_1, \ldots, x_a\}$, a collection of subsets, $S_1, \ldots, S_b \subseteq X$ and an integer $k < |X|$, the *set cover problem* consists in determining whether there exists a subset $Y \subseteq X$, such that $|Y| \leq k$ and

$$S_c \cap Y \neq \emptyset, \quad \forall c \text{ with } 1 \leq c \leq b.$$

We call such a set Y a set cover for X.

Theorem 2.1 (Pop [142]). *The Generalized Minimum Spanning Tree problem on trees is $\mathcal{NP}$-hard.*

Proof. In order to prove that the GMSTP on trees is *$\mathcal{NP}$-hard* it is enough to show that there exists an *$\mathcal{NP}$-complete* problem that can be polynomially reduced to a GMSTP.

We consider the set cover problem for a given finite set $X = \{x_1, \dots, x_a\}$, a collection of subsets of X, $S_1, \dots, S_b \subseteq X$ and an integer $k < |X|$.

We show that we can construct a graph $G = (V, E)$, having a tree structure, such that there exists a set cover $Y \subseteq X$, $|Y| \leq k$ if and only if there exists a generalized spanning tree in G with a cost of at most k.

The constructed graph G contains the following $m = a+b+1$ clusters $V_1, \dots, V_m$:

- V_1 consists of a single node denoted by r
- $V_2, \dots, V_{a+1}$ node sets (corresponding to $x_1, x_2, \dots, x_a \in X$) each of which has two nodes: one 'expensive' (see the construction of the edges), say $\overline{x_i}$, and one 'non-expensive', say $\hat{x}_i$, for $i = 2, \dots, a$, and
- b node sets, $V_{a+2}, \dots, V_m$ with $V_\nu = S_{\nu-(a+1)}$, for $\nu = a+2, \dots, m$.

Edges in G are constructed as follows:

(i) Each 'expensive node', say $\overline{x_t}$ of V_t, for all $t = 2, \dots, a+1$, is connected with r by an edge with a cost of 1 and each 'non-expensive' node, say $\hat{x}_t$ of V_t, for all $t = 2, \dots, a+1$, is connected with r by an edge with a cost of 0.

(ii) Choose any node $j \in V_t$ for any $t \in \{a+2, \dots, m\}$. Since $V_t \subset X$, then j coincides with a node in X, say $j = x_l$. We construct an edge between j and (the expensive node) $\overline{x_l} \in V_l$ with $l \in \{2, \dots, a\}$. The cost of the edges constructed this way is 0.

By design the graph $G = (V, E)$ has a tree structure.

Now suppose that there exists a generalized spanning tree in G, with a cost of at most k then, by choosing

$$Y := \{x_l \in X \mid \text{the expensive vertex } \overline{x_l} \in V_{l+1} \text{ corresponding to } x_l \text{is a vertex of the generalized spanning tree in } G\}$$

we see that Y is a set cover of X.

On the other hand, if there exists a set cover $Y \subseteq X$, $|Y| \leq k$ then, according to the construction of G, there exists a generalized spanning tree in G with a cost of at most k. □

2.2 An exact algorithm for the GMSTP

Let G' be the graph obtained from G after replacing all nodes of a cluster V_i with a supernode representing V_i. We will call the graph G' the global graph. For convenience, we identify V_i with the supernode that represents it. Edges of the graph G' are defined between each pair of the graph vertices $V_1, \dots, V_m$.

The local-global approach to the GMSTP aims at distinguishing between *global connections* (connections between clusters) and *local connections* (connections between nodes from different clusters). As we will see, having a global tree connection of the clusters makes it rather easy to find the least cost generalized spanning tree.

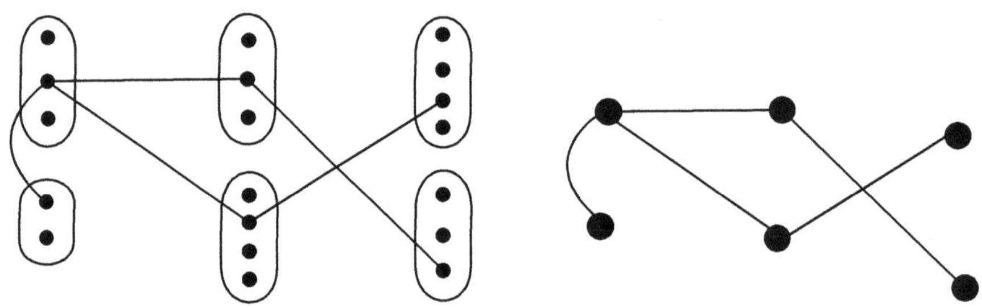

Figure 2.2. An example showing a generalized spanning tree corresponding to a global spanning tree.

To each global spanning tree there are several corresponding generalized spanning trees. Among these generalized spanning trees there exists one called the best generalized spanning tree (w.r.t. cost minimization), which can be determined either by dynamic programming or by solving a linear integer program.

In the following, based on the local-global approach and making use of dynamic programming, we will present an exact exponential time algorithm.

Given a spanning tree of the global graph G', which we shall refer to as a *global spanning tree*, we use dynamic programming in order to find the corresponding best (w.r.t. cost minimization) generalized spanning tree.

Fix an arbitrary cluster V_{root} as the root of the global spanning tree and orient all the edges away from vertices of V_{root}, according to the global spanning tree. A directed edge $\langle V_k, V_l \rangle$ of G', resulting from the orientation of edges of the global spanning tree, provides a natural definition of an orientation $\langle i, j \rangle$ of an edge $(i, j) \in E$, where $i \in V_k$ and $j \in V_l$. Let v be a vertex of cluster V_k for some $1 \le k \le m$. All such nodes v are potential candidates to be incident to an edge of the global spanning tree.

On the graph G, we denote by $T(v)$ the subtree rooted at such a vertex v from G. $T(v)$ includes all vertices reachable from v under the above orientation of the edges of G, based on the orientation of the edges of the global spanning tree. The *children* of $v \in V_k$, denoted by $C(v)$, are those vertices $u \in V_l$ which are heads of the directed edges $\langle u, v \rangle$ in the orientation. The leaves of the tree are those vertices without children.

Let $W(T(v))$ denote the minimum weight of a generalized subtree rooted at v. We want to compute

$$\min_{r \in V_{\text{root}}} W(T(r)).$$

We are now ready to present the dynamic programming recursion for solving the subproblem $W(T(v))$. The initialization is:

$$W(T(v)) = 0, \quad \text{if } v \in V_k \text{ and } V_k \text{ is a leaf of the global spanning tree.}$$

To compute $W(T(v))$ for a vertex $v \in V$ interior to a cluster, i.e., to find the optimal solution of the subproblem $W(T(v))$, we have to look at all vertices from the clusters V_l such that $C(v) \cap V_l \neq \emptyset$. If u denotes a child of the interior vertex v, then the recursion for v is as follows:

$$W(T(v)) = \sum_{l, C(v) \cap V_l \neq \emptyset} \min_{u \in V_l} [c(v, u) + W(T(u))].$$

Hence, for fixed v, we have to check at most n vertices. Consequently, for the given global spanning tree, the overall complexity of this dynamic programming algorithm is $O(n^2)$. Since, by Cayley's formula, the number of all distinct global spanning trees is m^{m-2}, we have established the following:

Theorem 2.2. *There exists a dynamic programming algorithm which provides an exact solution to the GMSTP in $O(m^{m-2}n^2)$ time, where n is the number of nodes and m is the number of clusters in the input graph.*

Clearly, the above is an exponential time algorithm, unless the number of clusters m is fixed.

A similar dynamic programming algorithm can provide an exact solution to the prize collecting generalized minimum spanning tree problem, with the difference that the dynamic programming recursion to solve the subproblem $W(T(v))$ for a node $v \in G$, should be considered as follows:

$$W(T(v)) = \sum l, C(v) \cap V_l \neq \emptyset \min_{u \in V_l} [c(v, u) + d(u) + W(T(u))].$$

where by $d(u)$ we denote the cost associated to the node u.

2.3 Mathematical models of the GMSTP

In order to formulate the GMSTP as an integer program, we introduce the following binary variables:

$$x_e = x_{ij} = \begin{cases} 1 & \text{if the edge } e = \{i, j\} \in E \text{ is included in the selected subgraph} \\ 0 & \text{otherwise} \end{cases},$$

$$z_i = \begin{cases} 1 & \text{if the node } i \in V \text{ is included in the selected subgraph} \\ 0 & \text{otherwise} \end{cases},$$

$$w_{ij} = \begin{cases} 1 & \text{if the arc } (i, j) \in A \text{ is included in the selected subgraph} \\ 0 & \text{otherwise} \end{cases}.$$

We use the vector notations $x = (x_{ij})$, $z = (z_i)$, $w = (w_{ij})$ and the notations $x(E') = \sum_{\{i,j\}\in E'} x_{ij}$, for $E' \subseteq E$, $z(V') = \sum_{i\in V'} z_i$, for $V' \subseteq V$ and $w(A') = \sum_{(i,j)\in A'} w_{ij}$, for $A' \subseteq A$.

2.3.1 Formulations based on tree properties

Generalized subtour elimination formulation (Myung et al. [128])
A feasible solution of the GMSTP can be regarded as a cycle free graph with $m - 1$ edges, connecting all the clusters, with one node selected from each cluster. Therefore, the GMSTP can be formulated as the following integer programming problem:

$$\begin{aligned}
\min \sum_{e\in E} c_e x_e & & & \\
\text{s.t.}\quad z(V_k) &= 1, & \forall\, k \in K = \{1, \dots, m\}, & \quad (2.1)\\
x(E(S)) &\le z(S - \{i\}), & \forall\, i \in S \subset V, 2 \le |S| \le n - 1, & \quad (2.2)\\
x(E) &= m - 1, & & \quad (2.3)\\
x_e &\in \{0, 1\}, & \forall\, e \in E, & \quad (2.4)\\
z_i &\in \{0, 1\}, & \forall\, i \in V. & \quad (2.5)
\end{aligned}$$

In the above formulation, constraints (2.1) guarantee that from every cluster we select exactly one node, constraints (2.2) eliminate all the subtours and finally constraint (2.3) guarantees that the selected subgraph has $m - 1$ edges. This formulation, introduced by Myung [128], is called the *generalized subtour elimination formulation* because constraints (2.2) eliminate all the cycles.

We denote the feasible set of the linear programming relaxation of this formulation by P_{sub}, where we replace the constraints (2.4) and (2.5) with $0 \le x_e, z_i \le 1$, for all $e \in E$ and $i \in V$.

Generalized cutset formulation (Myung et al. [128])
We may replace the subtour elimination constraints (2.2) by connectivity constraints, which results in the so-called *generalized cutset formulation* introduced in [128]:

$$\min \sum_{e \in E} c_e x_e$$
$$\text{s.t.} \quad (2.1), (2.3), (2.4), (2.5) \text{ and}$$
$$x(\delta(S)) \geq z_i + z_j - 1, \ \forall\, i \in S \subset V, j \notin S \tag{2.6}$$

where for all $S \subset V$ the cutset $\delta(S)$ is usually defined as:

$$\delta(S) = \{e = \{i, j\} \in E \mid i \in S, j \notin S\}.$$

We denote the feasible set of the linear programming relaxation of this formulation by P_{cut}. Myung et al. [128] proved that the following relation holds: $P_{\text{sub}} \subset P_{\text{cut}}$.

Generalized multicut formulation (Pop [142])
Our next model, the so-called *generalized multicut formulation,* is obtained by replacing simple cutsets by multicuts. Given a partition of the nodes $V = C_0 \cup C_1 \cup \ldots \cup C_k$, we define the multicut $\delta(C_0, C_1, \ldots, C_k)$ as the set of edges connecting different C_i and C_j. The generalized multicut formulation for the GMSTP is:

$$\sum_{e \in E} c_e x_e$$
$$\text{s.t.} \quad (2.1), (2.3), (2.4), (2.5) \text{ and}$$
$$x(\delta(C_0, C_1, \ldots, C_k)) \geq \sum_{j=0}^{k} z_{i_j} - 1,$$
$$\forall\, C_0, C_1, \ldots, C_k \text{ node partitions of } V \text{ and}$$
$$\forall\, i_j \in C_j \text{ for } j = 0, 1, \ldots, k. \tag{2.7}$$

Let P_{mcut} denote the feasible set of the linear programming relaxation of this model. Clearly, $P_{\text{mcut}} \subseteq P_{\text{cut}}$, in addition, it can be proved that the following result holds:

Proposition 2.3 (Pop [142]). $P_{\text{sub}} = P_{\text{mcut}}$.

Proof. Let $(x, z) \in P_{\text{sub}}$ and $C_0, C_1, \ldots, C_k$ a node partition of V.
Since

$$E = E(C_0) \cup E(C_1) \cup \cdots \cup E(C_k) \cup \delta(C_0, C_1, \ldots, C_k),$$

we get

$$x(\delta(C_0, C_1, \ldots, C_k) = x(E) - x(E(C_0)) - x(E(C_1)) - \cdots - x(E(C_k)).$$

For all $j = 0, 1, \dots, k$ and $i_j \in C_j$ we have:

$$x(E(C_j)) \leq z(C_j) - z_{i_j}.$$

Therefore we obtain:

$$x(\delta(C_0, C_1, \dots, C_k) \geq z(V) - 1 - \sum_{j=0}^{k} [z(C_j) - z_{i_j}] = \sum_{j=0}^{k} z_{i_j} - 1.$$

Conversely, let $(x, z) \in P_{\text{mcut}}$, $i \in S \subset V$ and consider the inequality (2.7) with $C_0 = S$ and with $C_1, \dots, C_k$ as singletons with union $V \setminus S$. Then

$$x(\delta(S, C_1, \dots, C_k)) \geq \sum_{j=0}^{k} z_{i_j} - 1 = z_i + z(V \setminus S) - 1,$$

where $i \in S \subset V$. Therefore

$$\begin{aligned} x(E(S)) &= x(E) - x(\delta(S, C_1, \dots, C_k)) \\ &\leq z(V) - 1 - z_i - z(V \setminus S) + 1 = z(S - i). \end{aligned}$$

□

Cluster subpacking formulation (Feremans [37])
We may strengthen the generalized subtour formulation of the GMSTP by replacing the subtour elimination constraints (2.2) with the *cluster subpacking constraints*

$$x(E(S)) \leq z(S \setminus V_k), \quad \forall S \subset V, 2 \leq |S| \leq n - 1, k \in K$$

We observe that these constraints are dominated by:

$$x(E(S')) \leq z(S' \setminus V_k) = z(S') - 1$$

where $S' = S \cup V_k$. Therefore, we arrive at the *cluster subpacking formulation* of the GMSTP, introduced by Feremans [37]:

$$\min \sum_{e \in E} c_e x_e$$

s.t. (2.1), (2.3), (2.4), (2.5) and

$$x(E(S)) \leq z(S) - 1, \ \forall \ S \subset V, 2 \leq |S| \leq n - 1, |\{k : V_k \subseteq S\}| \neq 0 \quad (2.8)$$

The cluster subpacking constraints (2.8) guarantee that the number of edges selected from any subset of nodes S with $S \subset V, 2 \leq |S| \leq n - 1$, cannot be greater than the number of nodes selected from that set minus 1. We will denote by P_{spack} the feasible set of the linear programming relaxation of the cluster subpacking formulation.

2.3.2 Formulations based on arborescence properties

Consider the directed graph $D = (V, A)$, obtained by replacing each edge $e = (i, j) \in E$ with the opposite arcs (i, j) and (j, i) in A, having the same weight as the edge $(i, j) \in E$. The directed version of the GMSTP introduced by Myung et al. [128], called *the generalized minimum spanning arborescence problem* is defined on the directed graph $D = (V, A)$, rooted at a given cluster, here chosen to be V_1 without loss of generality, and consists of determining a minimum cost arborescence which includes exactly one node from every cluster.

The next two formulations that we will present in this section, were introduced by Feremans et al. [37].

Directed generalized cutset formulation (Feremans [37])
We first consider a *directed generalized cutset formulation* of the GMSTP. In this model we consider the directed graph $D = (V, A)$ with the cluster V_1 chosen as a root without loss of generality, and we denote with $K_1 = K \setminus \{1\}$.

$$\begin{aligned}
\min \sum_{e \in E} & c_e x_e \\
\text{s.t. } z(V_k) &= 1, && \forall\, k \in K = \{1, \dots, m\} \\
x(E) &= m - 1 \\
w(\delta^-(S)) &\geq z_i, && \forall\, i \in S \subseteq V \setminus V_1 && (2.9)\\
w_{ij} &\leq z_i && \forall\, i \in V_1, j \notin V_1 && (2.10)\\
w_{ij} + w_{ji} &= x_e && \forall\, e = (i, j) \in E && (2.11)\\
x, z, w &\in \{0, 1\}. &&&& (2.12)
\end{aligned}$$

In this model, the constraints (2.9) and (2.10) guarantee existence of a path from the selected root node to any other selected node, which includes only the selected nodes. Let P_{dcut} denote the projection of the feasible set of the linear programming relaxation of this model into the (x, z)-space.

Another possible directed generalized cutset formulation, considered by Myung et al. in [128], was obtained by replacing (2.3) with the following constraints:

$$w(\delta^-(V_1)) = 0 \qquad (2.13)$$

$$w(\delta^-(V_k)) \leq 1, \quad \forall\, k \in K_1. \qquad (2.14)$$

Directed subpacking formulation (Feremans [37])
We now introduce a formulation of the GMSTP based on branchings. Consider, as in the previous formulation, the digraph $D = (V, A)$, with V_1 chosen as the cluster root.

The *directed subpacking formulation* of the GMSTP is defined as follows:

$$
\begin{aligned}
\min \sum_{e\in E} & c_e x_e \\
\text{s.t.}\quad z(V_k) &= 1, && \forall\, k \in K = \{1,\dots,m\} \\
x(E) &= m-1 \\
w(A(S)) &\le z(S-i), && \forall\, i \in S \subset V, 2 \le |S| \le n-1 && (2.15)\\
w(\delta^-(j)) &= z_j, && \forall\, j \in V \setminus V_1 && (2.16)\\
w_{ij} + w_{ji} &= x_e, && \forall\, e=(i,j) \in E \\
x, z, w &\in \{0,1\}.
\end{aligned}
$$

Let P_{dsub} denote the projection of the feasible set of the linear programming relaxation of this model into the (x,z)-space. Obviously, $P_{\text{dsub}} \subseteq P_{\text{sub}}$.

The following result was established by Feremans [37]: $P_{\text{dsub}} = P_{\text{dcut}} \cap P_{\text{sub}}$. Here, we now present a different proof of this result, see Pop [142].

Proposition 2.4. $P_{\text{branch}} = P_{\text{dcut}} \cap P_{\text{sub}}$.

Proof. First, we prove that $P_{\text{dcut}} \cap P_{\text{sub}} \subseteq P_{\text{branch}}$.

Let $(x,z) \in P_{\text{dcut}} \cap P_{\text{sub}}$. Using the relations (2.13) and (2.14), it is easy to see that the constraint (2.16) is satisfied. Therefore $(x,z) \in P_{\text{branch}}$.

We show that $P_{\text{branch}} \subseteq P_{\text{dcut}} \cap P_{\text{sub}}$.

It is obvious that $P_{\text{branch}} \subseteq P_{\text{sub}}$. Therefore, it remains to show $P_{\text{branch}} \subseteq P_{\text{dcut}}$.

Let $(x,z) \in P_{\text{branch}}$. For all $i \in V_1$ and $j \notin V_1$ take $S = \{i,j\} \subset V$. Then by (2.14) we have $w_{ij} + w_{ji} \le z_i$, which implies (2.9).

We now show that $w(\delta^-(l)) \le z_l$, for $l \in V_k, k \in K_1$.

Take $V^l = \{i \in V | (i,l) \in \delta^-(l)\}$ and $S^l = V^l \cup \{l\}$, then $w(\delta^-(l)) = w(A(S^l))$ and choose $i_l \in V^l$.

$$
\begin{aligned}
1 = \sum_{l\in V_k} w(\delta^-(l)) \qquad & = \sum_{l\in V_k} w(A(S^l)) \le \sum_{l\in V_k} z(S^l \setminus \{i_l\}) \\
= \sum_{l\in V_k} z_l + \sum_{l\in V_k} \sum_{j\in V^l\setminus\{i_l\}} z_j \qquad & = 1 + \sum_{l\in V_k} \sum_{j\in V^l\setminus\{i_l\}} z_j .
\end{aligned}
$$

Therefore, for all l, there is only one $i_l \in V^l$ with $z_{i_l} \neq 0$ and

$$
w(\delta^-(l)) = w(A(S^l)) \le z(S^l \setminus \{i_l\}) = z_l .
$$

For every $i \in S \subset V \setminus V_1$

$$
w(A(S)) = \sum_{i\in S} w(\delta^-(i)) - w(\delta^-(S)) \le z(S - \{i\}),
$$

which implies

$$
\begin{aligned}
w(\delta^-(S)) &\geq \sum_{i\in S} w(\delta^-(i)) - z(S) + z_i \\
&= \sum_{i\in S}\left[1 - \sum_{l\in V_k\setminus\{i\}} w(\delta^-(l))\right] - z(S) + z_i \\
&\geq \sum_{i\in S}\left[1 - \sum_{l\in V_k\setminus\{i\}} z_l\right] - z(S) + z_i = z(S) - z(S) + z_i = z_i.
\end{aligned}
$$

□

Directed cluster subpacking formulation (Feremans [37])
The directed counterpart of the cluster subpacking formulation of the GMSTP is given by:

$$
\begin{aligned}
\min \sum_{e\in E} & c_e x_e \\
\text{s.t.}\quad z(V_k) &= 1, && \forall\, k \in K = \{1,\dots,m\} \\
x(E) &= m-1 \\
w(A(S)) &\leq z(S) - 1, && \forall\, S \subset V, 2 \leq |S| \leq n-1, |\{k : V_k \subseteq S\} \neq 0 \\
w(\delta^-(j)) &= z_j && \forall\, j \in V \setminus V_1 \\
w_{ij} + w_{ji} &= x_e, && \forall\, e = (i,j) \in E \\
x, z, w &\in \{0,1\}.
\end{aligned}
$$

Let P_{dspack} denote the projection of the feasible set of the linear programming relaxation of this model into the (x,z)-space. Obviously, $P_{\text{dspack}} \subseteq P_{\text{spack}}$ and $P_{\text{spack}} = P_{\text{dspack}}$ (see [37]). This means that, even if there are fewer variables, a simple, undirected formulation can be as tight as a directed formulation. This is also the case for the MST problem and Steiner Tree Problem.

2.3.3 Flow based formulations

All the formulations described so far have an exponential number of constraints. The formulations that we will consider next have only a polynomial number of constraints but an additional number of variables. In order to provide compact formulations of the GMSTP, one possibility is to introduce 'auxiliary' flow variables beyond the natural binary edge and node variables.

We wish to send a flow between the nodes of the network and regard the edge variable x_e as indicating whether the edge $e \in E$ is able to carry any flow or not. We consider four such flow formulations: a single commodity model, a multicommodity model, a bidirectional flow model and a flow cut formulation. In each of these models, the flow variables will be directed, although the edges are undirected. That is, for each edge $\{i, j\} \in E$, we will have flow in both directions i to j and j to i.

Single commodity flow formulation Pop [142]
In the *single commodity model*, the source cluster V_1 sends one unit of flow to every other cluster. Let f_{ij} denote the flow on edge $e = \{i, j\}$ in the direction i to j. This leads to the following formulation:

$$\min \sum_{e \in E} c_e x_e$$

$$\text{s.t.} \quad z(V_k) = 1, \qquad \forall\, k \in K = \{1, \dots, m\}$$

$$x(E) = m - 1$$

$$\sum_{e \in \delta^+(i)} f_e - \sum_{e \in \delta^-(i)} f_e = \begin{cases} (m-1)z_i & \text{for} \quad i \in V_1 \\ -z_i & \text{for} \quad i \in V \setminus V_1 \end{cases} \tag{2.17}$$

$$f_{ij} \le (m-1)x_e, \qquad \forall\, e = \{i, j\} \in E \tag{2.18}$$

$$f_{ji} \le (m-1)x_e, \qquad \forall\, e = \{i, j\} \in E \tag{2.19}$$

$$f_{ij}, f_{ji} \ge 0, \qquad \forall\, e = \{i, j\} \in E \tag{2.20}$$

$$x, z \in \{0, 1\}.$$

In this model, the mass balance equations (2.17) imply that the network defined by any solution (x, z) must be connected. Since the constraints (2.1) and (2.3) state that the network defined by any solution contains $m - 1$ edges and one node from every cluster, every feasible solution must be a generalized spanning tree. Therefore, when projected into the space of the (x, z) variables, this formulation correctly models the GMSTP. We let P_{sflow} denote the projection of the feasible set of the linear programming relaxation of this model into the (x, z)-space.

Multicommodity flow formulation (Myung et al. [118])
A stronger relaxation is obtained by considering multicommodity flows. This *directed multicommodity flow model* was introduced by Myung et al. in [128]. In this model, every node set $k \in K_1$ defines a commodity. One unit of commodity k originates from V_1 and must be delivered to node set V_k. Letting f_{ij}^k be the flow of commodity k in arc (i, j), we obtain the following formulation:

$$\min \sum_{e \in E} c_e x_e$$

$$\text{s.t.} \quad z(V_k) = 1, \qquad \forall\, k \in K = \{1, \dots, m\}$$

$$x(E) = m - 1$$

$$\sum_{a\in\delta^+(i)} f_a^k - \sum_{a\in\delta^-(i)} f_a^k = \begin{cases} z_i, & i \in V_1 \\ -z_i, & i \in V_k,\ k \in K_1 \\ 0, & i \notin V_1 \cup V_k \end{cases} \tag{2.21}$$

$$f_{ij}^k \le w_{ij}, \qquad \forall\, a = (i,j) \in A,\ k \in K_1 \tag{2.22}$$

$$w_{ij} + w_{ji} = x_e, \qquad \forall\, e = \{i,j\} \in E$$

$$f_a^k \ge 0, \qquad \forall\, a = (i,j) \in A,\ k \in K_1 \tag{2.23}$$

$$x, z \in \{0,1\}.$$

In [128], Myung et al. presented a branch and bound procedure for solving the GMSTP. The computational efficiency of such a procedure greatly depends upon how quickly it generates good lower and upper bounds. Their lower bounding procedure was based on the directed multicommodity flow model, since its linear programming relaxation not only provides a tight lower bound, but also has a nice structure, on the basis of which an efficient dual ascent algorithm can be constructed. They also developed a heuristic algorithm, which finds a feasible solution for the GMSTP by using the obtained dual solution.

We let P_{mcflow} denote the projection of the feasible set of the linear programming relaxation of this model into the (x,z)-space. Pop [142] showed the the following relation between the feasible sets of the linear programming relaxation of the single commodity flow model and the multicommodity flow model:

Proposition 2.5. $P_{\text{mcflow}} \subseteq P_{\text{sflow}}$.

Proof. Let $(w,x,z,f) \in P_{\text{mcflow}}$, then

$$0 \le \sum_{k\in K_1} f_{ij}^k \le |K_1| w_{ij} \le (m-1)x_e.$$

With $f_{ij} = \sum_{k\in K_1} f_{ij}^k$ for every $e = \{i,j\} \in E$, we find

$$\sum_{e\in\delta^+(i)} f_a^k - \sum_{e\in\delta^-(i)} f_a^k = \sum_{k\in K_1} \Big(\sum_{a\in\delta^+(i)} f_a^k - \sum_{a\in\delta^-(i)} f_a^k \Big)$$

$$= \begin{cases} (m-1)z_i & \text{for } i \in V_1 \\ -z_i & \text{for } i \in V \setminus V_1. \end{cases}$$

□

Bidirectional flow formulation (Pop [142])

We obtain a closely related formulation by eliminating the variables w_{ij}. The resulting formulation consists of the constraints (2.1), (2.3), (2.21), (2.23) plus

$$f_{ij}^h + f_{ij}^k \le x_e, \ \forall\, h,k \in K_1 \quad \text{and} \quad \forall\, e \in E. \tag{2.24}$$

We refer to this model as the *bidirectional flow formulation* of the GMSTP and let P_{bdflow} denote its set of feasible solutions in (x, z)-space. Note that, since we have eliminated the variables w_a when constructing the bidirectional flow formulation, this model is defined on the undirected graph $G = (V, E)$, even though for each commodity k we permit flows f_{ij}^k and f_{ji}^k in both directions on edge $e = \{i, j\}$.

In the bidirectional flow formulation, the constraints (2.24), which we are called the bidirectional flow inequalities, link the flow of different commodities flowing in different directions on the edge $\{i, j\}$. These constraints model the fact that in any feasible generalized spanning tree, if we eliminate edge $\{i, j\}$ and divide the nodes in two sets, any commodity whose associated node is in the same set as the root node set, does not flow on edge $\{i, j\}$. Additionally, any two commodities whose associated nodes are both in the set without the root, both flow on edge $\{i, j\}$ in the same direction. So, whenever two commodities h and k both flow on edge $\{i, j\}$, they both flow in the same direction and therefore one of f_{ij}^h and f_{ij}^k equals zero.

Proposition 2.6 (Pop [142]). $P_{\text{mcflow}} = P_{\text{bdflow}}$.

Proof. If $(w, x, z, f) \in P_{\text{mcflow}}$ then, using the constraints (2.11) we have

$$f_{ij}^h + f_{ji}^k \le w_{ij} + w_{ji} = x_e, \quad \forall\, e = \{i, j\} \in E \quad \text{and} \quad \forall\, h, k \in K_1.$$

On the other hand, assume that $(x, z, f) \in P_{\text{bdflow}}$. From the constraints (2.24), we get

$$\max_h f_{ij}^h + \max_k f_{ji}^k \le x_e \quad \forall\, e = \{i, j\} \in E.$$

Hence, we can choose w such that $\max_h f_{ij}^h \le w_{ij}$ and $x_e = w_{ij} + w_{ji}$ for all $e = \{i, j\} \in E$. Take for example

$$w_{ij} = \frac{1}{2}(x_e + \max_h f_{ij}^h - \max_k f_{ji}^k).$$

Clearly, $(w, x, z, f) \in P_{\text{mcflow}}$. □

We obtain another formulation, referred to to as the *undirected multicommodity flow model*, by replacing the inequalities (2.22) with the weaker constraints:

$$f_{ij}^k \le x_e \quad \text{for every} \quad k \in K_1 \quad \text{and} \quad e \in E.$$

Flow cut formulation (Feremans [37])

The flow cut formulation is closely related to the multicommodity flow formulation. The main difference between the two is that the flow balance constraints (2.21) and constraints (2.22) are replaced with a directed cutset constraint using Gale's [44] necessary and sufficient feasibility condition for the existence of a feasible flow.

$$
\begin{aligned}
\min \sum_{e \in E} & c_e x_e \\
\text{s.t.} \quad & z(V_k) = 1, && \forall\, k \in K = \{1,\dots,m\} \\
& w_{ij} + w_{ji} = x_e, && \forall\, e = \{i,j\} \in E \\
& w(\delta^-(V_1)) = 0 && && (2.25) \\
& w(\delta^-(V_k)) \le 1, && \forall\, k \in K_1 && (2.26) \\
& w(\delta^-(S)) \ge z(V_k \cap S) - z(V_1 \cap S), && \forall\, S \subset V, 1 \le |S| \le n-1, \\
& && k \in K_1 && (2.27) \\
& x, z \in \{0,1\}.
\end{aligned}
$$

The directed cutset constraint (2.27), together with constraints (2.1), (2.11), (2.25) and (2.26), guarantees tree design and connectivity between all clusters. We refer to this model as the *flow cut formulation* of the GMSTP and let P_{fcut} denote its set of feasible solutions in (x,z)-space.

The relationship between the polyhedron defined by the linear relaxation of the flow cut formulation of the GMSTP and some of the polyhedrons of the linear relaxations of the corresponding formulations is (see [37], [142], [192]):

$$P_{\text{spack}} = P_{\text{dspack}} = P_{\text{fcut}} = P_{\text{mcflow}} = P_{\text{bdflow}}$$

2.3.4 A model based on Steiner tree properties

Raghavan [175] and Duin et al. [31] have transformed the GMSTP into a Steiner tree problem, which can be solved with any of the specialized algorithms developed for this problem.

Steiner tree formulation (Raghavan [175])
The *Steiner tree formulation*, introduced by Raghavan [175], is a multicommodity flow formulation of the GMSTP, after the GMSTP has been transformed into a Steiner tree problem with degree constraints. The transformation of the GMSTP to a Steiner tree problem is achieved in the following steps: let s_r be an artificial root node, and $T = \{t_1, t_2, \dots, t_k\}$ be a set of artificial sink nodes, which we require to be in the Steiner tree. Artificial node s_r is connected to each node $i \in V$ with the zero cost arc (s_r, i) and all nodes $i \in V_k$ of a given cluster $k = 1,\dots,m$ are connected with the zero cost arc (i, t_k) to the corresponding artificial node t_k. In this new graph, a set of commodities H is also defined, such that there is exactly one commodity $h \in H$, originating in node s_r and terminating in node t_k for each $k = 1,\dots,m$. The Steiner

tree formulation of the GMSTP is described as follows:

$$\min \sum_{e \in E} c_e x_e$$

$$\text{s.t.} \quad \sum_{(i,j)\in A: i \notin V_k, j \in V_k} w_{ij} = 1, \qquad \forall\, k \in K_1 \tag{2.28}$$

$$w_{ij} + w_{ji} \le 1, \quad \forall (i,j) \in A \tag{2.29}$$

$$\sum_{a \in \delta^+(i)} f_a^h - \sum_{a \in \delta^-(i)} f_a^h = \begin{cases} -1, & \text{if } i = O(h) \\ 1, & \text{if } i = D(h) \\ 0, & \text{otherwise} \end{cases} \quad \forall\, i \in V \text{ and } h \in H \tag{2.30}$$

$$0 \le f_{ij}^h \le w_{ij}, \qquad \forall (i,j) \in A, h \in H \tag{2.31}$$

$$w \in \{0,1\}.$$

The Steiner tree formulation of the GMSTP then is the standard multicommodity flow model for the Steiner tree problem with additional degree constraints (2.28) that ensure an in-degree of 1 for each of the clusters ($k \in K$). This formulation ensures the design of a minimum-cost directed Steiner tree, containing the root node s_r, all the nodes in T, and exactly one node from each cluster.

We denote by P_{stree} the set of feasible solutions in (x, z)-space of this formulation. In [175], Raghavan has shown that the Steiner tree formulation is equivalent to the multicommodity flow formulation, where the node variables were used to indicate whether a node is selected to be in the generalized spanning tree.

2.3.5 Local-global formulation of the GMSTP

Our last formulation was introduced by Pop [142] and aims at distinguishing between *global*, i.e. inter-cluster connections, and *local* ones, i.e. connections between nodes from different clusters.

We introduce variables y_{ij} ($i, j \in \{1, \ldots, m\}$) to describe the global connections. Hence, $y_{ij} = 1$ if cluster V_i is connected to cluster V_j and $y_{ij} = 0$ otherwise, and we assume that y represents a spanning tree.

The convex hull of all these y-vectors is generally known as the spanning tree polytope (on the contracted graph with vertex set $\{V_1, \ldots, V_m\}$, which we assume to be complete).

Following Yannakakis [205], this polytope, denoted by P_{MST}, can be represented by the following polynomial number of constraints:

$$\sum_{\{i,j\}} y_{ij} = m - 1$$

$$y_{ij} = \lambda_{kij} + \lambda_{kji}, \qquad \text{for} \quad 1 \le k, i, j \le m \quad \text{and} \quad i \ne j \tag{2.32}$$

$$\sum_{j} \lambda_{kij} = 1, \qquad \text{for} \quad 1 \le k, i, j \le m \quad \text{and} \quad i \ne k \tag{2.33}$$

$$\lambda_{kkj} = 0, \qquad \text{for} \quad 1 \le k, j \le m \tag{2.34}$$

$$y_{ij}, \lambda_{kij} \ge 0, \qquad \text{for} \quad 1 \le k, i, j \le m.$$

where the variables λ_{kij} are defined for every triplet of nodes k, i, j, with $i \ne j \ne k$ and their value for a spanning tree is:

$$\lambda_{kij} = \begin{cases} 1 & \text{if } j \text{ is the parent of } i, \text{ when we root the tree at } k \\ 0 & \text{otherwise.} \end{cases}$$

The constraints (2.32) mean that an edge (i, j) is in the spanning tree if and only if either i is the parent of j or j is the parent of i. The constraints (2.33) mean that, if we root a spanning tree at k, every node other than node k has a parent. Finally, the constraints (2.34) mean that the root k has no parent.

If the vector y describes a spanning tree on the contracted graph, the corresponding best (w.r.t. minimization of costs) *local solution* $x \in \{0, 1\}^{|E|}$ can be obtained via one of the following two methods, described by Pop [142] using dynamic programming (see Section 2.2) or integer linear programming.

If the 0-1 vector y describes a spanning tree on the contracted graph, the corresponding local solution $x \in \{0, 1\}^{|E|}$ minimizing the costs can be obtained by solving the following integer linear programming problem:

$$\min \sum_{e \in E} c_e x_e$$

$$\begin{aligned} \text{s.t.} \quad z(V_k) &= 1 && \forall\, k \in K = \{1, \dots, m\}, \\ x(V_l, V_r) &= y_{lr} && \forall\, l, r \in K = \{1, \dots, m\},\ l \ne r, \\ x(i, V_r) &\le z_i && \forall\, r \in K, \forall\, i \in V \setminus V_r, \\ x_e, z_i &\in \{0, 1\} && \forall\, e = \{i, j\} \in E,\ \forall\, i \in V, \end{aligned}$$

where $x(V_l, V_r) = \sum_{i \in V_l, j \in V_r} x_{ij}$ and $x(i, V_r) = \sum_{j \in V_r} x_{ij}$.

For given y, we denote the feasible set of the linear programming relaxation of this program by $P_{\text{local}}(y)$. The following result holds (Pop et al. [150]):

Proposition 2.7. *If y is the 0-1 incidence vector of a spanning tree of the contracted graph, then the polyhedron $P_{\text{local}}(y)$ is integral.*

Proof. Suppose that the 0-1 vector y describes a spanning tree T of the contracted graph. Then, in order to prove that the polyhedron $P_{\text{local}}(y)$ is integral, it suffices to show that every solution of the linear programming relaxation can be written as a convex combination of solutions corresponding to spanning trees:

$$(x, z) = \sum \lambda_T (x_T, z_T).$$

The proof is by induction, on the support of x, denoted by supp(x) and defined as follows:

$$\text{supp}(x) := \{x_e \mid x_e \neq 0,\ e \in E\}.$$

Suppose that there is a global connection between the clusters V_l and V_r (i.e. $y_{lr} = 1$) then

$$1 = x(V_l, V_r) = \sum_{i \in V_l} x(i, V_r) \leq \sum_{i \in V_l} z_i = 1,$$

which implies that $x(i, V_r) = z_i$.

We claim that supp(x) $\subseteq E$ contains a tree connecting all clusters. Assume the opposite is true and let $T^1 \subseteq E$ be a maximal tree in supp(x). Since T^1 does not connect all clusters, there is some edge (l, r) with $y_{lr} = 1$, such that T^1 has some vertex $i \in V_l$ but no vertex in V_r.

Then $z_i > 0$, and thus $x(i, V_r) = z_i > 0$, so T^1 can be extended by some $e = (i, j)$ with $j \in V_r$, which is a contradiction.

Now, let x^{T^1} be the incidence vector of T^1 and let

$$\alpha := \min\{x_e \mid e \in T^1\}.$$

If $\alpha = 1$, then $x = x^{T^1}$ and we are done.

Otherwise, let z^{T^1} be the vector with $z_i^{T^1} = 1$ if T^1 covers $i \in V$ and $z_i^{T^1} = 0$ otherwise. Then

$$(\widehat{x}, \widehat{z}) := ((1-\alpha)^{-1}(x - \alpha x^{T^1}), (1-\alpha)^{-1}(z - \alpha z^{T^1}))$$

is again in $P_{\text{local}}(y)$ and can be written, by induction, as a convex combination of tree solutions. Our claim now follows. □

A similar argument shows that the polyhedron $P_{\text{local}}(y)$ is integral, even in the case when the 0-1 vector y describes a cycle free subgraph in the contracted graph. If the 0-1 vector y contains a cycle of the contracted graph, then $P_{\text{local}}(y)$ is in general not integral. In order to show that this affirmation holds, we consider the following example, depicted in Figure 2.3.

If the lines drawn in the figure (i.e., $\{1,3\}$, $\{2,4\}$ etc.) have cost 1 and all the other lines (i.e., $\{1,4\}$, $\{2,3\}$ etc.) have cost $M \gg 1$, then $z \equiv \frac{1}{2}$ and $x \equiv \frac{1}{2}$ on the drawn lines is an optimal solution of $P_{\text{local}}(y)$, showing that the polyhedron $P_{\text{local}}(y)$ is not integral.

Local-global formulation (Pop [142])

The observations presented so far lead to our final formulation, called *local-global formulation* of the GMSTP as a 0-1 mixed integer programming problem, where only

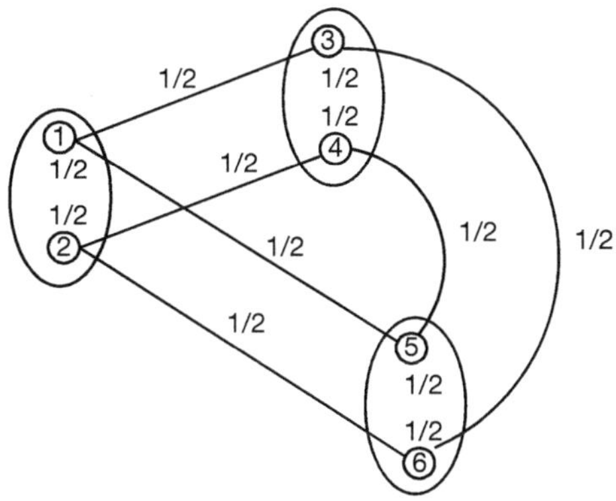

Figure 2.3. An example showing that the polyhedron $P_{\text{local}}(z)$ may have fractional extreme points.

the global variables y are forced to be integral:

$$
\begin{aligned}
\min \sum_{e \in E} c_e x_e & \\
\text{s.t.} \quad z(V_k) &= 1, && \forall\, k \in K = \{1, \dots, m\}, \\
x(E) &= m - 1, && \\
x(V_l, V_r) &= y_{lr}, && \forall\, l, r \in K = \{1, \dots, m\},\ l \neq r, \\
x(i, V_r) &\leq z_i, && \forall\, r \in K, \forall\, i \in V \setminus V_r, \\
y_{ij} &= \lambda_{kij} + \lambda_{kji}, && \forall\, 1 \leq k, i, j \leq m \quad \text{and} \quad i \neq j, \\
\sum_j \lambda_{kij} &= 1, && \forall\, 1 \leq k, i, j \leq m \quad \text{and} \quad i \neq k, \\
\lambda_{kkj} &= 0, && \forall\, 1 \leq k, j \leq m, \\
\lambda_{kij} &\geq 0, && \forall\, 1 \leq k, i, j \leq m, \\
x_e, z_i &\geq 0, && \forall\, e = (i, j) \in E,\ \forall\, i \in V, \\
y_{lr} &\in \{0, 1\}, && \forall\, 1 \leq l, r \leq m.
\end{aligned}
$$

This new formulation of the GMSTP is obtained by incorporating the constraints characterizing P_{MST}, with $y \in \{0, 1\}$, into $P_{\text{local}}(y)$. The local-global formulation of the GMSTP uses three types of variables to define the connections between clusters in the graph. The first group of variables, which we have already used in the previous formulations, are indicator edge variables, x_{ij}. These variables define the use of specific edges between nodes in the graph. On the other hand, the other two groups of variables, y_{lr} and λ_{klr}, define connections between clusters in the graph. The y_{lr} variables indicate whether the solution includes any of the edges that directly connect the clusters l and r. The λ_{klr} variables are used to define $|K|$ directed trees, one for each of the clusters in the graph.

This new model of the GMSTP has several nice properties, that can be used to develop exact solution procedures [142, 150], metaheuristic algorithms [76, 77], heuristic algorithms [143] and also heuristic and exact procedures for the PC-GMSTP [192, 54]. Among the described integer programming formulations, the local-global formulation is the most compact, in terms of number of variables and number of constraints.

Note that, although the presented formulations are defined for the GMSTP, these formulations are also valid for the PC-GMSTP. We only need to add the price of selecting a node from a cluster to the objective function.

The relationships between the polyhedrons, defined by the linear relaxations of the described GMSTP formulations, are depicted in the following figure:

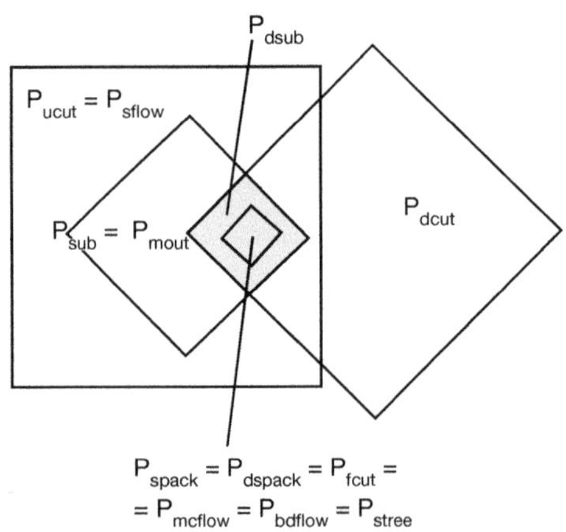

Figure 2.4. Relationship between the polyhedrons defined by the linear relaxations of the corresponding GMSTP formulations.

2.4 Approximation results for the GMSTP

Throughout this section, we will discriminate between so-called *positive results*, which establish the existence of some approximation algorithms, and so-called *negative results* which disprove the existence of good approximating results for some combinatorial optimization problems, under the assumption that $\mathcal{P} \neq \mathcal{NP}$.

We start with some basic results of approximation theory that follow the descriptions of Goemans [57] and Schuurman and Woeginger [183].

2.4.1 Introduction

Many of the optimization problems that we would like to solve are *$\mathcal{NP}$-hard*. Therefore, it is very unlikely that these problems can be solved by a polynomial-time algorithm. However, these problems still have to be solved in practice. In order to do so, we have to relax some of the requirements. In general, there are three different possibilities.

- **Superpolynomial-time algorithms:** Even though an optimization problem is *$\mathcal{NP}$-hard*, there are "good" and "not so good" algorithms for solving it exactly. The "not so good" algorithms certainly include most of the simple enumeration methods, where one enumerates all the feasible solutions and chooses the one with the optimal value of the objective function. Such methods have a very high time complexity. The "good" algorithms includes methods like *branch-and-bound*, where an analysis of the problem at hand is used to discard most of the feasible solutions, before they are even considered. These approaches allow one to obtain exact solutions of reasonably large instances of a problem, but their running time still depends exponentially on the size of the problem.
- **Average-case polynomial-time algorithms:** For some problems, it is possible to have algorithms which require superpolynomial-time on only a few instances whereas the other instances run in polynomial time. A famous example is the *Simplex Method* for solving problems in linear programming.
- **Approximation Algorithms:** We may also relax the requirement to obtain an exact solution of the optimization problem and be content with a solution that is "not too far away" from the optimum. This is partially justified by the fact that, in practice, it usually suffices to obtain a slightly sub-optimal solution.

Clearly, there are good and bad approximation algorithms. We need some means of determining the quality of an approximation algorithm and a way of comparing different algorithms. There are a few criteria to consider:

Average-case performance: Here, we consider some probability distribution on the set of all possible instances of a given problem. Based on this assumption, an expectation of the performance may then be found. Results of this kind strongly depend on the choice of the initial distribution and do not provide us with any information about the performance on a particular instance.

Experimental performance: This approach is based on running the algorithm on a few "typical" instances. It has been mostly used to compare the performance of several approximation algorithms. Of course the result depends on the choice of the "typical" instances and may vary from experiment to experiment.

Worst-case performance: This is usually done by establishing upper and lower bounds for approximate solutions, in terms of the optimum value. In case of minimization problems, we try to establish upper bounds, in case of maximization problems, we wants to find lower bounds.

The advantage of the worst-case bounds on the performance of approximation algorithms is that, given any instance of the optimization problem, we are guaranteed that the approximate solution stays within these bounds. It should also be noted that approximation algorithms usually find solutions much closer to the optimum than the worst-case bounds suggest. Thus, it is of interest in and of itself, to see how tight the bounds on the performance of each algorithm are, that is, how bad the approximate solution can really get. This is usually done by providing examples of specific instances, for which the approximate solution is very far from the optimum solution.

Even for simple algorithms, establishing worst-case performance bounds often requires a very deep understanding of the problem at hand and the use of powerful theoretical methods from areas like linear programming, combinatorics, graph theory, probability theory, etc.

We consider an $\mathcal{NP}$*-hard* optimization problem, for which, as we have seen, it is difficult to find the exact optimal solution within polynomial time. At the expense of reducing the quality of the solution by relaxing some of the requirements, we can often get a considerable speed-up in the complexity. This leads us to the following definition:

Definition 2.8 (Approximation algorithms). Let X be a minimization problem and $\alpha > 1$. An algorithm APP is called an α-approximation algorithm for problem X if, for all instances I of X, it delivers a feasible solution, in polynomial time, with objective value APP(I) such that

$$\text{APP}(I) \leq \alpha \text{OPT}(I), \tag{2.35}$$

where by APP(I) and OPT(I) we denote the values of an approximate solution and that of an optimal solution for instance I, respectively.

The value α is called the *performance guarantee* or the *worst case ratio* of the approximation algorithm APP. The closer α is to 1 the better the algorithm is.

2.4.2 Positive results: the design of the approximation algorithms

With regard to approximations, positive results involve the design and analysis of polynomial time approximation algorithms. This section concentrates on such positive results; we will outline the main strategy.

Assume that we need to find an approximation algorithm for an $\mathcal{NP}$*-hard* optimization problem X. How shall we proceed? In what follows, we will concentrate on minimization problems, but the ideas apply equally well to maximization problems.

In what follows, given an instance I of an optimization problem and some algorithm for its approximation, we will denote by OPT $=$ OPT(I) the value of an optimum solution and by APP $=$ APP(I) the value of the solution produced by the approximation algorithm. We are interested in an α-approximation algorithm, i.e. an algorithm APP such that

$$\text{APP} \leq \alpha \text{OPT}.$$

Directly relating APP to OPT can be difficult. Therefore, we try instead to find a lower bound for the optimal solution, denoted by LB, satisfying the following conditions:

$$\begin{aligned} &1. \quad \text{LB} \leq \text{OPT}, \\ &2. \quad \text{APP} \leq \alpha \text{LB}. \end{aligned}$$

Consider now the following optimization problem:

$$\text{OPT} = \min_{x \in S} f(x).$$

A lower bound on OPT can be obtained by so-called *relaxation*. Consider the following related optimization problem:

$$\text{LB} = \min_{x \in R} g(x).$$

Definition 2.9 (Lower bound). LB is called a lower bound on OPT and the optimization problem is called a relaxation of the original problem, if the following two conditions hold:

$$\begin{aligned} &S \subseteq R \\ &g(x) \leq f(x) \quad \text{for all} \quad x \in S. \end{aligned}$$

The above conditions imply the following:

$$\text{LB} = \min_{x \in R} g(x) \leq \min_{x \in S} f(x) = \text{OPT}.$$

which shows that LB is a lower bound on OPT.

Many combinatorial optimization problems can be formulated as integer or mixed integer programs and the linear programming relaxations of these programs provide a lower bound for the optimal solution. Later, we will show how to use a linear programming relaxation of an integer programming formulation of the GMSTP to get an α-approximation algorithm for the GMSTP with bounded cluster size.

There exist two main techniques for deriving an approximately optimal solution from a solution of the relaxed problem:

1. Rounding

The idea is to solve the linear programming relaxation and then convert the obtained fractional solution into an integral solution. The approximation guarantee is established by comparing the costs of the integral and fractional solutions.

Therefore, we find an optimal solution x^* to the relaxation, and we "round" $x^* \in R$ to an element $x' \in S$. Then, we prove that $f(x') \leq \alpha g(x^*)$, which implies that

$$f(x') \leq \alpha \text{LB} \leq \alpha \text{OPT}.$$

Often, randomization is helpful. In this case $x^* \in R$ is randomly rounded to some element $x' \in S$, such that $E[f(x')] \leq \alpha g(x^*)$. These algorithms can sometimes be derandomized, in which case one finds an x'' such that $f(x'') \leq E[f(x')]$.

An example of the application of the rounding technique is provided by the Minimum Weight Vertex Cover Problem, with the result due to Hochbaum [75].

2. Primal-Dual

We consider the linear programming relaxation of the primal program. This method iteratively constructs an integer solution to the primal program and a feasible solution to the dual program, $max\,\{h(y) : y \in D\}$. The approximation guarantee is established by comparing the two solutions.

The weak duality property of the relaxation implies:

$$\max\{h(y) : y \in D\} \leq \min\{g(x) : x \in R\}.$$

We construct $x \in S$ from $y \in D$ such that:

$$f(x) \leq \alpha h(y) \leq \alpha h(y_{\max}) \leq \alpha g(x_{\min}) \leq \alpha \text{OPT}.$$

Notice that y can be any element of D, and must not necessarily be an optimal solution to the dual.

More details about the primal-dual technique can be found in [7] and [203].

2.4.3 A negative result for the GMSTP

For some hard combinatorial optimization problems it is possible to show that they don't have an approximation algorithm, unless $\mathcal{P} = \mathcal{NP}$. In order to get a result of this form it is enough to show that the existence of an α-approximation algorithm would allow one to solve some decision problem, known to be $\mathcal{NP}$*-complete*, in polynomial time.

Applying this scheme to the GMSTP, we obtain an in-approximability result. This result is a different formulation, in terms of approximation algorithms, of a result provided by Myung et al. [128], which says that even finding a near optimal solution for the GMSTP is $\mathcal{NP}$*-hard*. Our proof is slightly different from the proof provided in [128].

Theorem 2.10. *Under the assumption $\mathcal{P} \neq \mathcal{NP}$, there is no α-approximation algorithm for the GMSTP.*

Proof. Assume that there exists an α-approximation algorithm APP for the GMSTP, where α is a real number greater than or equal to 1. This means that

$$\text{APP}(I) \leq \alpha \text{OPT}(I),$$

for every instance I, where $\text{OPT}(I)$ and $\text{APP}(I)$ are the values of the optimal solution and of the solution found by the algorithm APP, respectively.

Then, we can show that APP also solves the node-cover problem for a given graph $G = (V, E)$ and an integer k such that $k < |V|$. This result contradicts the assumption that $\mathcal{P} \neq \mathcal{NP}$.

We construct a graph $G' = (V', E')$ and the edge cost function such that the algorithm APP finds a feasible solution with a value no greater than α times the optimal cost, if and only if G contains C, where C is a node cover of G, i.e. a subset of V, such that all the edges of G are adjacent to at least one node of C.

The graph G' contains the following $m = |E| + k + 1$ node sets (clusters), $V'_1, \dots, V'_m$:

- V'_1 consists of a single node, denoted by r,
- $V'_2, \dots, V'_{k+1}$ are identical node sets, each of which has $|V|$ nodes, corresponding to the nodes of V, and
- $|E|$ node sets, $V'_{k+2}, \dots, V'_m$, each of which contains a single node corresponding to an edge $e \in E$.

Edges in G' are constructed as follows:

(i) For all $t = 2, \dots, k + 1$, each node of V'_t is connected to r by an edge. The set consisting of these edges is denoted by E'_1.

(ii) Let i be a node of V'_t for any $t \in \{2, \dots, k + 1\}$ and let j be a node of V'_t for any $t \in \{k + 2, \dots, m\}$. Then, an edge is constructed between i and j if the edge of G corresponding to j is incident to the node of G corresponding to i and we denote the set of those edges by E'_2.

(iii) We also construct an edge between i and j, even though the edge of G corresponding to j is not incident to the node of G corresponding to i, and we let E'_3 denote the set of those edges.

We let $E' = E'_1 \cup E'_2 \cup E'_3$.

The cost of each edge is defined as follows:

$$c_{ij} = \begin{cases} 0 & \text{for all} \quad i, j \in E'_1 \\ 1 & \text{for all} \quad i, j \in E'_2 \\ (|E| + 1)\alpha & \text{for all} \quad i, j \in E'_3 \end{cases}.$$

We claim that G contains C if and only if

$$\text{APP}(I) \leq \alpha |E|,$$

where the instance I corresponds to G' and its cost function.

Note that there always exists a generalized spanning tree in G': all clusters different from the identical clusters $V'_2, \ldots, V'_{k+1}$ have only one node and if we select k nodes from $V'_2, \ldots, V'_{k+1}$, where we take one node from each cluster such that each node of C is included, then these k nodes, together with the remaining nodes selected, always uniquely form a generalized spanning tree of G' with edges in $E'_1 \cup E'_2$, by the definition of G'.

Suppose now that G' contains a generalized spanning tree and let C be a set of distinct nodes selected from the clusters, $V'_2, \ldots, V'_{k+1}$ in the tree, then C is a node cover of G.

Therefore, we have shown that a generalized spanning tree, only using edges in $E'_1 \cup E'_2$, exists in G' if and only if G contains a node cover C.

If G contains C, then $\text{OPT}(I) = |E|$. Moreover, if G does not contain C, then any generalized spanning tree of G' should use at least one edge in E'_3 and thus

$$\text{OPT}(I) \geq |E| - 1 + \alpha(|E| + 1) > \alpha |E|.$$

In particular, if G contains a node cover C, then the approximation algorithm APP will produce a solution with value

$$\text{APP}(I) \leq \alpha |E| = \alpha \text{OPT}(I),$$

i.e. the solution does not use any edge from E'_3 and APP identifies a node cover. □

In the next section however, after making some further assumptions, we will present a positive result for the GMSTP.

2.4.4 An approximation algorithm for the GMSTP with bounded cluster size

As we have seen in the previous section, under the assumption $\mathcal{P} \neq \mathcal{NP}$, there exists no α-approximation algorithm for the GMSTP.

However under the following assumptions:

A1: the graph has bounded cluster size, i.e. $|V_k| \le \rho$, for all $k = 1, \dots, m$,

A2: the cost function is strict positive and satisfies the triangle inequality, i.e. $c_{ij} + c_{jk} \ge c_{ik}$ for all $i, j, k \in V$,

a polynomial approximation algorithm for the GMSTP is possible.

Under the above assumptions, we present, in this section, an approximation algorithm for the GMSTP with performance ratio 2ρ. The approximation algorithm is constructed following the ideas of Slavik [187], which deals with the Generalized Traveling Salesman Problem and Group Steiner Tree Problem.

An integer programming formulation of the GMSTP

We consider the following integer programming formulation of the GMSTP:

Problem IP1:

$$Z_1 = \min \sum_{e \in E} c_e x_e,$$

$$\begin{aligned}
\text{s.t.}\quad & z(V_k) = 1, && \forall\, k \in K = \{1, \dots, m\}, && (2.36)\\
& x(\delta(S)) \ge z_i, && \forall\, i \in S,\ \forall S \subset V \text{ such that for some} && (2.37)\\
& && \text{cluster } V_k, \text{ with } k \in K,\ S \cap V_k = \emptyset, &&\\
& x(E) = m - 1, && && (2.38)\\
& x_e \in \{0, 1\}, && \forall\, e \in E, && (2.39)\\
& z_i \in \{0, 1\}, && \forall\, i \in V. && (2.40)
\end{aligned}$$

Here, we used the standard notations introduced in Section 2.3.

In the above formulation, condition (2.36) guarantees that a feasible solution contains exactly one vertex from every cluster. Condition (2.37) guarantees that any feasible solution induces a connected subgraph. Condition (2.38) simply assures that any feasible solution has $m - 1$ edges and, because the cost function is strict positive this constraint is redundant.

Now, we consider the linear programming relaxation of the integer programming formulation of the GMSTP. To do that, we simply replace the conditions (2.39) and (2.40) in IP1 with new conditions:

$$0 \le x_e \le 1, \quad \text{for all } e \in E, \tag{2.41}$$

$$0 \le z_i \le 1, \quad \text{for alll } i \in V. \tag{2.42}$$

We assume that assumptions A1 and A2 hold.

The algorithm for approximating the optimal solution of the GMSTP is as follows:

Algorithm 2.11 ("Approximate the GMSTP").

Input: A complete graph $G = (V, E)$, with strictly positive cost function on the edges, satisfying the triangle inequality, and with the nodes partitioned into the clusters $V_1, \dots, V_m$ with bounded size, $|V_k| \leq \rho$.

Output: A tree $T \subset G$, spanning some vertices $W' \subset V$, which includes exactly one vertex from every cluster, and approximates the optimal solution to the GMSTP.

1. Solve the linear programming relaxation of the problem IP1 and let $(z^*, x^*, Z_1^*) = ((z_i^*)_{i=1}^n, (x_e^*)_{e \in E}, Z_1^*)$ be the optimal solution.
2. Set $W^* = \{i \in V \mid z_i^* \geq \frac{1}{\rho}\}$ and consider $W' \subset W^*$, with the property that W' contains exactly one vertex from each cluster, and find a minimum spanning tree $T \subset G$ on the subgraph G' generated by W'.
3. Output APP $=$ cost(T) and the generalized spanning tree T.

Even though the linear programming relaxation of the problem IP1 has an exponential number of constraints, it can still be solved in polynomial time by using either the ellipsoid method, with a min-cut max-flow oracle [59] or Karmarkar's algorithm [96], because the linear programming relaxation can be formulated "compactly" (the number of constraints polynomially bounded) using flow variables see Section 2.3.

Auxiliary results

We start this subsection with a result established by Goemans and Bertsimas [58] and called the *parsimonious property*.

Given a complete undirected graph $G = (V, E)$, we associate with each edge $(i, j) \in E$ a cost c_{ij} and for any pair (i, j) of vertices, let r_{ij} be the connectivity requirement between i and j (r_{ij} is assumed to be symmetric, i.e. $r_{ij} = r_{ji}$). A network is called *survivable*, if it has at least r_{ij} edge disjoint paths between any pair (i, j) of vertices.

The *survivable network design problem* consists in finding the minimum cost survivable network. This problem can be formulated as the following integer program:

$$
\begin{aligned}
(\mathrm{IP}_\emptyset(r)) \quad & IZ_\emptyset(r) = \min \sum_{e \in E} c_e x_e & & \\
\text{s.t.} \quad & x(\delta(S)) \geq \max_{(i,j) \in \delta(S)} r_{ij}, & & S \subset V,\ S \neq \emptyset, \qquad (2.43)\\
& 0 \leq x_e, & & e \in E,\\
& x_e \quad \text{integral}, & & e \in E.
\end{aligned}
$$

We denote by $IZ_\emptyset(r)$ the optimal value of the above integer program. Let $(P_\emptyset(r))$ denote the linear programming relaxation of $(\mathrm{IP}_\emptyset(r))$, obtained by dropping the integrality restrictions and let $Z_\emptyset(r)$ be its optimal value.

By definition, the degree of vertex $i \in V$ is $d_x(i) = x(\delta(i))$, for any feasible solution x, to either $(\mathrm{IP}_\emptyset(r))$ or to $(P_\emptyset(r))$. Because of the constraints (2.43), for $S = \{i\}$, the degree of vertex i is at least equal to $\max_{j \in V\setminus\{i\}} r_{ij}$. If $d_x(i) = \max_{j \in V\setminus\{i\}} r_{ij}$, we say that x is *parsimonious* at vertex i. If we impose the requirement of the solution x to be parsimonious at all the vertices of a set $D \subseteq V$, we get some interesting variations of $(\mathrm{IP}_\emptyset(r))$ and $(P_\emptyset(r))$, denoted by $(\mathrm{IP}_D(r))$ and $(P_D(r))$ respectively. The formulation of $(\mathrm{IP}_D(r))$ as an integer program is:

$$
\begin{aligned}
(\mathrm{IP}_D(r)) \quad & IZ_D(r) = \min \sum_{e \in E} c_e x_e, & \\
\text{s.t.} \quad & x(\delta(S)) \geq \max_{(i,j) \in \delta(S)} r_{ij}, & S \subset V,\ S \neq \emptyset, \\
& x(\delta(i)) = \max_{j \in V\setminus\{i\}} r_{ij}, & i \in D,\ D \subseteq V, \\
& 0 \leq x_e, & e \in E, \\
& x_e \ \text{ integral}, & e \in E.
\end{aligned}
$$

We denote by $IZ_D(r)$ the optimal value of the above integer program. Let $(P_D(r))$ denote the linear programming relaxation of $(IP_D(r))$, obtained by dropping the integrality restrictions and let $Z_D(r)$ be its optimal value.

Theorem 2.12 (parsimonious property, Goemans and Bertsimas [58]). *If the costs c_{ij} satisfy the triangle inequality, then*

$$Z_\emptyset(r) = Z_D(r)$$

for all subsets $D \subseteq V$.

The proof of this theorem is based on a result on connectivity properties of Eulerian multigraphs, from Lovász [114]. Now let $W \subset V$ and consider the following linear program:

Problem LP2:

$$
\begin{aligned}
& Z_2^*(W) = \min \sum_{e \in E} c_e x_e, & & \\
\text{s.t.} \quad & x(\delta(S)) \geq 1, & S \subset V,\ \text{s.t. } W \cap S \neq \emptyset \neq W \setminus S, & \quad (2.44) \\
& x(\delta(i)) = 0, & i \in V \setminus W, & \quad (2.45) \\
& 0 \leq x_e \leq 1, & e \in E. & \quad (2.46)
\end{aligned}
$$

Replacing the constraints (2.46) with the integrality constraints $x_e \in \{0, 1\}$, we obtain the formulation of the minimum tree, spanning the subset of nodes $W \subset V$.

Consider the following relaxation of the problem LP2.

Problem LP3:

$$Z_3^*(W) = \min \sum_{e \in E} c_e x_e$$

$$\text{s.t.} \quad x(\delta(S)) \geq 1, \qquad S \subset V, \text{ s.t. } W \cap S \neq \emptyset \neq W \setminus S,$$

$$0 \leq x_e, \qquad e \in E. \tag{2.47}$$

Thus, we have omitted constraint (2.45) and relaxed constraint (2.46).

The following result is a straightforward consequence of the parsimonious property, if we choose $r_{ij} = 1$, if $i, j \in W$, and 0 otherwise, and $D = V \setminus W$.

Lemma 2.13. *The value of the optimal solutions to problems LP2 and LP3 are the same, that is*

$$Z_2^*(W) = Z_3^*(W).$$

Consider the following problem:

Problem IP4:

$$Z_4 = \min \sum_{e \in E} c_e x_e,$$

$$\text{s.t.} \quad x(\delta(S)) \geq 1, \qquad S \subset V, \text{ s.t. } S \neq \emptyset \neq V, \tag{2.48}$$

$$x_e \in \{0, 1\}, \qquad e \in E. \tag{2.49}$$

Clearly, this is the integer programming formulation of the MST (minimum spanning tree) problem. Let LP4 be the LP relaxation of this formulation, that is, we simply replace the constraint (2.49) by the constraint $0 \leq x_e \leq 1$, for all $e \in E$.

Denote by Z_4^* the value of the optimal solution of the LP4. The following known result for minimum spanning trees holds:

Proposition 2.14.

$$L^T(V) \leq \left(2 - \frac{2}{|V|}\right) Z_4^*,$$

where $L^T(V)$ denotes the cost of the minimum spanning tree on V.

Proof. See for example [187]. □

Let $W \subset V$, then Proposition 2.14 can be easily modified to obtain:

Proposition 2.15.

$$L^T(W) \leq \left(2 - \frac{2}{|W|}\right) Z_2^*(W).$$

Proof. Let (x_e) be a feasible solution to LP2. If $e \notin E(W) = \{(i,j) \mid i,j \in W\}$, this implies $x_e = 0$ and using Proposition 2.14 we prove the inequality. □

Performance Bounds

Let $(z^*, x^*, Z_1^*) = ((z_i^*)_{i=1}^n, (x_e^*)_{e \in E}, Z_1^*)$ be the optimal solution to the linear programming relaxation of the GMSTP. Define

$$\widehat{x}_e = \rho x_e^*$$

$$\widehat{z}_i = \begin{cases} 1 & \text{if} \quad z_i^* \geq \frac{1}{\rho} \\ 0 & \text{otherwise.} \end{cases}$$

$W^* = \left\{ i \in V \mid z_i^* \geq \dfrac{1}{\rho} \right\}$ $=\{i \in V \mid \widehat{z}_i = 1\}$. Because we need only one vertex from every cluster, we delete extra vertices from W^* and consider $W' \subset W^*$, such that $|W'|$ = m and W' consists of exactly one vertex from every cluster.

Since LP1 is the linear programming relaxation of the problem IP1, we have

$$Z_1^* \leq Z_1.$$

Now, let us show that $(\widehat{x}_e)_{e \in E}$ is a feasible solution to LP3 with $W = W'$. Indeed, $\widehat{x}_e \geq 0$ for all $e \in E$, hence condition (2.47) is satisfied. Let $S \subset V$ be such that $W' \cap S \neq \emptyset \neq W' \setminus S$ and choose some $i \in W' \cap S$. Hence $\widehat{z}_i = 1$ and $z_i^* \geq \frac{1}{\rho}$. Then we have

$$\widehat{x}(\delta(S)) = \sum_{e \in \delta(S)} \widehat{x}_e = \rho \sum_{e \in \delta(S)} x_e^* \geq \rho z_i^* \geq \rho \frac{1}{\rho} = 1,$$

by definition of $\widehat{x}_e$ and the fact that the (x_e^*) solve LP1. Hence the $(\widehat{x}_e)$ satisfy constraint (2.44) in LP3. Therefore, for the approximation algorithm of the GMSTP APP, the following holds:

$$\begin{aligned}
\text{APP} = L^T(W') \quad & \leq \left(2 - \frac{2}{|W'|}\right) Z_2^*(W') = \left(2 - \frac{2}{|W'|}\right) Z_3^*(W') \\
\leq \left(2 - \frac{2}{|W'|}\right) \sum_{e \in E} c_e \widehat{x}_e \quad & = \left(2 - \frac{2}{|W'|}\right) \rho \sum_{e \in E} c_e x_e^* = \left(2 - \frac{2}{|W'|}\right) \rho Z_1^* \\
\leq \left(2 - \frac{2}{|W'|}\right) \rho Z_1 \quad & = \left(2 - \frac{2}{|W'|}\right) \rho \text{OPT}.
\end{aligned}$$

And since $W' \subset V$, that is, m = $|W'| \leq |V|$ = n, we have proved the following.

Theorem 2.16. *The performance ratio of the algorithm "Approximate GMSTP" for approximating the optimum solution to the GMSTP satisfies:*

$$\frac{\text{APP}}{\text{OPT}} \leq \left(2 - \frac{2}{n}\right)\rho.$$

In the following, we generalize the "Approximate GMSTP" algorithm, and its analysis, to the case where, in addition to the cost c_{ij} associated with each edge $e = (i, j) \in E$, there is a cost, say d_i , associated with each vertex $i \in V$.

In this case, the GMSTP can be formulated as the following integer program:

$$\begin{aligned} \text{OPT} = \min \sum_{e \in E} c_e x_e + \sum_{i \in V} d_i z_i \\ \text{s.t.} \quad (2.36)\text{–}\,(2.40). \end{aligned}$$

Suppose that $(\overline{x}, \overline{z})$ is an optimal solution. Then, the optimal value OPT of this integer program consists of two parts:

$$L_{\text{OPT}} := \sum_{e \in E} c_e \overline{x_e} \quad \text{and} \quad V_{\text{OPT}} := \sum_{i \in V} d_i \overline{z_i}.$$

Under the same assumptions A1 and A2, the algorithm for approximating the optimal solution of the GMST problem, in this case, is as follows:

1. Solve the linear programming relaxation of the previous integer program and let $(z^*, x^*) = ((z_i^*)_{i=1}^n, (x_e^*)_{e \in E})$ be the optimal solution.

2. Set $W^* = \{i \in V | z_i^* \geq \frac{1}{\rho}\}$ and consider $W' \subset W^*$, with the property that W' contains exactly one vertex from each cluster, and find a minimum spanning tree $T \subset G$ on the subgraph G' generated by W'.

3. Output APP = ecost(T) + vcost(T) and the generalized spanning tree T.

where by ecost(T) and vcost(T) we denoted the cost of the tree T with respect to the edges, respectively to the nodes.

Regarding the performance bounds of this approximation algorithm, using the same auxiliary results, and defining $(\widehat{x}_e, \widehat{z}_i)$ as we did at the beginning of this subsection, the following inequalities hold:

$$\begin{aligned} L^T(W') &\leq \rho\left(2 - \frac{2}{n}\right) L_{\text{OPT}}, \\ V^T(W') &\leq \rho V_{\text{OPT}}, \end{aligned}$$

where $L^T(W')$, $V^T(W')$ denote the cost of the tree T spanning the nodes of W'

with respect to the edges and to the nodes, respectively, and, as before, $W' \subset W^* = \{i \in V \mid z_i^* \geq \frac{1}{\rho}\}$, such that $|W'| = \mathrm{m}$ and W' consists of exactly one vertex from every cluster.

For the approximation algorithm proposed in this case, the following holds:

$$\begin{aligned} \text{APP} = L^T(W') + V^T(W') &\leq \rho\left(2 - \frac{2}{n}\right) L_{\text{OPT}} + \rho V_{\text{OPT}} \\ \leq \rho\left(2 - \frac{2}{n}\right)(L_{\text{OPT}} + V_{\text{OPT}}) &= \rho\left(2 - \frac{2}{n}\right) \text{OPT}. \end{aligned}$$

In the case of the at least generalized minimum spanning tree problem, Feremans et al. [41] considered a geometric case of the problem, where the graph is complete, all vertices are situated in the plane and the Euclidean distance between the vertices defines the edge cost. This special case of the problem, with the clustering structured as specified, is called *grid clustering*.

For the case of the grid clustering, the same authors proved that the problem is strongly NP-hard, they provided an exact exponential time dynamic programming algorithm and, based on this algorithm, they developed a polynomial time approximation scheme, namely if T^{APPX} is the generalized spanning tree, constructed in polynomial time, using the dynamic programming algorithm, then the following result holds:

Theorem 2.17 (Feremans et al. [41]).

$$\frac{c(T^{\text{APPX}}) - c(T^{\text{OPT}})}{c(T^{\text{OPT}})} \leq \epsilon.$$

2.5 Solving the GMSTP

Myung et al. [128] used a branch and bound procedure in order to solve the GMSTP. Their lower procedure is a heuristic method, which approximates the linear programming relaxation associated with the dual of the multicommodity flow formulation of the GMSTP. They also developed a heuristic algorithm, which finds a primal feasible solution for the GMSTP, using the obtained dual solution and they reported the exact solution of instances with up to 100 vertices. The GMSTP was solved to optimality for nodes up to 200 by Feremans [37], using a branch-and-cut algorithm. More recently, Pop et al. [150] have proposed a new integer programming formulation of the GMSTP, based on a distinction between local and global variables and a solution procedure that finds an optimal solution on instances with up to 240 vertices.

The difficulty of obtaining optimum solutions for the GMSTP has led to the development of several metaheuristics. The first such algorithms were the tabu search (TS) heuristic of Feremans [37] and the simulated annealing (SA) heuristic of Pop [142]. An improved version of the SA was described in [154]. Two variants of the TS heuristic and of a four variable neighborhood search (VNS) based heuristic were later devised by Ghosh [49]. Another VNS algorithm for the GMSTP was proposed by Hu et al. [76]. The authors report that their VNS approach can produce solutions that are comparable to those obtained by means of the second variant of the TS heuristic of Ghosh [49]. The same authors proposed an improved VNS algorithm, by combining it with integer linear programming [77]. Golden et al. [53] have devised a local search heuristic (LSH) and a genetic algorithm (GA) for the GMSTP. Both algorithms have yielded improvements on TSPLIB instances with sizes between $198 \leq n \leq 225$. The LSH outperform the GA on none of these instances. Another TS heuristic for the GMSTP was devised by Wang et al. [206]. The authors note, that their TS heuristic produces solutions slightly better than those obtained by the GA of Golden et al. [53]. An attribute based tabu search heuristic, employing new neighborhoods, was proposed by Oncan et al. [135]. The authors mention that their TS based heuristic yields the best results for all instances. Recently, a memetic algorithm was described by Pop et al. [164], while Hu and Raidl [81] proposed an evolutionary algorithm with solution archive for solving the GMSTP in which the already considered solutions are stored in an appropriate data structure allowing a fast detection of duplicate solutions.

In this section we will describe four algorithmic methods for solving the generalized minimum spanning tree problem:

- a branch-and-cut algorithm, based on the undirected cluster subpacking formulation;
- a heuristic method, based on the local-global approach;
- a rooting procedure, based on the local-global integer programming formulation of the GMSTP;
- an improved version of the simulated annealing for solving the GMSTP.

2.5.1 A branch-and-cut algorithm for solving the GMSTP

Branch-and-cut is a combinatorial optimization method for solving integer linear programs. The method is a hybrid of branch and bound and cutting plane methods, where, after solving the LP relaxation, and having been unsuccessful in pruning the node on the basis of the LP solution, we try to find a violated cut. If one or more violated cuts are found, they are added to the formulation and the LP is solved again. If none are found, we branch. For more information on this method we refer to [123].

In the following, we present the branch-and-cut algorithm for solving the GMSTP, as developed by Feremans et al. [40]. This algorithm is based on the undirected cluster subpacking formulation of the GMSTP, that was described in subsection 2.3.1.

Algorithm 2.18 (The Branch-and-Cut Algorithm).

Step 1: (*Initialization*) The LP relaxation of the undirected cluster subpacking formulation of the GMSTP, containing only the constraints (2.1) and (2.3), is inserted into a list L. The best-known value, denoted by opt_{best}, is obtained by means of a tabu search heuristic, that explores the solution space.

Step 2: (*termination check and subproblem selection*) If L is empty, then stop, otherwise, extract one subproblem from the list according to a best-first rule.

Step 3: (*subproblem solution*) Solve the subproblem, using an LP-solver, and denote its optimal value by *opt*. If $opt \geq opt_{\text{best}}$, go to Step 2. Otherwise, if this solution is feasible, call a local improvement procedure. If this solution is fractional, then cal a rounding procedure, in order to improve it.

Step 4: (*separation procedure*) Search for violations of the generalized subtour eliminations constraints:

$$x(E(S)) \leq z(S) - 1, \ \forall \ S \subset V, \ 2 \leq |S| \leq n-1, \ |\{k : V_k \subseteq S\}| \neq 0$$

with the two special cases:

$$\begin{aligned} x(E(\{i\} : V_k)) &\leq z_i, & i &\in V \setminus (W \cup V_k), \\ x(\delta(i)) &\geq z_i, & i &\in V. \end{aligned}$$

where W denotes the set of vertices for which the cluster is a singleton and we defined for $S, T \subseteq V$

$$E(S : T) = \{e = \{i, j\} \in E \mid i \in S, \ j \in T\}. \tag{2.50}$$

If violations are identified, introduce the corresponding constraints into the subproblem and go to Step 3.

Step 5: (*separation procedure: other valid inequalities*) Search for violations of odd-cycle inequalities:

$$\sum_{k=1}^{m} (E(V_k^1 : V^2_{(k \, mod \, m)+1})) \leq \lfloor m/2 \rfloor \tag{2.51}$$

where we assumed that each cluster V_k, $k = 1, \dots, m$ is partitioned into two nonempty subsets V_k^1 and V_k^2 and, for violations of odd-hole inequalities,

$$x(F) \leq \lfloor |F|/2 \rfloor \tag{2.52}$$

where $F \subseteq E$, with the property that $|F|$ is odd, $|F| \geq 3$ and it induces a cordless cycle in the corresponding *conflict graph* $G_C = (V_C, E_C)$, where each edge of E

is a vertex of G_C and an edge of E_C is defined between e_1 and e_2 of E if and only if e_1 and e_2 are in *conflict*. Two edges incident to different vertices of the same cluster are said to be in conflict.

If violations are identified, introduce the corresponding constraints into the subproblem and go to Step 3.

Step 6: (*branching*) Create two new subproblems by branching on a constraint $z(V_k) = 1$. Add these two subproblems to the list L and go to Step 2.

Branching on a constraint was performed as in Williams [207]. Assuming that there exists at least one vertex $i \in V$ such that $z_i^* > 0$, where $G^* = (V^*, E^*)$ denotes the support graph associated with a fractional solution (x^*, z^*), with $V^* = \{i \in V \mid z_i^* > 0\}$ and $E^* = \{\{i, j\} \in E \mid x_{ij}^* > 0\}$, we can create two descendants of the current subproblem as follows: let V_k be the cluster with at least one vertex i such that $0 < z_i^* < \frac{1}{2}$. This cluster is divided into two sets V_k^1 and V_k^2 such that $z^*(V_k^1)$ and $z^*(V_k^2)$ are close to $\frac{1}{2}$. The two child subproblems are then generated from the current subproblem by adding $z^*(V_k^1) = 0$ and $z^*(V_k^2) = 1$ for the other.

The branch-and-cut algorithm just described was coded by Feremans et al. [40] in C++ and used the ABACUS 2.3 library (see Thinel [198]), with CPLEX 7.0 as LP-solver (see ILOG [85]). The tests were performed on a Generic sun4u sparc SUNW, Ultra-5.10 workstation (360 MHz, 128 MB RAM), running the SunOS 5.8 operating system. The algorithm was tested on randomly generated instances and on other instances. The obtained results are presented in Table 2.1.

2.5.2 A heuristic algorithm for solving the GMSTP

Simple constructive heuristic algorithms based on Prim's, Kruskal's and Sollin's algorithms for the minimum spanning tree problem have been proposed for the GMSTP by Feremans [37]. In this section, we will describe a better heuristic algorithm, based on the local-global approach.

We start with some notations: let F^{global} denote the current edge set in the complete global graph. S^{global} denotes the current node set in the complete global graph, F denotes the current edge set in the graph G and S denotes the current node set in the graph G.

Given a global subtree connecting some of the clusters, we denote by T the minimum generalized subtree corresponding to the given connection of the clusters. This minimum generalized subtree can be determined easily, using the dynamic programming method described in Section 2.2.

It is easy to observe that $|F| = |F^{\text{global}}|$ and $|S| = |S^{\text{global}}|$ hold.

Table 2.1. Computational results for non-Euclidean problems (average of five trials per type of instance).

Pb. size		Rooting procedure		Branch and cut		Myung's results		Heuristic results	
m	n	LB/OPT	CPU	LB/UB	CPU	LB/OPT	CPU	OPT/UB	CPU
8	24	100	0.0	100	0.0	100	0.0	–	–
	32	100	0.0	100	0.2	100	0.2	–	–
	48	100	0.2	100	1.4	94.3	3.2	–	–
	80	100	0.6	100	4.2	94.9	17.6	–	–
10	30	100	0.1	100	1.0	89.1	0.0	85.10	0
	40	100	0.7	100	1.0	–	–	92.74	0
	60	100	0.9	100	3.2	87.8	3.2	89.98	0.004
	100	100	3.5	100	8.8	91.3	17.6	96.34	0.008
12	36	100	0.1	100	1.8	89.6	6.0	90.06	0.002
	48	100	1.6	99.2	3.2	91.3	54.9	93.30	0.006
	72	100	5.6	100	6.8	100	6.8	94.17	0.006
	120	100	14.5	–	–	–	–	92.77	0.022
15	45	100	0.2	100	3.6	89.0	17.4	95.74	0.002
	90	100	5.9	100	21.4	–	–	89.64	0.010
	150	100	40.5	98.8	42.4	–	–	92.70	0.044
18	54	100	0.5	99.5	7.6	–	–	89.23	0.002
	108	100	9.4	–	–	–	–	93.38	0.016
	180	100	193.8	–	–	–	–	91.83	0.076
20	60	100	3.8	–	–	–	–	86.81	0.004
	120	100	11.4	96.3	39.8	–	–	89.05	0.024
	200	100	407.6	94.6	191.4	–	–	89.53	0.104
25	75	100	21.6	–	–	–	–	87.99	0.012
	150	100	25.1	88.3	178.8	–	–	88.99	0.084
	200	100	306.6	97.8	140.6	–	–	93.75	0.146
30	90	100	40.0	–	–	–	–	86.32	0.025
	180	100	84.0	96.6	114.6	–	–	90.63	0.032
	240	100	341.1	–	–	–	–	89.64	0.068
40	120	100	71.6	100	92.6	–	–	88.13	0.062
	160	100	1 713.2	94.2	288.6	–	–	89.68	0.080

The heuristic algorithm works as follows:

Algorithm 2.19 (Heuristic algorithm for solving the GMSTP).

$F^{\text{global}} := \emptyset$; $S^{\text{global}} := \{V_r\}$, where V_r is an arbitrary cluster;
$F := \emptyset$; $S := \{v_{i_0}\}$, where $v_{i_0} \in V_r$;
while $|S| < m$ do
From among all global edges with exactly one end in S^{global}, find a global edge $(V_k, V_l) \in E^{\text{global}}$ that minimizes the cost of the corresponding generalized tree;
$F^{\text{global}} = F^{\text{global}} \cup \{(V_k, V_l)\}$;
$S^{\text{global}} := S^{\text{global}} \cup \big(\{V_k, V_l\} \setminus S^{\text{global}}\big)$;
$F := \big\{(v_i, v_j) | (v_i, v_j) \in T\big\}$;
$S := \big\{v_i | v_i \in V_k \cap T,\ V_k \in S^{\text{global}}\big\}$;
end{while}

---

where the cost of the of the current generalized tree is given by the sum of costs of the selected edges plus the costs of the nodes selected.

The following result holds:

Theorem 2.20. *The previous algorithm yields a generalized spanning tree of G that approximates the optimal solution of the GMSTP and its running time is $O(m^2n^2)$, where m is the number of clusters and n is the number of nodes of G.*

The proof of this result was given by Pop and Zelina [145]. In the same paper some computational experiments were presented, that showed the efficiency of the heuristic algorithm, based on the local-global approach, in comparison to the constructive heuristic algorithms based on Prim's, Kruskal's and Sollin's algorithms.

2.5.3 Rooting procedure for solving the GMSTP

Based on the local-global formulation, we proposed a solution procedure called the *rooting procedure*, which is described in the following.

Instead of considering the 0-1 local-global mixed integer programming problem, we consider the constraints that characterize the polytope P_{MST} only for fixed k, $1 \leq k \leq m$. We then get a relaxation that we will denote by P^k.

Using the description of Yannakakis for the global spanning tree polytope, this situation corresponds to the case where we randomly choose one cluster V_k and root the global tree only at the root k.

$$
\begin{aligned}
\min \sum_{e\in E} & c_e x_e \\
\text{s.t.} \quad z(V_k) &= 1, && \forall\, k \in K = \{1,\ldots,m\}, \\
x(E) &= m-1, \\
x(V_l, V_r) &= y_{lr}, && \forall\, l,r \in K = \{1,\ldots,m\},\ l \neq r, \\
x(i, V_r) &\le z_i, && \forall\, r \in K, \forall\, i \in V \setminus V_r, \\
(P^k) y_{ij} &= \lambda_{kij} + \lambda_{kji}, && \forall\, 1 \le k,i,j \le m \text{ and } i \neq j,\ \underline{k \text{ fixed}}, \\
\sum_j \lambda_{kij} &= 1, && \forall\, 1 \le k,i,j \le m \text{ and } i \neq k,\ \underline{k \text{ fixed}}, \\
\lambda_{kkj} &= 0, && \forall\, 1 \le k,j \le m,\ \underline{k \text{ fixed}}, \\
\lambda_{kij} &\ge 0, && \forall\, 1 \le k,i,j \le m,\ \underline{k \text{ fixed}}, \\
x_e, z_i &\ge 0, && \forall\, e = (i,j) \in E,\ \forall\, i \in V, \\
y_{lr} &\in \{0,1\}, && \forall\, 1 \le l,r \le m.
\end{aligned}
$$

If the optimal solution of this relaxation (solved with CPLEX) produces a generalized spanning tree, we have given the optimal solution of the GMSTP. Otherwise, we obtain a subgraph containing at least one cycle and we add the corresponding constraints (from the characterization of P_{MST}), in order to break that cycle (i.e. root the global tree also in a second cluster, contained in the cycle) and proceed this way until we arrive at the optimal solution of the GMSTP. We call this procedure the *rooting procedure*.

Our algorithms have been coded in C and compiled with a HP-UX cc compiler. For solving the linear and mixed integer programming problems we used CPLEX 6.5. The computational experiments were performed on a HP 9000/735 computer with a 125 MHz processor and 144 MB memory.

According to the method of generating the edge costs, the generated problems are classified as one of two types: the Euclidean case and the non-Euclidian case. The clusters in both cases are random and we assume that every cluster has the same number of nodes.

For the instances in the structured Euclidean case, m squares (clusters) are "packed in a square" and in each of these m clusters n_c nodes are selected at random. The costs between nodes are given by the Euclidean distances between the nodes. So, in this model, the clusters may be interpreted as physical clusters. In the other models, such an interpretation is not valid.

For the unstructured Euclidean case, $n = mn_c$ nodes are randomly generated in $[0, 100]^2$ with costs given by the Euclidean distances. But then, the clusters are chosen randomly from among these points. Finally, in the non-Euclidean model the edge costs are generated at random on the interval $[0, 100]$.

For each type of instance we considered five trials.

Here, we compare the computational results for the non-Euclidian model, obtained when solving the problem, with use of our rooting procedure and the heuristic algorithm, with the computational results given by Myung et al. in [128] and by Feremans et al. [40], using their branch-and-cut algorithm. The computational results are presented in the next table.

The first two columns in the table specify the size of the problem: the number of clusters (m) and the number of nodes (n). The next columns describe the rooting procedure and contain the lower bounds, obtained as a percentage of the optimal value of the GMSTP (LB/OPT), and the computation times (CPU), in seconds, for solving the GMSTP. In addition, the second table shows the minimum number of roots, chosen by the rooting procedure, in order to get the optimal solution of the GMSTP. The last columns of the first table contain the lower bounds as a percentage of the upper bounds of the GMST problem (LB/UB), the computation times (CPU), in seconds, for solving the GMSTP with the branch and cut algorithm, the lower bounds as a percentage of the optimal value of the GMSTP (LB/OPT), and the computation times (CPU) obtained by Myung, the upper bounds as a percentage of the optimum value, and the computational results obtained using the proposed heuristic algorithm. In the table the sign '–' means that the corresponding information was not provided.

As can be seen, in all instances we considered, for graphs with a number of nodes up to 240, the optimal solution of the GMSTP was found by using our rooting procedure. It is worth mentioning that for the instances considered in the table, the maximum number of clusters, chosen as roots, in order to get the optimal solution of the problem, was 5. We can also see from the table, that the proposed heuristic algorithm provides us with good suboptimal solutions to the GMSTP with very good computation times for all the instances considered.

2.5.4 Solving the GMSTP with Simulated Annealing

Simulated annealing (SA) is a generic probabilistic metaheuristic for the global optimization problem in applied mathematics, which is to locate a good approximation to the global minimum of a given function in a large search space. It is inspired by Monte Carlo methods in statistical mechanics. The name and inspiration come from annealing in metallurgy, which is a technique that involves heating and controlled cooling of a material, to increase the size of its crystals and reduce their defects. The heat causes the atoms to become unstuck from their initial positions (a local minimum of the internal energy) whereupon they start to wander randomly through states of higher energy. The slow cooling gives them more chances of finding configurations with lower internal energy than the initial one. It is often used when the search space is discrete (e.g., all tours that visit a given set of cities).For certain problems, simulated annealing may be more effective than exhaustive enumeration, provided that the goal is merely to find an acceptable solution within a fixed amount of time, rather than the best possible solution.

The ideas that form the basis of simulated annealing were first published by Metropolis et al. [121] as an algorithm to simulate the cooling of a material in a heat bath, a process known as annealing.

Kirkpatrick et al. [97] and Cerny [16] suggested that this type of simulation could be used to search for feasible solutions of an optimization problem. Their approach can be regarded as a variant of the local search technique.

A central problem for local search is the occurrence of *local optima*, i.e. nodes in the search space where no neighbor is a strict improvement over the current node, in terms of the cost function, but which are not global optima. In the simulated annealing heuristic non-improving local moves are allowed in order to overcome the local optima, but their frequency is governed by a probability function which changes as the algorithm progresses. It has been shown (see [52]) that with a "large enough" initial temperature and a proper cooling schedule, SA guarantees a globally optimal solution. However, this theoretical result is particularly helpful, since the annealing time needed to ensure a significant probability of success will usually exceed the time required for a complete search of the solution space.

The annealing algorithm for a minimization problem with solution space S, objective function c and neighborhood structure N can be stated as follows:

Algorithm 2.21.

Select a starting solution $s_0 \in S$, an initial temperature $T_0 > 0$, a temperature reduction function α and a number of iterations per temperature L;

Repeat
    Repeat
        Randomly select $s \in N(s_0)$ and compute $\delta = c(s) - c(s_0)$;
        If $\delta < 0$
        then $s_0 = s$
        else generate random p, uniformly distributed over the range $(0, 1)$;
        if $p < \exp(-\delta/T)$ then $s_0 = s$
    Until *iterationcount* $= L$;
Set $T = \alpha(T)$;
Until the stopping condition is true.

s_0 is the approximation of the optimal solution.

Here, L represents the number of repetitions at each temperature, i.e. the number of evaluated, randomly chosen candidate solutions, in the neighborhood of the current solution. Therefore, at each stage, we evaluate L randomly chosen candidate solutions in the neighborhood of the current solution. If a candidate solution improves on the current solution, it is accepted. Otherwise, it is accepted with a probability $p(T, \delta) = \exp(-\delta/T)$, which depends on the control parameter T (the temperature in the physical equivalent) and the amount δ by which the move worsens the current solution. This relation ensures that the probability of accepting a move to a very poor solution is very small. After completion of each stage, the temperature is reduced. The way the temperature is reduced is known as the *cooling schedule*. Given a relatively high temperature T at the beginning of the process, the probability of accepting non-improving moves is fairly high. As the process continues, the temperature decreases and non-improving moves become less likely.

The search is continued until there is some evidence that suggests that the probability of improving on the best solution found so far is very low. At this stage, the system is said to be frozen.

Given a neighborhood structure, simulated annealing can be viewed as an algorithm that continuously attempts to transform the current configuration into one of its neighbors.

Comparing iterative improvement and simulated annealing, it is apparent that the situation where the control parameter in the simulated annealing algorithm is set to 0 corresponds to a version of iterative improvement. Conversely, simulated annealing is a generalization of iterative improvement in that it accepts, deterioration of the cost function, with non-zero but gradually decreasing probability. It is not clear, however, whether it performs better than repeated application of iterative improvements (for a number of different initial configurations): both algorithms converge asymptotically to a globally minimal configuration of the problem at hand.

In order to implement the simulated annealing algorithm for the GMSTP, a number of decisions must be made. These can be divided into two categories: *generic decisions*, which are concerned with parameters of the annealing algorithm itself, including factors such as the initial temperature, the cooling schedule (governed by the parameter L, i.e. number of repetitions and the choice of the temperature reduction function α), and the stopping condition, and *problem specific decisions*, involving the choice of the solution space, the form of the cost function, an initial solution (initialization) and the employed neighborhood structure.

In our implementation of the simulated annealing algorithm for the GMSTP, we chose the following cooling schedule:

1. **Initial value of the control parameter, T_0.**
   Following Johnson et al. [90], we determine T_0 by calculating the average increase in cost δ^+, for a number of random transitions:
   $$T_0 = \frac{\delta^+}{\ln(\chi_0^{-1})},$$
   where $\chi_0 = 0.8$ is a given acceptance rate.

2. **Final value of the control parameter.**
The stopping criteria that we used, determining the final value of the control parameter, was specified by fixing the number of values T_k, for which the algorithm is to be executed, ending on a suitably low temperature.

3. **Number of repetitions at each temperature.**
The number of iterations at each temperature, which is related to the size of the neighborhoods, may vary from temperature to temperature. It is important to stay at lower temperatures for a long time, to ensure possible local optima have been fully explored. Therefore, we increased the value of L arithmetically (by adding a constant factor).

4. **Decrementing the control parameter.**
We used the following decrement rule:
$$\alpha(T) = rT$$
where r is a constant smaller than but close to 1, called the cooling rate.
As in [90], we used $r = 0.95$. This corresponds to a fairly slow cooling process.

In the following, we present the problem specific decisions.

Solution space
We define the solution space as the set of all the feasible solutions. Suppose that the graph $G = (V, E)$ consists of n nodes which are partitioned into m clusters $V_1, \ldots, V_m$, all having the same number of nodes, i.e.
$$|V_1| = \ldots = |V_m| = \frac{n}{m}.$$
By Cayley's formula, we know that the total number of global spanning trees (i.e. trees connecting the clusters) is equal to m^{m-2}. Given a global spanning tree there are $(\frac{n}{m})^m$ possible generalized spanning trees (feasible solutions of the GMSTP).

Therefore, the solution space of the GMSTP contains
$$m^{m-2} \cdot \left(\frac{n}{m}\right)^m = n^m \cdot m^{-2}$$
elements. The number of elements in the solution space is exponentially large, so that they cannot be explored exhaustively.

Objective function
We will consider the objective function of the integer programming formulations of the GMSTP, described in Section 2.3.

Initialization
Simulated Annealing is an improvement heuristic, which requires an initial solution. It is desirable to start with a relatively good solution (as compared with, for example,

a randomly generated one), otherwise a great deal of effort will be spent on reaching the first local minimum.

In our case, an initial feasible solution for the GMSTP was found with the following greedy algorithm:

Algorithm 2.22.

Input: A connected graph $G = (V, E)$, with nodes partitioned into m clusters and edges defined between nodes from different clusters with positive cost.

Output: A generalized spanning tree $T = (W, F)$ of G.

$F := \emptyset$ (F is the edge set of the current tree T)
$W := \{v_i\}$, $v_i \in V_k$, $1 \leq k \leq m$ (W is the node set of T)

while $|S| < m$ **do**

From among all edges with exactly one end in W, find an edge (v_i, v_j) with minimum cost and, whenever we choose a node from a cluster, delete all the other nodes from that cluster and all edges adjacent to the deleted nodes;
$F := F \cup (v_i, v_j)$
$W := W \cup (\{v_i, v_j\} \setminus W)$

end; (while).

An algorithm similar to the greedy algorithm may be applied in order to find a feasible solution for the GMSTP (i.e. a generalized spanning tree of G), if we start with a generalized tree $T = (W, F)$ of G ($|W| < m$), instead of randomly selecting a node from a random cluster. We will call this algorithm the *local greedy algorithm.*

Neighborhood structure
The neighborhood is one of the most important factors in designing a simulated annealing algorithm. It should be designed such that the structures common to good solutions are preserved after the neighborhood operation is applied.

In the case of the GMSTP, we defined a neighborhood structure as follows: The *neighborhood* of any given generalized spanning tree T is defined as the set of all the generalized spanning trees obtained with the local greedy algorithm, for any subtree of T. The size of the neighborhood of any given generalized spanning tree T is given by the number of subtrees. The maximum number of subtrees of a tree with m nodes is reached for trees with nodes distributed on exactly one level, except the root, which has degree $m - 1$. Therefore, in this case, the maximum size of the neighborhood is $2^{m-1} + m - 1$, see [195]. Given a current solution of the GMSTP, i.e. a generalized spanning tree T, for any subtree U of T, using the local greedy algorithm, we obtain a candidate solution for the simulated annealing in the neighborhood of the current solution. The next figure shows a current solution and a modified solution.

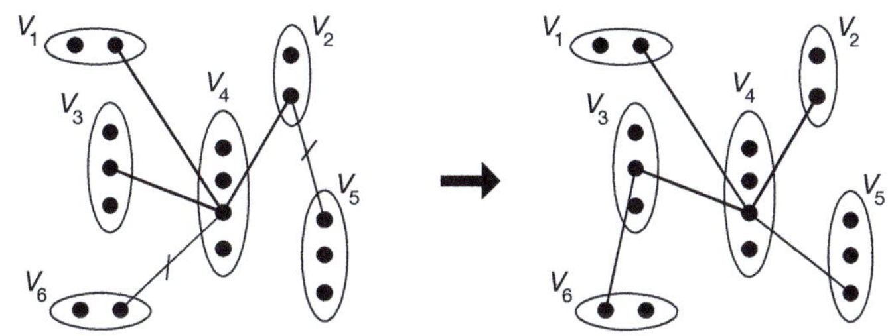

Figure 2.5. A current solution and a modified solution of the GMSTP.

In the previous figure, we considered a graph G whose nodes are partitioned into 6 clusters and a current solution T. Randomly selecting a subtree U of T (for example by discarding the edges connecting nodes from clusters V_4 with V_5 and V_2 with V_5) we apply the local greedy algorithm, in order to obtain the best modified solution in the neighborhood of T, given the subtree U.

Implementation details and computational results

Our computational experiments were performed on a Dual Intel Pentium III Processor – 1000 MHz computer with 512 MB RAM. The procedures for the combined Simulated Annealing and local greedy scheme were written in Java.

After some preliminary tests, we have chosen the following cooling schedule: we start with a temperature $T_0 = 100$, the cooling rate $r = 0.95$ will be used to reduce the temperature in each step, and in each stage $L = 30$ moves are evaluated.

We considered an undirected graph $G = (V, E)$ having n nodes, partitioned into m clusters, such that each cluster contains the same number of nodes. Edges are defined between all the nodes from different clusters and the costs of the edges were randomly generated in the interval [1,100]. For each type of instance (n, m) we randomly generated five different problems, i.e. with different costs of the edges. The instances that we used are similar to those considered by Myung [128] and Feremans [37].

In the next table, we compare the computational results obtained for solving the GMSTP, using the combined simulated annealing, with a local greedy algorithm, with the computational results given by Feremans [37] and Pop [142, 150].

The first two columns in the table list the size of the problem as the number of clusters (m) and the number of nodes (n). The next two columns describe the simulated annealing procedure and contain the upper bounds, obtained as a percentage of the optimal value of the GMSTP (UB/OPT), and the computation times (CPU) in seconds for solving the GMSTP. The last columns contain the lower bounds as a percentage of the optimal value of the GMST problem (UB/OPT), the computation times (CPU) in seconds for solving the GMSTP with the rooting procedure (see [142, 150]), the lower bounds as a percentage of the upper bound of the GMSTP (LB/UB), and the computation times (CPU) in seconds for solving the GMSTP with a branch and cut algorithm (see [37]). In the last two columns '–' means that for those instances, [37] did not provide computational results.

It is important to notice that we were able to obtain the optimal solution of the GMSTP, using our simulated annealing combined with a local greedy algorithm, for all instances considered,.

From Table 2.2, we observe that, for instances up to 140 nodes, in the case of simulated annealing, computation times were longer, compared with the computation times corresponding to the other methods. However, for larger instances, while the computation times for the rooting procedure increase exponentially, the computation times in the case of simulated annealing are reasonable.

For medium and large instances of the GMSTP (graphs which contain more than 50 clusters and 200 nodes), we compare the results of our SA to Variable Neighborhood Search (VNS) (Hu, Leiner and Raidl [77]), Tabu Search with recency and frequency based memory (TS) (Ghosh [49]), and, in the case of TSPlib instances, also to the Genetic Algorithm (GA) from Golden, Raghavan and Stanojevic [53].

The algorithms have been tested on two kinds of instances. One was introduced by Ghosh [49], who created three types of benchmark instances called grouped Euclidean, random Euclidean, and non-Euclidean random instances, and the other one applies geographical clustering (see Fischetti at al. [43]) on TSPlib instances. In the case of the grouped Euclidean instances, squares with side length *span* were associated with clusters and were regularly distributed on a grid of size $col \times row$. The nodes of each cluster are randomly distributed, within the corresponding square. For the second type of benchmark instances, the nodes belonging to the same cluster are not necessarily close to each other, since they are randomly generated within a square of size 1000×1000 and randomizing the cluster assignment. In these two cases, the cost of the edges is given by the Euclidean distance between the nodes. Finally, in the case of the non-Euclidean random benchmark instances, the cost of the edges was randomly chosen in the integer interval $[0, 1000]$. All considered graphs have a complete set of edges. The details about these benchmark instances are listed in the Tables 2.3 and 2.4.

In the next two tables, Table 2.5 and 2.6, we present the names of the instance, number of nodes, (average) number of nodes per cluster, and (average) objective values of the final solutions, as obtained by the different algorithms. The best values are printed bold. In case of SA and VNS, we also provide corresponding standard deviations, denoted by σ, and objective values. best values are printed bold. In case of SA and VNS, we also provide corresponding standard deviations of objective values.

In Table 2.5, we compare our SA to TS, VNS and VNDS on grouped Euclidean instances, random Euclidean instances, and non-Euclidean instances.

Analyzing the results in Tables 2.5 and 2.6, we observe that, among the considered approaches, VNS and TS are overall the most powerful algorithms.

Table 2.2. Computational results for solving the GMSTP.

Pb. size		*SA algorithm*		*Rooting procedure*		*Branch and cut*	
m	n	UB/OPT	CPU	LB/OPT	CPU	LB/UB	CPU
8	24	100	2.6	100	0.0	100	0.0
	32	100	3.7	100	0.0	100	0.2
	48	100	5.6	100	0.3	100	1.4
	80	100	10.2	100	2.2	100	4.2
10	30	100	4.8	100	0.2	100	1.0
	40	100	6.7	100	0.3	100	1.0
	60	100	10.8	100	0.7	100	3.2
	100	100	20.6	100	1.9	100	8.8
12	36	100	8.3	100	0.7	100	1.8
	48	100	11.6	100	0.9	99.2	3.2
	72	100	19.1	100	3.7	100	6.8
	120	100	39.0	100	9.3	–	–
15	45	100	16.6	100	1.3	100	3.6
	60	100	21.6	100	1.7	97.1	6.2
	90	100	39.7	100	9.6	100	21.4
	150	100	80.2	100	34.1	98.8	42.4
18	54	100	27.2	100	2.6	99.5	7.6
	108	100	68.1	100	13.8	–	–
	180	100	143.8	100	150.4	–	–
20	60	100	37.5	100	7.2	–	–
	120	100	181.9	100	33.2	96.3	39.8
	200	100	219.8	100	203.8	94.6	191.4
25	75	100	80.8	100	17.5	–	–
	150	100	221.7	100	266.4	88.3	178.8
	200	100	331.1	100	284.5	97.8	140.6
30	90	100	150.4	100	60.2	–	–
	180	100	412.0	100	721.8	96.6	114.6
	240	100	627.4	100	1 948.9	–	–
40	120	100	382.1	100	142.6	100	92.6
	160	100	561.0	100	572.3	94.2	288.6
	280	100	691.2	100	25 672.3	–	–

Table 2.3. Benchmark instances adopted from Ghosh [49]. Each instance has a constant number of nodes per cluster.

Instance set	$\|V\|$	$\|E\|$	r	$\frac{\|V\|}{r}$	col	row	sep	$span$
Grouped Eucl 125	125	7750	25	5	5	5	10	10
Grouped Eucl 500	500	124 750	100	5	10	10	10	10
Grouped Eucl 600	600	179 700	20	30	5	4	10	10
Grouped Eucl 1 280	1 280	818 560	64	20	8	8	10	10
Random Eucl 250	250	31 125	50	5	–	–	–	–
Random Eucl 400	400	79 800	20	20	–	–	–	–
Random Eucl 600	600	179 700	20	30	–	–	–	–
Non-Eucl 200	200	19 900	20	10	–	–	–	–
Non-Eucl 500	500	124 750	100	5	–	–	–	–
Non-Eucl 600	600	179 700	20	30	–	–	–	–

Table 2.4. TSPlib instances with geographical clustering, from Feremans [37]. Numbers of nodes varies for each cluster.

Name of instance	$\|V\|$	$\|E\|$	r	$\frac{\|V\|}{r}$	$d_{\min}$	$d_{\max}$
gr137	137	9316	28	5	1	12
kroa150	150	11 175	30	5	1	10
d198	198	19 503	40	5	1	15
krob200	200	19 900	40	5	1	8
gr202	202	20 301	41	5	1	16
ts225	225	25 200	45	5	1	9
pr226	226	25 425	46	5	1	16
gil262	262	34 191	53	5	1	13
pr264	264	34 716	54	5	1	12
pr299	299	44 551	60	5	1	11
lin318	318	50 403	64	5	1	14
rd400	400	79 800	80	5	1	11
fl417	417	86 736	84	5	1	22
gr431	431	92 665	87	5	1	62
pr439	439	96 141	88	5	1	17
pcb442	442	97 461	89	5	1	10

Table 2.5. Results on instance sets from Ghosh [49].

Instances				TS	VNDS	SA		VNS	
Set	$\|V\|$	r	$\|V\|/r$	$C(T)$	$\overline{C(T)}$	$\overline{C(T)}$	σ	$\overline{C(T)}$	σ
Grouped Eucl 125	125	25	5	*141.1*	*141.1*	152.3	0.52	*141.1*	0.00
	125	25	5	*133.8*	*133.8*	150.9	0.74	*133.8*	0.00
	125	25	5	143.9	145.4	156.8	0.00	*141.4*	0.00
Grouped Eucl 500	500	100	5	*566.7*	577.6	642.3	0.00	567.4	0.57
	500	100	5	*578.7*	584.3	663.3	1.39	585.0	1.32
	500	100	5	*581.6*	588.3	666.7	1.81	583.7	1.82
Grouped Eucl 600	600	20	30	85.2	87.5	93.9	0.00	*84.6*	0.11
	600	20	30	*87.9*	90.3	99.5	0.28	*87.9*	0.00
	600	20	30	88.6	89.4	99.2	0.17	*88.5*	0.00
Grouped Eucl 1280	1280	64	20	327.2	329.2	365.1	0.46	*315.9*	1.91
	1280	64	20	322.2	322.5	364.4	0.00	*318.3*	1.78
	1280	64	20	332.1	335.5	372.0	0.00	*329.4*	1.29
Random Eucl 250	250	50	5	*2285.1*	2504.9	2584.3	23.82	2300.9	40.27
	250	50	5	*2183.4*	2343.3	2486.7	0.00	2201.8	23.30
	250	50	5	*2048.4*	2263.7	2305.0	16.64	2057.6	31.58
Random Eucl 400	400	20	20	*557.4*	725.9	665.1	3.94	615.3	10.8
	400	20	20	724.3	839.0	662.1	7.85	*595.3*	0.00
	400	20	20	604.5	762.4	643.7	14.54	*587.3*	0.00
Random Eucl 600	600	20	30	541.6	656.1	491.8	7.83	*443.5*	0.00
	600	20	30	540.3	634.0	542.8	25.75	*573.0*	10.2
	600	20	30	627.4	636.5	469.5	2.75	*469.0*	11.9
Non-Eucl 200	200	20	10	*71.6*	94.7	76.9	0.21	*71.6*	0.00
	200	20	10	*41.0*	76.6	41.1	0.02	*41.0*	0.00
	200	20	10	*52.8*	75.3	86.9	5.38	*52.8*	0.00
Non-Eucl 500	500	100	5	*143.7*	203.2	200.3	4.44	152.5	3.69
	500	100	5	*132.7*	187.3	194.3	1.20	148.6	4.27
	200	20	10	*162.3*	197.4	205.6	0.00	166.1	2.89
Non-Eucl 600	600	20	30	*14.5*	59.4	22.7	1.49	15.6	1.62
	20	30	5	17.7	23.7	22.0	0.82	*16.1*	1.24
	600	20	30	*15.1*	29.5	22.1	0.44	16.0	1.66

Table 2.6. Results on TSPlib instances with geographical clustering.

TSPlib Instances				TS	VNDS	SA		GA	VNS	
Name	$\lvert V \rvert$	r	time	$C(T)$	$\overline{C(T)}$	$\overline{C(T)}$	std. dev.	$\overline{C(T)}$	$\overline{C(T)}$	std. dev.
gr137	137	28	150 s	*329.0*	330.0	352.0	0.00	*329.0*	*329.0*	0.00
kroa150	150	30	150 s	*9815.0*	*9815.0*	10 885.6	25.63	*9815.0*	*9815.0*	0.00
d198	198	40	300 s	7062.0	7169.0	7468.73	0.83	*7044.0*	*7044.0*	0.00
krob200	200	40	300 s	11 245.0	11 353.0	12 532.0	0.00	*11 244.0*	*11 244.0*	0.00
gr202	202	41	300 s	*242.0*	249.0	258.0	0.00	243.0	*242.0*	0.00
ts225	225	45	300 s	62 366.0	63 139.0	67 195.0	34.49	62 315.0	*62 268.5*	0.51
pr226	226	54	300 s	*55 515.0*	*55 515.0*	56 286.6	40.89	*55 515.0*	*55 515.0*	0.00
gil262	262	53	300 s	*942.0*	979.0	1022.8	0.00	–	942.3	1.02
pr264	264	54	300 s	*21 886*	22 115.0	23 445.8	68.27	–	21 886.5	1.78
pr299	299	60	450 s	20 399.0	20 578.0	22 989.4	11.58	–	*20 322.6*	14.67
lin318	318	64	450 s	18 521.0	18 533.0	20 268.0	0.00	–	*18 506.8*	11.58
rd400	400	80	600 s	*5943.0*	6056.0	6440.8	3.40	–	5943.6	9.69
fl417	417	84	600 s	1034.0	1036.0	1080.5	0.51	–	*1033.0*	0.18
pr439	439	88	600 s	51 852.0	52 104.0	55 694.1	45.88	–	*51 847.9*	40.92
pcb442	442	89	600 s	*19 621.0*	19 961.0	21 515.1	5.15	–	19 702.8	52.11

2.6 Notes

The generalized minimum spanning tree problem (GMSTP) has been introduced by Myung et al. [128] in 1995. Since then, several research papers have appeared which deal with different aspects of the problem, such as complexity, approximation results, integer programming formulations, and solving the problem with exact, heuristic, metaheuristic and hybrid algorithms. The problem finds several important applications in designing metropolitan area networks [47] and regional area networks [173], and in determining the location of the regional service centers [142], energy transportation [135], agricultural irrigation [30], etc.

Within this chapter the following results were presented:

- a survey of the (mixed) integer programming formulations of the GMSTP;
- a local-global approach to the problem, which has lead to the development of exact solution procedures, heuristic and metaheuristic algorithms for GMSTP, and other generalized network design problems;
- an exact algorithm for the GMSTP
- an approximation algorithm for the problem, in the case where the size of the clusters is bounded;
- four algorithmic methods: a branch-and-cut algorithm, based on the the undirected cluster subpacking formulation of the GMSTP, a heuristic algorithm, based on the local-global approach, a rooting procedure, based on the local-global integer programming formulation, and a simulated annealing heuristic for solving the GMSTP.

The material presented in this chapter is mainly based on the results published by Pop [142, 143, 151], Feremans et al. [40] and Pop et al. [149, 154].

Chapter 3

The Generalized Traveling Salesman Problem (GTSP)

The traveling salesman problem (TSP) certainly is one of the classical and most intensively studied and analyzed representatives of combinatorial optimization problems, with a lot of solution methodologies and several applications in, e.g. logistics, transportation, planning, microchips manufacturing, DNA sequencing, etc.

Some mathematical problems related to the TSP were treated in the 1800s by W.R. Hamilton and the TSP already appeared in the literature in the early 19th century. It was published in Vienna, in the 1930s, by the mathematician and economist Karl Menger [120] and later was promoted by H. Whitney and M. Flood in the circles of Princeton, being named the *traveling salesman problem.*

Comparing the TSP to other combinatorial optimization problems, the main difference is that very powerful problem-specific methods, such as the Lin–Kernighan algorithm [113] and effective branch and bound methods are available, that are able to achieve a global optimal solution in very high problem dimensions.

Several extensions of the TSP have been considered in the literature: the time-dependent TSP, the bottleneck TSP, TSP with profits, TSP with mixed deliveries and collections, TSP with backhauls, one commodity pickup and delivery TSP, the multiple TSP, etc.

In this chapter, we are concerned with an extension of the traveling salesman problem (TSP) called the the generalized traveling salesman problem (GTSP). Given a complete undirected graph whose nodes are partitioned into a number of subsets (clusters) and with nonnegative costs associated to the edges, the GTSP consists of finding a minimum-cost Hamiltonian tour including exactly one node from each cluster. Therefore, the TSP is a special case of the GTSP, where each cluster consists of exactly one node.

A variant of the GTSP is the problem of finding a minimum cost Hamiltonian tour including *at least* one vertex from each cluster. This problem was introduced by Laporte and Nobert [103] and by Noon and Bean [133].

The GTSP has several applications to location problems, telecommunication problems, railway optimization, planning, postal routing, manufacture of microchips, logistics, computer file sequencing, etc. More information on the problem and its applications can be found in Fischetti, Salazar and Toth [42, 43], Laporte, Asef-Vaziri and Sriskandarajah [105], Laporte and Nobert [103], etc.

3.1 Definition and complexity of the GTSP

The *generalized traveling salesman problem* (GTSP) seeks to find a minimum-cost tour H, spanning a subset of nodes, such that H contains exactly one node from each cluster V_i, $i \in \{1, \ldots, m\}$.

We will call such a cycle a *generalized Hamiltonian tour*. An example of a generalized Hamiltonian tour, for a graph with nodes partitioned into 6 clusters, is presented in Figure 3.1.

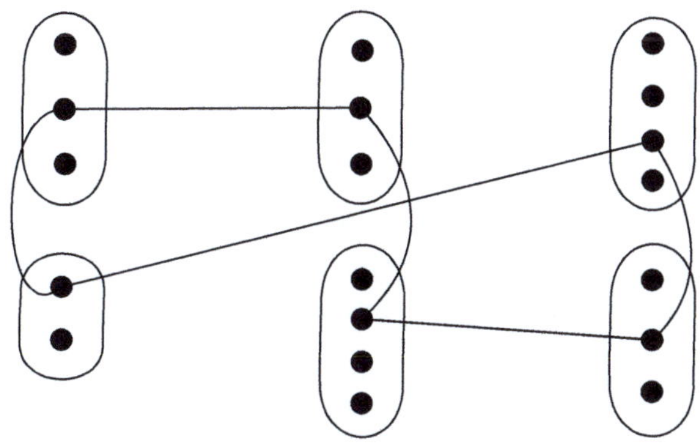

Figure 3.1. Example of a generalized Hamiltonian tour.

The GTSP involves two related decisions:

- choosing a node subset $S \subseteq V$, such that $|S \cap V_k| = 1$, for all $k = 1, \ldots, m$.
- finding a minimum cost Hamiltonian cycle in the subgraph of G, induced by S.

Both the GTSP and the at least version of the GTSP are $\mathcal{NP}$-hard, since they reduce to the traveling salesman problem, when each cluster consists of exactly one node.

3.2 An efficient transformation of the GTSP into the TSP

Because of the complexity of generalized network design problems, efficient transformations of these problems into classical combinatorial optimization problems seem to be an appropriate approach. In addition, for the classical combinatorial optimization problems, there exist many heuristic and optimal solution methods.

Motivated by the fact that there exists a large variety of exact and heuristic algorithms for TSP, see for example [64, 109], several efficient transformations of the generalized traveling salesman problem into TSP have been developed:

- the first transformation of the GTSP was described by Noon and Bean [134]. They transformed the GTSP into a standard asymmetric TSP over the same number of nodes, using a two stages approach: in the first stage the GTSP was transformed into a clustered TSP, which was then transformed into a standard asymmetric TSP. Ben-Arieh et al. [10] investigated the Noon and Bean transformation in computational experiments.

- Lien, Ma and Wah [112] provided a transformation of the GTSP into classical TSP, in which the number of nodes of the transformed TSP was quite large, in fact more than three times the number of nodes in the associated GTSP.

- Dimitrijevic and Saric [26] developed another transformation, that decreased the size of the corresponding TSP. In their method, the number of nodes of the TSP was twice the number of nodes of the original GTSP.

- Behzad and Modarres [8] provided an efficient transformation, in which the number of nodes in the transformed TSP does not exceed the number of nodes in the original GTSP.

In the following, we will describe the efficient transformation of the GTSP into the TSP proposed by Behzad and Modarres [8], which is a modification of the transformation developed by Dimitrijevic and Saric [26]. It should be mentioned that, although in the transformation, the underlying TSP is represented by a directed graph, the graph associated with the GTSP is not necessarily a directed one.

Let v_i^r denote the i-th node of the cluster V_r. We define the TSP on a directed graph G' associated to G as follows:

1. The set of nodes of G and G' are identical.

2. All nodes of each cluster are connected by arcs into a cycle in G'. We denote by $v_{i(s)}^r$ the node following v_i^r in the cycle.

3. The costs of the arcs of the transformed graph G' are defined as:

$$c'(v_i^r, v_{i(s)}^r) = 0$$

and

$$c'(v_i^r, v_j^t) = c(v_{i(s)}^r, v_j^t) + M, \quad r \neq t$$

where M must be a sufficiently large number, for example $\sum_{\{i,j\}\in E} c(i,j)$.

On the directed graph G' a path is called a *cluster path*, if it consists of all nodes of the cluster. A Hamiltonian cycle with cost less than $(m + 1)M$ is called a *clustered Hamiltonian cycle*.

Lemma 3.1. *(Behzad and Modarres [8]) Every clustered Hamiltonian cycle in G' enters and leaves each cluster exactly once.*

Proof. From the definition of the costs in the directed graph G', we see that the costs of moving from one cluster to another is at least M. Since every Hamiltonian cycle visits all the m clusters, its cost is at least mM. If the clustered Hamiltonian cycle in G' does not traverse along the cluster paths, this implies that it enters and leaves each of the m clusters more than once. Therefore, the cost of the tour cannot be less than $(m + 1)M$, which is a contradiction. □

We can now define the one-to-one correspondence between clustered Hamiltonian cycles in G' and generalized Hamiltonian tours in G:

1. Consider a clustered Hamiltonian cycle in G' and connect the first nodes of its cluster paths together in the order of their corresponding clusters, then the result is a generalized Hamiltonian tour in G.
2. Consider a generalized Hamiltonian tour in G that includes the following nodes $\cdots \to v_i^r \to v_j^t \to \cdots$, $r \neq t$. Replacing the node v_i^r with the V_r-th cluster path starting with v_i^r and then connecting the last node of this path to the next cluster path starting with v_j^t, we obtain a clustered Hamiltonian cycle in G'.

It is obvious that the cost of a generalized Hamiltonian tour in G visiting m clusters is equal to the cost of the corresponding clustered Hamiltonian cycle in G' minus mM.

In Figure 3.2 we present an optimal solution of the GTSP, defined on the graph G, and the corresponding TSP defined on the graph G'. The nodes of each cluster are cycled in G' by solid arcs, as depicted in the figure. The optimal solution of the TSP as well as its corresponding solution in the GTSP are illustrated by dash-dot arcs.

Even though the idea of transforming the GTSP into the TSP seems to be very promising, this approach has very limited application, because it requires exact solutions of the obtained TSP instances. Even a near-optimal solution of such a TSP may correspond to an infeasible GTSP solution.

3.3 An exact algorithm for the Generalized Traveling Salesman Problem

In this section, we present an algorithm that finds an exact solution to the generalized traveling salesman problem.

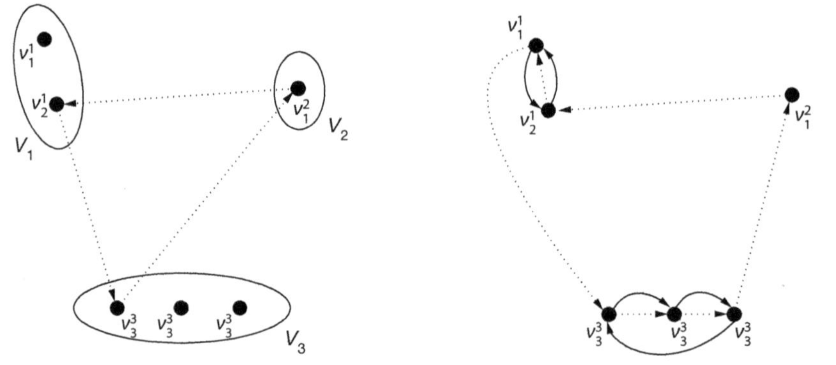

Figure 3.2. Illustration of the GTSP and the associated TSP with their optimal solutions.

Let G' be the graph obtained from G after replacing all nodes of a cluster V_i with a supernode representing V_i. For convenience, we identify V_i with the supernode that represents it. We assume that G', with vertex set $\{V_1, \dots, V_m\}$, is complete.

Given a sequence in which the clusters are visited $(V_{k_1}, \dots, V_{k_m})$, we want to find the best feasible Hamiltonian tour H^* (w.r.t. cost minimization), visiting the clusters according to the given sequence. This can be done in polynomial time, by solving $|V_{k_1}|$ shortest path problems as we will describe below.

We construct a layered network, depicted in Figure 3.3, having $m + 1$ layers corresponding to the clusters $V_{k_1}, \dots, V_{k_m}$. In addition, we duplicate the cluster V_{k_1}. The layered network contains all nodes of G plus some extra nodes v' for each $v \in V_{k_1}$. There is an edge $\{i, j\}$ for each $i \in V_{k_l}$ and $j \in V_{k_{l+1}}$ $(l = 1, \dots, m-1)$, with cost c_{ij}. Moreover, there is an edge $\{i, j'\}$ for each $i \in V_{k_m}$ and $j' \in V_{k_1}$, with cost $c_{ij'}$.

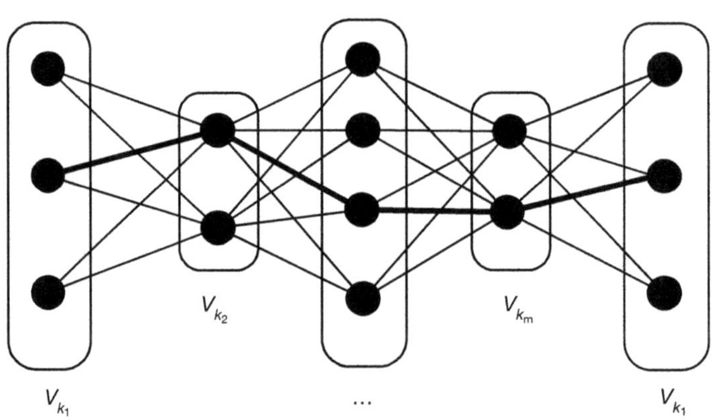

Figure 3.3. Example of a generalized Hamiltonian tour in the constructed layered network LN.

For any given $v \in V_{k_1}$, we consider paths from v to v', $w' \in V_{k_1}$, that visit exactly one node from each cluster $V_{k_2}, \dots, V_{k_m}$, hence giving it a feasible generalized Hamiltonian tour.

Conversely, each generalized Hamiltonian tour visiting the clusters according to the sequence $(V_{k_1}, \dots, V_{k_m})$ corresponds to a path in the layered network from a certain node $v \in V_{k_1}$ to $w' \in V_{k_1}$.

Therefore, it follows that the best generalized Hamiltonian tour H^* (w.r.t. cost minimization) visiting the clusters in a given sequence, can be found by determining all shortest paths from each $v \in V_{k_1}$ to the corresponding $v' \in V_{k_1}$, that visit exactly one node from each of the clusters $(V_{k_2}, \dots, V_{k_m})$.

The overall time complexity then is $|V_{k_1}|O(|E| + \log n)$, i.e. $O(n|E| + nlogn)$ in the worst case. We can reduce the time by choosing $|V_{k_1}|$ as the cluster with minimum cardinality.

Notice that the above procedure leads to an $O((m-1)!(n|E| + nlogn))$ time exact algorithm for the GTSP, obtained by trying all the $(m-1)!$ possible cluster sequences. So, we have established the following result:

Theorem 3.2. *The above procedure provides an exact solution to the generalized traveling salesman problem in $O((m-1)!(n|E| + nlogn))$ time, where n is the number of nodes, $|E|$ is the number of edges and m is the number of clusters in the input graph.*

Clearly, the presented algorithm is an exponential time algorithm, unless the number of clusters m is fixed.

3.4 Integer programming formulations of the GTSP

3.4.1 Formulations based on the properties of Hamiltonian tours

In order to formulate the GTSP problem as an integer program, we introduce the binary variables $x_e \in \{0, 1\}$, $e \in E$ and $z_i \in \{0, 1\}$, $i \in V$, to indicate whether an edge e respectively a node i is contained in the Hamiltonian tour. A feasible solution to the GTSP can be seen as a cycle with m edges, connecting all clusters and exactly one node from each cluster. Therefore the GTSP can be formulated as the following integer programming problem:

$$\min \sum_{e \in E} c_e x_e$$

$$\text{s.t.} \quad z(V_k) = 1, \qquad \forall\, k \in K = \{1, \dots, p\}, \tag{3.1}$$

$$x(\delta(i)) = z_i, \qquad \forall\, i \in V, \tag{3.2}$$

$$x(E(S)) \le z(S - \{i\}), \qquad \forall\, i \in S \subset V, 2 \le |S| \le n-2, \tag{3.3}$$

$$x_e \in \{0, 1\}, \qquad \forall\, e \in E, \tag{3.4}$$

$$z_i \in \{0, 1\}, \qquad \forall\, i \in V. \tag{3.5}$$

Here, we used the same notations as in Chapter 2 for the case of the generalized minimum spanning tree problem.

In this formulation, the objective function clearly describes the cost of an optimal generalized tour. Constraints (3.1) guarantee that we select exactly one node from each cluster, constraints (3.2) are *degree constraints*. They specify that, if a node of G is selected, it is entered and left exactly once. The constraints (3.3) are *subtour elimination constraints*, that prohibit the formation of subtours, i.e. tours on subsets of less than n nodes. Because of the degree constraints, subtours over one node (and hence, over $n-1$ nodes) cannot occur. Therefore, it is valid to define constraints (3.3) only for $2 \le |S| \le n-2$. Finally, the constraints (3.4) and (3.5) impose binary conditions on the variables.

This formulation, introduced by Fischetti et al. [43], is called the *generalized subtour elimination formulation*, since the constraints (3.3) eliminate all cycles on subsets with less than n nodes.

Since, in the GTSP, exactly one node from each cluster must be visited, we can drop the intra-cluster edges. In view of this reduction, the constraints (3.2) are equivalent to

$$x(\delta(V_k)) = 2, \quad \forall k \in K = \{1, \ldots, m\},$$

and the constraints (3.3) are equivalent to

$$x(E(S)) \le r - 1, \quad \forall S = \cup_{j=1}^{r} V_j \quad \text{and} \quad 2 \le r \le m-2.$$

We may replace the subtour elimination constraints (3.3) with connectivity constraints, resulting in the *generalized cutset formulation*, introduced in [43] as well

$$\begin{aligned} &\min \sum_{e \in E} c_e x_e, \\ \text{s.t.}\quad &(3.1), (3.2), (3.4), (3.5), \\ &x(\delta(S)) \ge 2(z_i + z_j - 1), \quad \forall S \subset V, \quad 2 \le |S| \le n-2, \\ &\qquad\qquad i \in S, \quad j \notin S. \end{aligned} \tag{3.6}$$

Both formulations that we have described so far, have an exponential number of constraints. The formulations that we will consider next will only have a polynomial number of constraints, but we will add a number of variables.

3.4.2 Flow based formulations

In order to give compact formulations of the GTSP, one possibility is to introduce 'auxiliary' flow variables, beyond the natural binary edge, and node variables.

We wish to send a flow between the nodes of the network and view an edge variable x_e as indicating whether the edge $e \in E$ is able to carry a flow or not. We consider three such flow formulations: a single commodity model, a multicommodity model, and a bidirectional flow model. In each of these models, although the edges are undirected, the flow variables will be directed. That is, for each edge $\{i, j\} \in E$, we will have flow in both the directions i to j and j to i.

In the *single commodity model*, the source cluster V_1 sends one unit of flow to every other cluster. Let f_{ij} denote the flow on edge $e = \{i, j\}$ in the direction i to j. This leads to the following formulation:

$$\min \sum_{e \in E} c_e x_e$$

$$\text{s.t.} \quad (3.1), (3.2),$$

$$\Delta(f_e) = \mu, \tag{3.7}$$

$$f_{ij}, f_{ji} \leq (m-1)x_e, \quad \forall e = (i, j) \in E, \tag{3.8}$$

$$f_{ij}, f_{ji} \geq 0, \tag{3.9}$$

$$x, z \in \{0, 1\},$$

where $\Delta(f_e) = \sum_{e \in \delta^+(i)} f_e - \sum_{e \in \delta^-(i)} f_e$ and

$$\mu = \begin{cases} (m-1)z_i & \text{for} \quad i \in V \setminus V_1 \\ -z_i & \text{if} \quad i \in V_1 \end{cases}.$$

In this model, the constraints (3.7) restrict $m - 1$ units of a single commodity flow to the cluster V_1 and 1 unit of flow out of each of the other clusters. These constraints are called *mass balance equations* and imply that the network defined by any solution (x, z) must be connected. Since the constraints (3.1) and (3.2) state that the network defined by any solution contains one node from every cluster and satisfies the degree constraints, every feasible solution must be a generalized Hamiltonian tour. Therefore, when projected into the space of the (x, z) variables, this formulation correctly models the GTSP.

We let P_{flow} denote the projection of the feasible set of the linear programming relaxation of this model into the (x, z)-space.

A stronger relaxation is obtained by considering multicommodity flows. In this model every node set $k \in K_1 = \{2, \dots, m\}$ defines a commodity. One unit of commodity k originates from V_1 and must be delivered to node set V_k. Letting f_{ij}^k be the flow of commodity k in arc (i, j) we obtain the following formulation:

$$\min \sum_{e \in E} c_e x_e,$$

$$\text{s.t.} \quad (3.1), (3.2),$$

$$\Delta(f_a^k) = \mu_k, \qquad \forall\, k \in K_1, \tag{3.10}$$

$$f_{ij}^k \leq w_{ij}, \qquad \forall\, a = (i, j) \in A, k \in K_1, \tag{3.11}$$

$$w_{ij} + w_{ji} = x_e, \qquad \forall\, e = (i,j) \in E, \tag{3.12}$$

$$f_a^k \geq 0, \qquad \forall\, a = (i,j) \in A, k \in K_1, \tag{3.13}$$

$$x, z \in \{0,1\},$$

where

$$\mu_k = \begin{cases} z_i & \text{for} \quad i \in V_1 \\ -z_i & \text{if} \quad i \in V_k \\ 0 & \text{if} \quad i \notin V_1 \cup V_k \end{cases} .$$

and the binary variables w_{ij}, $(i,j) \in A$ indicate whether an arc (i,j) is contained in the Hamiltonian tour. The directed graph $D = (V, A)$ is obtained by replacing each edge $e = \{i,j\} \in E$ with the opposite arcs (i,j) and (j,i) in A, with the same weight as the edge $\{i,j\} \in E$.

We let P_{mcflow} denote the projection of the feasible set of the linear programming relaxation of this model into the (x,z)-space.

Proposition 3.3. $P_{\text{mcflow}} \subseteq P_{\text{flow}}$.

Proof. Let $(w,x,z,f) \in P_{\text{mcflow}}$, then

$$0 \leq \sum_{k \in K_1} f_{ij}^k \leq |K_1| w_{ij} \leq (m-1) x_e.$$

With $f_{ij} = \sum_{k \in K_1} f_{ij}^k$, for every $e = \{i,j\} \in E$, we find

$$\Delta(f_a^k) = \sum_{k \in K_1} \left(\sum_{a \in \delta^+(i)} f_a^k - \sum_{a \in \delta^-(i)} f_a^k \right) = \mu.$$

□

We obtain a closely related formulation by eliminating the variables w_{ij}. The resulting formulation consists of the constraints (3.1), (3.2), (3.4), (3.5), (3.10), (3.13) plus

$$f_{ij}^h + f_{ji}^k \leq x_e, \quad \forall\, k,h \in K_1 \quad \text{and} \quad \forall\, e \in E. \tag{3.14}$$

We refer to this model as the *bidirectional flow formulation* of the GTSP with P_{bdflow} denoting its set of feasible solutions in (x,z)-space.

In the bidirectional flow formulation, the constraints (3.14), which are called the *bidirectional flow inequalities*, link the flow of different commodities flowing in different directions on the edge $\{i,j\}$. These constraints model the following fact: in any feasible generalized spanning tree, if we eliminate the edge $\{i,j\}$ and divide the nodes in two sets, any commodity whose associated node lies in the same set as the root node set, does not flow on edge $\{i, j\}$. Any two commodities, whose associated nodes both lie in the set without the root, both flow on edge $\{i, j\}$ in the same direction. So, whenever two commodities h and k both flow on edge $\{i, j\}$, they flow in the same direction and hence, one of f_{ij}^h and f_{ij}^k equals zero.

Proposition 3.4. $P_{\text{mcflow}} = P_{\text{bdflow}}$.

Proof. If $(w, x, z, f) \in P_{\text{mcflow}}$, using the constraints (3.11) and (3.12), we have

$$f_{ij}^h + f_{ji}^k \leq w_{ij} + w_{ji} = e,$$

for all $e = \{i, j\} \in E$ and for all $h, k \in K_1$.

If, on the other hand, we assume that $(x, z, f) \in P_{\text{bdflow}}$, then, using the constraints (3.14), we get

$$\max_h f_{ij}^h + \max_k f_{ji}^k \leq x_e, \quad \forall e = \{i, j\} \in E.$$

Hence, we can choose w such that $\max_h f_{ij}^h \leq w_{ij}$ and $x_e = w_{ij} + w_{ji}$ for all $e = \{i, j\} \in E$. For example, take

$$w_{ij} = \frac{1}{2}\big(x_e + \max_h f_{ij}^h + \max_k f_{ji}^k\big).$$

Clearly, we obtain $(w, x, z, f) \in P_{\text{mcflow}}$. □

3.4.3 A local-global formulation

Our last model arises from discriminating between global variables and local ones. Recall the construction of the graph G', obtained by shrinking each cluster of G into a single node.

We introduce the variables y_{ij} ($i, j \in \{1, \dots, m\}$) to describe the inter-cluster (global) connections. Hence, $y_{ij} = 1$ if the cluster V_i is connected to the cluster V_j and $y_{ij} = 0$ otherwise. We assume that y represents a Hamiltonian tour. The convex hull of all these z-vectors is generally known as the Hamiltonian tour polytope on the global graph G'.

Following Miller et al. [122] this polytope, denoted by P_{TSP}, can be represented by the following polynomial number of constraints:

$$\sum_{\substack{j=1 \\ j \neq i}}^{m} y_{ij} = 1, \qquad \forall i \in \{1, \dots, m\},$$

$$\sum_{\substack{i=1 \\ i \neq j}}^{m} y_{ij} = 1, \qquad \forall j \in \{1, \dots, m\},$$

$$u_i - u_j + (m-1)y_{ij} \le m-2, \qquad \forall i,j \in \{2,\dots,m\} \quad \text{and} \quad i \neq j, \tag{3.15}$$
$$1 \le u_i \le m-1, \qquad \forall i \in \{2,\dots,m\}. \tag{3.16}$$

where the additional variables u_i represent the sequence in which city i is visited, with $i \neq 1$.

The constraints (3.15) ensure that the solution contains no subtour on a set of nodes S with $|S| \le m-1$ and hence, that no subtour involves less than m nodes. The constraints (3.16) ensure that the u_i variables are uniquely defined for any feasible tour.

If the vector y describes a Hamiltonian tour on the global graph G', the corresponding best *generalized Hamiltonian tour* (w.r.t. cost minimization) $(x,z) \in \{0,1\}^{|E|\times|V|}$ can be obtained either by determining all shortest paths from each $v \in V_{k_1}$ to each $w' \in V_{k_1}$ that visit exactly one node from clusters $(V_{k_2},\dots,V_{k_m})$, or by solving the following 0 - 1 programming problem:

$$\begin{aligned}
&\min \sum_{e\in E} c_e x_e, \\
&\text{s.t.} \quad z(V_k) = 1, && \forall\, k \in K = \{1,\dots,m\}, \\
&\quad x(V_l, V_r) = y_{lr}, && \forall\, l,r \in K = \{1,\dots,m\}, \\
&\quad x(i, V_r) \le z_i, && \forall\, r \in K, \forall\, i \in V \setminus V_r, \\
&\quad x_e, z_i \in \{0,1\}, && \forall e = \{i,j\} \in E, \forall i \in V.
\end{aligned}$$

For a given y, we denote the feasible set of the linear programming relaxation of this program by $P_{\text{local}}(y)$.

Pop et al. [150] proved that if z is the 0 - 1 incidence vector of a spanning tree of the contracted graph, the polyhedron $P_{\text{local}}(y)$ is integral. But, as we will show in the next example a similar result does not hold when y is the incidence vector of the Hamiltonian tour, because if y is the 0 - 1 incidence vector of a Hamiltonian tour on the contracted graph, $P_{\text{local}}(y)$ may not be integral.

If the lines drawn in Figure 3.4 (i.e., $\{1,3\}$, $\{2,4\}$, etc.) have cost 1 and all other lines (i.e., $\{1,4\}$, $\{2,3\}$, etc.) have cost $M \gg 1$, then $y \equiv \frac{1}{2}$ and $x \equiv \frac{1}{2}$, on the drawn lines, is an optimal solution of $P_{\text{local}}(z)$, showing that the polyhedron $P_{\text{local}}(y)$ is not integral.

The observations presented so far lead to our final formulation, called the local-global formulation of the GTSP, as a 0-1 mixed integer programming problem:

$$\begin{aligned}
&\min \sum_{e\in E} c_e x_e, \\
&s.t. \quad z(V_k) = 1, && \forall\, k \in K = \{1,\dots,m\}, \\
&\quad x(V_l, V_r) = y_{lr}, && \forall\, l,r \in K = \{1,\dots,m\},
\end{aligned}$$

$$\begin{aligned} x(i, V_r) &\le z_i, && \forall\, r \in K, \forall\, i \in V \setminus V_r,\\ y &\in P_{\mathrm{TSP}}, &&\\ y_{lr} &\in \{0, 1\}, && 1 \le l, r \le m. \end{aligned}$$

This new formulation of the GTSP was obtained by incorporating the constraints characterizing P_{TSP}, with $y \in \{0, 1\}$, into $P_{\mathrm{local}}(y)$.

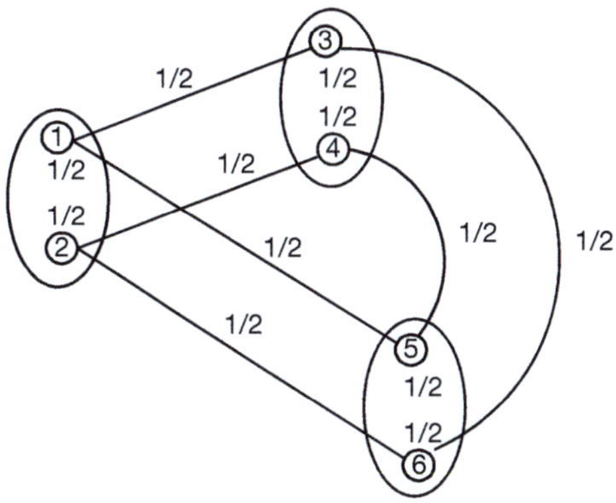

Figure 3.4. Example, showing that $P_{\mathrm{local}}(y)$ may have fractional extreme points if y is the is the 0 - 1 incidence vector of a Hamiltonian tour on the contracted graph.

3.5 Solving the Generalized Traveling Salesman Problem

The researchers paid a lot of attention to solving the GTSP. As we have seen in the previous section, several transformations of the GTSP into TSP have been proposed. Two approximation algorithms for the GTSP have been developed. Slavik [187] developed a $3\rho/2$-approximation algorithm for the GTSP, where ρ is the number of nodes in the largest cluster, that is, $\rho = \max_{i=1,\dots,m}\{|V_i|\}$ and with the additional assumption that the distance matrix satisfies the triangle inequality. As ρ may be quite large, the worst-case bound may be relatively weak. Garg and Konjevod [46] described an approximation algorithm for the group Steiner problem, which provided an $O(\log^2(n)\log(\log(n))\log(m))$ approximation algorithm for GTSP.

An efficient exact algorithm has been proposed by Fischetti et al. [43], based on a branch-and-bound algorithm that solved the problem to optimality for instances with up to 89 clusters and 442 nodes, etc. This method is the state-of-the-art exact algorithm for solving the GTSP. Noon and Bean [133] described a Lagrangian relaxation approach, in order to compute a lower bound on the total cost of an optimal solution of the GTSP.

The difficulty of obtaining optimum solutions for the GTSP has led to the development of several heuristic and metaheuristic algorithms. Noon [132] described several heuristics for the GTSP, the most promising of which is an adaptation of the well known nearest-neighbor heuristic for the TSPg. Fischetti et al. [43] propose similar adaptations of the farthest-insertion, nearest-insertion, and cheapest-insertion heuristics. Renaud and Boctor [178] developed the most sophisticated heuristic published to date, which they call GI^3 (Generalized Initialization, Insertion, and Improvement). It consists of three phases and it is a generalization of the I^3 heuristic proposed by Renaud et al. [179]. Recently, Cacchiani et al. [14] proposed a multistart heuristic, which starts iteratively with a randomly chosen set of vertices and applies a decomposition approach, combined with improvement procedures. Karapetyan and Gutin [95] investigated specific neighborhoods for GTSP and algorithms that can explore these neighborhoods quickly.

Several metaheuristic approaches have been proposed for solving the GTSP, including a genetic algorithm that uses a random-key encoding, which assures that the constructed solutions by crossover and mutation operators are feasible, and which was introduced by Snyder and Daskin [188]. Pintea et al. [139] developed a metaheuristic, based on an ant colony system, called the reinforcing ant colony system (RACS) algorithm, because it uses new pheromone rules and pheromone evaporation techniques. A genetic algorithm with several new features, including isolated initial populations and a new reproduction mechanism, based upon the TSP ordered crossover operator, was described by Silberholz and Golden [184]. An ensemble of discrete differential evolution algorithms, with parallel populations, was developed by Tasgetiren et al. [197]. Particle Swarm Optimization based algorithms have been developed by Shi et al. [185] and by Tasgetiren et al. [196]. Gutin et al. [65, 66]. Bontoux et al. [13] proposed memetic algorithms, which combine the power of genetic algorithms with local search procedures.

Recently, two hybrid algorithms have been developed by Pop et al. [161, 166], in which a genetic algorithm, respectively a consultant-guided search technique are combined with a local-global approach, in order to efficiently solve the GTSP. The memetic algorithm obtained by Gutin and Karapetyan [65] by combining a genetic algorithm with a powerful local search procedure can be considered the most competitive algorithm published to date.

In this section we will describe two algorithmic methods for solving the generalized traveling salesman problem:

- a reinforcing ant colony system
- a hybrid heuristic technique obtained by combining the consultant-guided search technique with a local-global approach to the GTSP.

3.5.1 Reinforcing ant colony system for solving the GTSP

In this section, we describe a new algorithm called the reinforcing ant colony system (RACS), based on the Ant Colony System (ACS) for solving the Generalized Traveling Salesman Problem.

The new algorithm enforces construction of a valid solution, used in (ACS), with an update pheromone rule as in [140] and pheromone evaporation as in Stüzle and Hoos, [194].

Let $V_k(v)$ denote the node v from the cluster V_k. The RACS algorithm for the GTSP works as follows:

• Initially, the ants are placed in the nodes of the graph, choosing random *clusters* and randomly selecting a node from the chosen cluster.

• At iteration $t+1$, each ant moves to a new node in an unvisited *cluster*, and the parameters controlling the algorithm are updated.

• Each edge is labeled by a trail intensity. Let $\tau_{ij}(t)$ represent the trail intensity of edge (i, j) at time t. When an ant decides which node it moves to next, it does so with a probability based on the distance to that node (i.e. cost of the edge) and the amount of trail intensity on the connecting edge.

The inverse of the distance from a node to the next node, is known as the *visibility*, $\eta_{ij} = \frac{1}{c_{ij}}$.

• During each unit of time, evaporation takes place, to stop the intensity trails from increasing unbounded. The rate evaporation is denoted by ρ, and its value is between 0 and 1.

In order to stop ants from visiting the same *cluster* in the same tour a tabu list is maintained. This prevents ants from visiting *clusters* they have previously visited. The ant tabu list is cleared after each completed tour.

• To favor selection of an edge with high pheromone value, τ, and high visibility η, a probability function ${p^k}_{iu}$ is considered. Denoting by ${J^k}_i$ the neighbors of node i unvisited by ant k and letting $u \in {J^k}_i, u = V_k(v)$, being the node v from the unvisited cluster V_k, this probability function is defined as follows:

$$ {p^k}_{iu}(t) = \frac{[\tau_{iu}(t)][\eta_{iu}(t)]^\beta}{\sum\limits_{o \in {J^k}_i} [\tau_{io}(t)][\eta_{io}(t)]^\beta}, \tag{3.17} $$

where β is a parameter used for tuning the relative importance of the edge cost in selecting the next node.

If $q > q_0$, ${p^k}_{iu}$ is the probability of choosing $j = u$, where $u = V_k(v)$ is the next node (the current node is i).

If $q \le q_0$ the next node j is chosen as follows:

$$ j = \operatorname{argmax}_{u \in J_i^k} \{\tau_{iu}(t)[\eta_{iu}(t)]^\beta\} \tag{3.18} $$

where q is a random variable uniformly distributed in the interval $[0, 1]$, and q_0 is a parameter similar to the temperature in simulated annealing, with $0 \le q_0 \le 1$.

The parameter q_0 is playing an important role, having the chance to improve the algorithm. In the RACS algorithm, the value of the parameter q_0 is changed at each iteration. For example, for an initial value of q_0 closer to 1, the value of q_0 is reduced by a small value c_0, for example $c_0 = 0.1$. If $q_0 \leq c_0$ then q_0 is modified.

• After each transition the trail intensity is updated, using the correction rule from [140]:

$$\tau_{ij}(t+1) = (1-\rho)\tau_{ij}(t) + \rho\frac{1}{n \cdot L^+}. \tag{3.19}$$

where L^+ is the cost of the best tour.

• In ant colony systems only the ant that generates the best tour is allowed to *globally* update the pheromone. The global update rule is applied to the edges belonging to the *best tour*. The correction rule is

$$\tau_{ij}(t+1) = (1-\rho)\tau_{ij}(t) + \rho\Delta\tau_{ij}(t), \tag{3.20}$$

where $\Delta\tau_{ij}(t)$ is the inverse cost of the best tour.

• In order to avoid stagnation, we used the pheromone evaporation technique introduced by Stüzle and Hoos, [194]. When the pheromone trail is above an upper bound $\tau_{\max}$, the pheromone trail is re-initialized.

The pheromone evaporation is used after the global pheromone update rule.

The RACS algorithm computes a sub-optimal solution for a given time $time_{\max}$ and can be stated as follows:

Algorithm 3.5 (Reinforcing Ant Colony System algorithm for the GTSP).

For every edge (i, j) do $\tau_{ij}(0) = \tau_0$
  for $k = 1$ to m do
    place ant k on a randomly chosen node
    from a randomly chosen cluster
  let T^+ be the shortest tour found and L^+ its length
  while ($t <$ time$_{\max}$) do
    for $k = 1$ to m do
      build tour $T^k(t)$ by applying **nc**-1 times
        choose the next node j from an unvisited cluster,

$$j = \begin{cases} \operatorname{argmax}_{u \in J_i^k}[\tau_{iu}(t)] \cdot [\eta_{iu}]^\beta & \text{if} \quad q \leq q_0 \\ J & \text{if} \quad q > q_0 \end{cases}$$

where $J \in J_i^k$ is chosen with probability:

$$p^k{}_{ij}(t) = \frac{[\tau_{ij}(t)] \cdot [\eta_{ij}(t)]^\beta}{\Sigma_{l \in J^k{}_i} [\tau_{ij}(t)] \cdot [\eta_{ij}(t)]^\beta}$$

and where i is the current node

apply the new local update rule

$$\tau_{ij}(t+1) = (1-\rho) \cdot \tau_{ij}(t) + \rho \cdot \frac{1}{nL^+}$$

end for

for $k = 1$ to m do

compute the length $L^k(t)$ of the tour $T^k(t)$

if an improved tour is found then update $T^k(t)$ and $L^k(t)$

for every edge $(i,j) \in T^+$ do

update pheromone trails by applying the rule:

$$\tau_{ij}(t+1) = (1-\rho) \cdot \tau_{ij}(t) + \rho \cdot \Delta\tau_{ij}(t)$$

where $\Delta\tau_{ij}(t) = \frac{1}{L^+}$

if $(\tau_{ij}(t) > \tau_{\max})$ then $\tau_{ij}(t) = \tau_0$

end for

end while

Print the shortest tour T^+ and its length L^+

3.5.2 Computational results

To evaluate the performance of our algorithm, the RACS was compared to three other heuristics from the literature. For numerical experiments, we used problems from the *TSPLIB* library [177], which provides optimal objective values for each of the problems. Several problems involving Euclidean distances have been considered.

Originally, in these problems, the sets of nodes are not divided into clusters. To divide them into subsets, we used the procedure, proposed by Fischetti et al. [43], called *Clustering*. This procedure sets the number of clusters $nc = [n/5]$, identifies the nc nodes farthest from each other, called centers, and assigns each remaining node to its nearest center. Obviously, some real world problems may have different cluster structures, but the solution procedure presented in this paper is able to handle any cluster structure.

The initial value of all pheromone trails, τ_0 is

$$\tau_0 = \frac{1}{n \cdot L_{nn}}, \tag{3.21}$$

where L_{nn} is the result of *Nearest Neighbor, (NN)* algorithm. *NN* is perhaps the most natural heuristic for the *TSP*. In the *NN* algorithm, the rule is to always go to the nearest as-yet-unvisited location. The corresponding tour traverses the nodes in the order they were constructed.

For the pheromone evaporation phase, let $\tau_{\max}$ denote the upper bound

$$\tau_{\max} = \frac{1}{1-\rho} \cdot \frac{1}{L_{nn}}. \tag{3.22}$$

Given a randomly value q_0 from the interval [0, 1], the value of the parameter q_0 is modified as follows:

$$q_0 = q_0 - 0.1; \quad \text{if} \quad q_0 < 0.1 \quad \text{then} \quad q_0 = 0.9. \tag{3.23}$$

As in all other ant systems, the choice of parameters in the algorithm is critical. Until now no mathematical analysis has been developed, which would give the optimal parameter in every situation.

In the proposed RACS algorithm, based on preliminary experiments, the values of the parameters were chosen as follows: $\beta \in \{2, 5\}$, $\rho \in \{0.1, 0.5\}$, $q_0 \in \{0.9, 0.5\}$, $c_0 = 0.1$ and $m = 2$.

In the next table, Table 3.1, we compare the computational results for solving the GTSP, using the RACS algorithm, with the computational results obtained using the classical Nearest Neighbor (NN) heuristic [178], a composite heuristic GI^3 [178] and a random key-Genetic Algorithm [188].

All solutions are the average of five successive runs of the algorithm, for each problem. The termination criterion is given by the maximal computing time $\text{time}_{\max}$, set by the user. In our case $\text{time}_{\max} = 5$ minutes.

The columns in the table are as follows:

- **Instance**: The name of the test problem. The digits at the beginning of the name give the number of clusters (nc). Those at the end give the number of nodes (n);
- **Optimal value**: The optimal objective value for the problem, as given in [188].
- **RACS, NN, GI^3, GA**: The objective value returned by RACS, the classical Nearest Neighbor heuristic from [178], the GI^3 heuristic from [178] and a random-key Genetic Algorithm from [188].

From Table 3.1, we can see that the Reinforcing Ant Colony System algorithm performs well, in many cases finding the optimal solution. The Reinforcing System finds good solutions within the pre-specified time interval: in 18 out of 36 instances,

Table 3.1. Reinforcing Ant Colony System (RACS) versus other algorithms for solving the GTSP.

Instance	Optimal Value	RACS	NN	GI^3	GA
11EIL51	174	174	181	174	174
14ST70	316	316	326	316	316
16EIL76	209	209	234	209	209
16PR76	64 925	64 925	76 554	64 925	64 925
20RAT99	497	497	551	497	497
20KROA100	9 711	9 711	10 760	9 711	9 711
20KROB100	10 328	10 328	10 328	10 328	10 328
20KROC100	9 554	9 554	11 025	9 554	9 554
20KROD100	9 450	9 450	10 040	9 450	9 450
20KROE100	9 523	9 523	9 763	9 523	9 523
20RD100	3 650	3 650	3 966	3 653	3 650
21EIL101	249	249	260	250	249
21LIN105	8 213	8 213	8 225	8 213	8 213
22PR107	27 898	27 898	28 017	27 898	27 898
22PR124	36 605	36 607.2	38 432	36 762	36 605
26BIER127	72 418	72 418	83 841	76 439	72 418
28PR136	42 570	42 570	47 216	43 117	42 570
29PR144	45 886	45 925.6	46 746	45 886	45 886
30KROA150	11 018	11 021.2	11 712	11 018	11 018
30KROB150	12 196	12 197.2	13 387	12 196	12 196
31PR152	51 576	51 588.4	53 369	51 820	51 576
32U159	22 664	22 707	26 869	23 254	22 664
39RAT195	854	854	1 048	854	854.2
40D198	10 557	10 559.2	12 038	10 620	10 557
40KROA200	13 406	13 407.2	16 415	13 406	13 406
40KROB200	13 111	13 114.2	17 945	13 111	13 113.4
45TS225	68 345	69 056.2	72 691	68 756	68 435.2
46PR226	64 007	64 028	68 045	64 007	64 007
53GIL262	1 013	1 021.2	1 152	1 064	1 016.2
53PR264	29 549	29 549.6	33 552	29 655	29 549
60PR299	22 615	22 726.2	27 229	23 119	22 631
64LIN318	20 765	20 790.2	24 626	21 719	20 836.2
80RD400	6 361	6 416.2	7 996	6 439	6 509
84FL417	9 651	9 730.2	10 553	9 697	9 653
88PR439	60 099	60 898	67 428	62 215	60 316.8
89PCB442	21 657	22 022.8	26 756	22 936	22 134

we obtained the known optimal values for the GTSP, and for the remaining instances the percentage gap between the known optimal values and the values obtained using our RACS approach was at most 1.5%.

It is possible to improve the results by choosing a longer execution time or by making better choices for the parameters there. The RACS algorithm for the Generalized Traveling Salesman Problem may be improved if finely tuned parameter values are used. Also, an efficient combination with other algorithms might improve the results.

3.5.3 A hybrid heuristic approach for solving the GTSP

We start with a brief description of the Consultant-Guided Search (CGS) metaheuristic. We refer the reader to [86] for a detailed presentation.

CGS is a recent swarm intelligence technique for solving hard combinatorial optimization problems, which takes inspiration from the way real people make, decisions based on advice received from consultants. It has been successfully applied to the classical Traveling Salesman Problem (TSP) [87] and to the Quadratic Assignment Problem [88].

CGS is a population-based method. An individual of the CGS population is a virtual person, who can act simultaneously as both a client and a consultant. As a client, a virtual person constructs a solution to the problem in each iteration. As a consultant, a virtual person provides advice to clients, in accordance with its *strategy*. Usually, at each step of the construction of the solution, there are several variants which a client can choose from. The variant recommended by the consultant has a higher probability for being chosen, but the client may opt for one of the other variants, which will be selected based on some heuristic.

At the beginning of each iteration, a client chooses a consultant based on his *personal preference* and on the consultant's *reputation*. The reputation of a consultant increases with the number of successes achieved by his clients. A client achieves a success, if he constructs a solution which is better than all others found by any client guided by the same consultant up to that point. Each time a client achieves a success, the consultant adjusts his strategy, in order to reflect the sequence of decisions made by the client. Because the reputation fades over time, a consultant needs his clients to constantly achieve successes, in order to maintain his reputation. If the consultant's reputation sinks below a minimum value, he will take a *sabbatical leave*, during which he will stop offering advice to clients and will instead start searching for a new strategy for future use.

In this subsection, we describe an algorithm for the GTSP that combines the consultant-guided search technique with a local-global approach and improves the solutions by using a local search procedure. The GTSP instances of most practical importance are symmetric problems with Euclidean distances, where the clusters are composed of nodes that are spatially close one to the other. We design our algorithm to take advantage of the structure of these instances.

The algorithm

At each iteration, a client constructs a global tour, that is, a Hamiltonian cycle in the global graph. The strategy of a consultant is also represented by a global tour, which the consultant advertises to his clients. The algorithm applies a local search procedure, in order to improve the global tour that represents either the global solution of a client, or the strategy of a consultant in sabbatical mode.Then, using the cluster optimization procedure described in Section 3.2, the algorithm finds the best generalized tour, which corresponds to the global tour returned by the local search procedure.

In order to compare the strategies constructed during the sabbatical leave, a consultant uses the cost of the generalized tour corresponding to each strategy. Similarly, the success of a client is evaluated based on the cost of the generalized solution. The pseudocode of our algorithm is shown in Algorithm 3.6.

A virtual person may be in either *normal* or *sabbatical* mode. During the initialization phase (lines 2–5), virtual people are created and placed in sabbatical mode. Based on his mode, on each iteration of the algorithm (lines 7–31), a virtual person constructs either a global solution to the problem (line 19) or a global consultant strategy (line 9). Later in this subsection, we describe the operations involved in the construction of a global solution or strategy, as well as the method used by a client in order to choose a consultant for the current iteration (line 17).

Global strategies and global solutions are improved by applying a local search procedure (lines 10 and 20). The *clusterOptimization* procedure, described in Section 3.2, is then used to find the best generalized strategy (line 11) that corresponds to the current global strategy or to find the best generalized solution (line 21) that corresponds to the current global solution. After constructing a global strategy, a virtual person in sabbatical mode checks if the corresponding generalized strategy is the best generalized strategy found since the beginning of the sabbatical (lines 12–15). Similarly, after constructing a global solution, a client checks the corresponding generalized solution, in order to decide if he has achieved a success and if so, he updates the strategy of his consultant (lines 22–26). At the end of each iteration, the reputation and action mode of each virtual person are updated (lines 30–31).

Algorithm 3.7 details how consultants' reputations are updated, based on the successes achieved by their clients.

Reputations fade over time with a constant rate, given by the parameter *fadingRate* (line 4). The reputation of a consultant is incremented with each success achieved by one of his clients (line 5) and he receives an additional bonus of 10 for finding a best-so-far solution (lines 6–9).

The reputation of a consultant cannot exceed a maximum value (lines 10–12) and the algorithm prevents the reputation of the best consultant, that is, the consultant that has found the best-so-far solution, from sinking below a given value (lines 13–17).

The constant parameter *initialReputation* represents the reputation assigned to a consultant at the end of the sabbatical leave.

Algorithm 3.8 details how the action mode of each virtual person is updated. Consultants whose reputations have sunk below the minimum level are placed in sabbatical mode, while consultants whose sabbatical leave has finished are placed in normal mode.

Algorithm 3.9 shows the actions taken to place a virtual person in sabbatical or in normal action mode.

Algorithm 3.6 (The CGS-GTSP algorithm).

```
procedure CGS-GTSP()
   create the set 𝒫 of virtual persons
   foreach p ∈ 𝒫 do
      setSabbaticalMode(p)
   end foreach
   while (termination condition not met) do
      foreach virtual person p do
         if actionMode[p] = sabbatical then
            currStrategy[p] ← constructStrategy(p)
              applyLocalSearch(currStrategy[p])
              genStrategy ← clusterOptimization(currStrategy[p])
              if cost(genStrategy) < bestStrategyCost then
                 bestStrategy[p] ← currStrategy[p]
                 bestStrategyCost[p] ← cost(currStrategy[p])
              end if
         else
              c ← chooseConsultant(p)
              if c ≠ null then
               currSol[p] ← constructSolution(p,c)
               applyLocalSearch(currSol[p])
               (currGenSol[p]) ← clusterOptimization(currSol[p])
               if currGenSol[p] is better than all solutions found by a client
                     of c since last sabbatical then
                  successCount[c] ← successCount[c] + 1
                  strategy[c] ← currSol[p]
               end if
              end if
         end if
      end foreach
      updateReputations()
      updateActionModes()
   end while
end procedure
```

Algorithm 3.7 (Procedure to update the reputations).

```
procedure updateReputations()
  foreach p ∈ 𝒫 do
    if actionMode[p] = normal then
      rep[p] ← rep[p] * (1 - fadingRate)
      rep[p] ← rep[p] + successCount[p]
      if cost(currGenSol[p]) < cost(bestSoFarSol) then
        bestSoFarSol ← currGenSol[p]
        rep[p] ← rep[p] + 10              // reputation bonus
      end if
      if rep[p] > 10 * initialReputation then
        rep[p] ← 10 * initialReputation
      end if
      if p is the best consultant then
        if rep[p] < initialReputation then
          rep[p] ← initialReputation
        end if
      end if
    end if
  end foreach
end procedure
```

Algorithm 3.8 (Procedure to update the action modes).

```
procedure updateActionModes()
  foreach p ∈ 𝒫 do
    if actionMode[p] = normal then
      rep[p] < 1 then
        setSabbaticalMode(p)
      end if
    else
      sabbaticalCountdown ← sabbaticalCountdown - 1
      if sabbaticalCountdown = 0 then
        setNormalMode(p)
      end if
    end if
end procedure
```

Algorithm 3.9 (Procedures to set sabbatical and normal mode).

```
procedure setSabbaticalMode(p)
   actionMode[p] ← sabbatical
   bestStrategy[p] ← null
   bestStrategyCost[p] ← ∞
   sabbaticalCountdown ← 20
end procedure

procedure setNormalMode(p)
   actionMode[p] ← normal
   rep[p] ← initialReputation
   strategy(p) ← bestStrategy[p]
end procedure
```

Strategy and construction of the solution

The heuristic used during the sabbatical leave to build a new strategy is based on virtual distances between the supernodes in the global graph. We compute the virtual distance between two supernodes as the distance between the centers of mass of the two corresponding clusters. The choice of this heuristic is justified by the class of problems for which our algorithm is designed, which are symmetric instances with Euclidean distances, where the nodes of a cluster are spatially close one to the other.

By introducing virtual distances between clusters, we have the possibility to use candidate lists, in order to restrict the number of choices available at each construction step. For each cluster V_i, we consider a candidate list that contains the closest *cand* clusters, with *cand* a parameter. This way, the feasible neighborhood of a person k in cluster V_i represents the set of clusters in the candidate list of cluster V_i that this person k has not yet visited. Several heuristic algorithms for the TSP use candidate lists in the solution construction phase (see [13]) for examples of their use in Ant Colony Optimization algorithms), but candidate lists have not been widely used to construct solutions for the GTSP. Our algorithm uses candidate lists during the construction of both strategy solution.

The use of candidate lists may significantly improve the time required by an algorithm, but it could also lead to missing good solutions. Therefore, the choice of appropriate candidate list sizes and elements is critical for an algorithm to work. In the case of the TSP, candidate lists with size 20 are frequently used, but other values between 10 and 40 are also common [91]. For GTSP, instances with clusters composed of nodes spatially close to each other, the appropriate sizes of the candidate lists are considerably smaller. Our experiments show that values of 4 or 5 are adequate in this case.

During the sabbatical leave, a consultant uses a random proportional rule to decide which cluster to visit next. For a consultant k, currently at cluster V_i, the probability to choose cluster V_j is given by the following formula:

$$p_{ij}^k = \frac{1/d_{ij}}{\sum_{l \in \mathcal{N}_i^k}(1/d_{il})} \tag{3.24}$$

where $\mathcal{N}_i^k$ is the feasible neighborhood of person k when at cluster V_i and d_{il} is the virtual distance between clusters V_i and V_l.

As mentioned before, the feasible neighborhood $\mathcal{N}_i^k$ contains the set of clusters in the candidate list of cluster V_i that person k has not yet visited. If all clusters in the candidate list have already been visited, the consultant may choose one of the clusters not in the candidate list, by using a random proportional rule similar to the one given by formula (3.24).

Use of virtual distances between clusters as a heuristic during sabbatical leave, leads to reasonably good initial strategies. In general, however, after applying the cluster optimization procedure, a global tour that is optimum with respect to the virtual distances between clusters, does not produce the optimum generalized tour. Therefore, during the solution construction phase, the algorithm does not rely on the distances between clusters, although it still uses candidate lists in order to determine the feasible neighborhood of a cluster.

In each step, a client receives a recommendation for the next cluster to be visited. This recommendation is based on the global tour advertised by the consultant. Let V_i be the cluster visited by client k during a construction step in the current iteration. To decide which cluster to recommend for the next step, the consultant finds the position at which the cluster V_i appears in its advertised global tour and identifies the clusters preceding and succeeding V_i in this tour. If neither of these two clusters has already been visited by the client, the consultant randomly recommends one of them. If only one is unvisited, this one is chosen as the recommendation. Finally, if both clusters are already visited, the consultant is not able to make a recommendation for the next step.

The client does not always follow the consultant's recommendation. The rule used for choosing the next cluster V_j to move to, is given by the following formula:

$$V_j = \begin{cases} v & \text{if } v \neq null \wedge q \leq q_0, \\ \text{random}(\mathcal{N}_i^k) & \text{otherwise.} \end{cases} \tag{3.25}$$

where:

- V_p is the cluster recommended by the consultant for the next step;
- q is a random variable uniformly distributed in [0,1] and q_0 ($0 \leq q_0 \leq 1$) is a parameter;

- $\mathcal{N}_i^k$ is the feasible neighborhood of person k, when at cluster V_i;
- random is a function that randomly chooses one element from the set given as its argument.

Again, if all clusters in the candidate list have already been visited, the feasible neighborhood $\mathcal{N}_i^k$ is empty. In this case, a client that ignores the recommendation of its consultant may choose one of the clusters not in the candidate list, by using a random proportional rule similar to the one given by formula (3.24).

The personal preference of a client for a given consultant is computed as the inverse of the cost of the generalized tour that corresponds to the global tour advertised by the consultant. In conjunction with the reputation, personal preference is used by clients in order to compute the probability for choosing a given consultant k:

$$p_k = \frac{(\text{reputation}_k \times \text{preference}_k)^2}{\sum_{c \in \mathcal{C}} (\text{reputation}_c \times \text{preference}_c)^2} \tag{3.26}$$

where $\mathcal{C}$ is the set of all available consultants.

At each iteration, the consultant's k reputation fades with a constant rate r:

$$\text{reputation}_k \leftarrow \text{reputation}_k (1 - r) \tag{3.27}$$

A consultant enters sabbatical mode when his reputation sinks below a minimum level. In our algorithm, this minimum reputation value is set to 1 and the duration of the sabbatical leave is set to 20 iterations. At the end of the sabbatical leave, a consultant's reputation is reset to a given value, specified by the parameter *initialReputation.*

In addition, the algorithm prevents the reputation of the best consultant from sinking below this initial value. We designate as best consultant the consultant whose strategy has led to the construction of the best-so-far solution.

Each time a client achieves a success, the reputation of his consultant is incremented. Should a client construct a best-so-far solution, the consultant's reputation receives a supplementary *bonus*. The reputation of a consultant cannot exceed the maximum value given by the parameter *maxReputation*.

An variant of the algorithm, that uses confidence

Here, we propose a a variant of our algorithm, based on the approach introduced in [87], which correlates the recommendation of a consultant with a certain level of confidence. Each arc in the global tour, advertised by a consultant, has an associated strength. Strengths are updated each time the consultant adjusts his strategy. If an arc in the newly advertised tour was also present in the old advertised tour, its strength will be incremented. Otherwise, its strength is set to 0. The strength of an arc could be interpreted as the consultant's confidence in recommending this arc to a client. A client is more likely to accept recommendations made with greater confidence.

This idea is expressed in our algorithm by allowing the value of the parameter q_0 from formula (3.25) to vary in a given range, at each construction step:

$$q_0 = \begin{cases} q_{\min} + s \cdot \frac{q_{\max} - q_{\min}}{s_{\max}} & \text{if } s < s_{\max} \\ q_{\max} & \text{otherwise} \end{cases}, \tag{3.28}$$

where s is the strength of the recommended arc and $q_{\min}$, $q_{\max}$ and $s_{\max}$ are constant parameters. The use of confidence somewhat compensates for the absence of a heuristic during the solution construction phase.

Local search

The global tours built during the strategy construction and solution construction phases are improved by using a local search procedure of which a generic description is given in Algorithm 3.10.

Algorithm 3.10 (The local search procedure).

1 **procedure** applyLocalSearch(H_G)
2   $H \leftarrow$ *clusterOptimization*(H_G)
3   **foreach** $H'_G \in$ *tourNeighborhood*(H_G) **do**
4   **if** *quickCheck*(H'_G) **then**
5       $H' \leftarrow$ *partialClusterOptimization*(H'_G, H_G)
6   **if** $cost(H') < cost(H)$ **then**
7         $H_G \leftarrow H'_G$
8         $H \leftarrow H'$
9   **end if**
10   **end if**
11   **end foreach**
12 **end procedure**

H_G and H'_G denote global Hamiltonian tours, that is, tours in the graph of clusters, while H and H' denote generalized Hamiltonian tours. Our algorithm can be combined with any local search procedure that conforms to the above algorithmic structure. The working of the *clusterOptimization* function (line 2) was explained in Section 3.2.

The *cost* function (line 6) computes the cost of a generalized Hamiltonian tour. The other functions referred in Algorithm 3.10 are only specified in generic form and they must be implemented by each concrete instantiation of the local search procedure.

The *tourNeighborhood* function (line 3) should return a set of global tours that represent the neighborhood of the global tour H_G provided as its argument. The *quickCheck* function (line 4) is intended to speed up the local search by quickly rejecting a candidate global tour from the partial cluster optimization, if this tour is not likely to lead to an improvement.

The *partialClusterOptimization* function (line 5) starts with the generalized tour obtained by traversing the nodes of H in accordance with the ordering of clusters in the global tour H'_G. Then, it reallocates some vertices in the resulting generalized tour, and tries to improve its cost. Typically, this function considers only a limited number of vertices for reallocation and it usually has a lower complexity than the *clusterOptimization* function.

The generalized tour constructed by the function *partialClusterOptimization* is accepted only if its cost is better than the cost of the current generalized tour (lines 6–9).

We provide two instantiations of the generic local search procedure shown in Algorithm 3.10: one based on a 2-opt local search and one based on a 3-opt local search. We describe here only the 2-opt based variant. Except for the fact that it considers exchanges between 3 arcs, the 3-opt based local search is very similar to the 2-opt based variant.

In the 2-opt based local search, the *tourNeighborhood* function returns a set of global tours obtained by replacing a pair of arcs (C_α, C_β) and (C_γ, C_δ) in the original global tour with the pair of arcs (C_α, C_γ) and (C_β, C_δ). In order to reduce the number of exchanges that must be taken into consideration, the set returned by our *tourNeighborhood* function includes only tours for which γ is in the candidate list of α. In other words, a pair of arcs is considered for exchange, only if the center of mass of the cluster γ is close to the center of mass of the cluster α.

The *partialClusterOptimization* function used in this case is similar to the RP1 procedure introduced in [43]. Let (C_α, C_β) and (C_γ, C_δ) be the two arcs from the original global tour H_G that have been replaced with (C_α, C_γ) and (C_γ, C_δ) in the neighboring global tour H'_G, as shown in Figure 3.5:

The vertices in clusters C_α, C_β, C_γ and C_δ can then be reallocated, in order to minimize the cost of the generalized tour. For this purpose, we have to determine the two node pairs (u', w') and (v', z') such that:

$$\begin{aligned} d_{iu'} + d_{u'w'} + d_{w'h} &= \min\{d_{ia} + d_{ab} + d_{bh} \mid a \in C_\alpha, b \in C_\gamma\} \\ d_{jv'} + d_{v'z'} + d_{z'k} &= \min\{d_{ja} + d_{ab} + d_{bk} \mid a \in C_\beta, b \in C_\delta\} \end{aligned} \tag{3.29}$$

This computation requires $|C_\alpha||C_\gamma| + |C_\beta||C_\delta|$ comparisons. In order to speed up the local search, we first perform a quick rejection test. We compute the quantities from formula (3.29), only if the following inequality holds:

$$\begin{aligned} &d_{\min}(C_\rho, C_\alpha) + d_{\min}(C_\alpha, C_\gamma) + d_{\min}(C_\gamma, C_\sigma) + d_{\min}(C_\pi, C_\beta) + \\ &\quad d_{\min}(C_\beta, C_\delta) + d_{\min}(C_\delta, C_\tau) \; < \; d_{iu} + d_{uv} + d_{vj} + d_{hw} + d_{wz} + d_{zk} \end{aligned} \tag{3.30}$$

where $d_{\min}(C_\alpha, C_\beta)$ is the minimum distance between each pair of vertices from clusters C_α and C_β. These minimum distances are computed only once, when starting the algorithm.

Experimental setup

We have implemented our algorithm as part of a software package written in Java, which is available online at http://swarmtsp.sourceforge.net/. At this address we provide all information necessary to reproduce our experiments.

In all experiments, we used candidate lists of length 4. The other parameters of the algorithm have been tuned, using the paramILS configuration framework [83]. We have generated a set of 100 Euclidean TSP instances with n cities, uniformly distributed in the interval $[200, 500]$ and with coordinates uniformly distributed in a square of dimension $10\,000 \times 10\,000$.

These instances have then been converted to the GTSP by applying the CLUSTERING procedure introduced in [43]. This procedure sets the number of clusters $s = \lceil n/5 \rceil$, identifies the s nodes farthest from each other and assigns each remaining node to its nearest center. We have used the resulting GTSP instances as training data for paramILS.

Before starting the tuning procedure, we ran our algorithm 10 times on each instance in the training set, using a default configuration. Each run has been terminated after $n/10$ seconds and we have stored the best result obtained for each GTSP instance. During the tuning procedure, these best known results are used as termination condition for our algorithm.

Each time paramILS evaluates a parameter configuration with respect to a given instance, we determine the mean time (averaged over 10 trials) that our algorithm

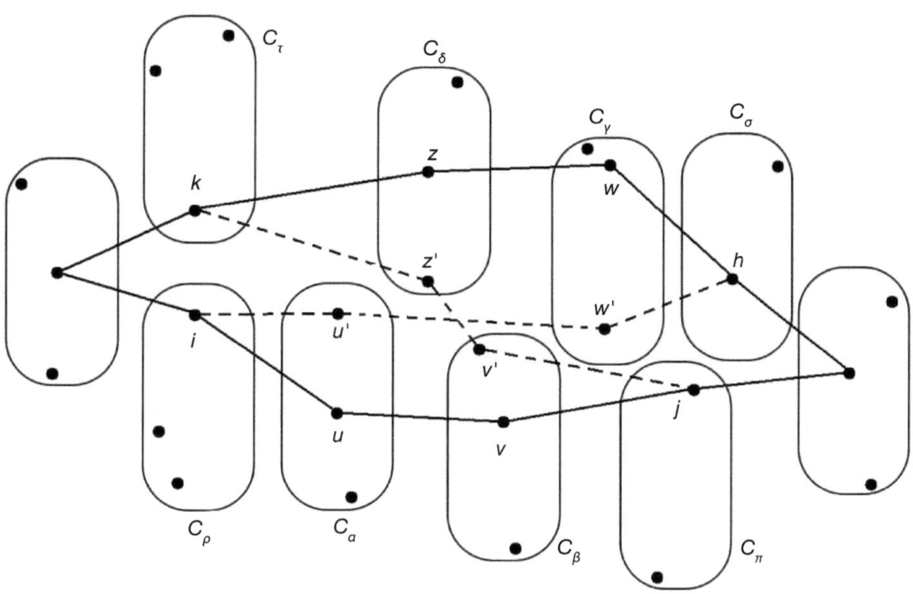

Figure 3.5. The generalized 2-opt exchange.

Table 3.2. Parameter configuration for the standard algorithm.

Parameter	Value	Description
m	8	number of virtual persons
q_0	0.8	see formula 3.25
initialReputation	6	reputation after sabbatical; see Algorithm 3.6 and Algorithm 3.8
reputationFadingRate	0.003	reputation fading rate; see Algorithm 3.6
candidateListSize	5	number of clusters in the candidate list

Table 3.3. Parameter configuration for the algorithm variant that uses confidence.

Parameter	Value	Description
$q_{\min}$	0.7	parameters used to compute
$q_{\max}$	0.98	the value of q_0;
$s_{\max}$	3	*see formulas* (3.25), (3.28)

needs to obtain a result at least as good as the best known result for this instance, using the given parameter configuration.

The best parameter configuration, found after 10 iterations of paramILS, is given in Table 3.2.

For the algorithm variant that uses confidence, we have used the same procedure as for the standard algorithm, but tuned only the values of the parameters $q_{\min}$, $q_{\max}$ and $s_{\max}$. For the parameters m, initialReputation, reputationFadingRate and candidateListSize we have used the values in Table 3.2.

The best parameter configuration found for the algorithm variant that uses confidence, after 10 iterations of paramILS, is given in Table 3.3.

3.5.4 Computational results

The performance of the proposed algorithm has been tested on 18 GTSP problems generated from symmetric Euclidean TSP instances. These TSP instances, containing between 198 and 442 nodes, are drawn from the TSPLIB [209] benchmark library. The corresponding GTSP problems are obtained by applying the CLUSTERING procedure introduced in [43]. For 16 of these GTSP instances, the optimum objective values have been determined by Fischetti et al. [43]. For the remaining 2 instances, the best known results are conjectured to be optimal.

Currently, the memetic algorithm of Gutin and Karapetyan [65] clearly outperforms all published GTSP heuristics. Therefore, we use this algorithm as a yardstick for evaluating the performance of the different variants of our algorithm. We use the following acronyms to identify the algorithms used in our experiments:

- GK: the memetic algorithm of Gutin and Karapetyan [65].
- CGS-2: the standard variant of our algorithm, combined with 2-opt local search.
- CGS-3: the standard variant of our algorithm, combined with 3-opt local search.
- CGS-C-2: the variant of our algorithm that uses confidence, combined with 2-opt local search.
- CGS-C-3: the variant of our algorithm that uses confidence, combined with 3-opt local search.

For each GTSP instance, we run each algorithm 25 times and report the average time needed to obtain the optimal solution. For the GK algorithm, we use the C++ implementation offered by its authors. The running times for GK differ from the values reported in [65], because we run our experiments on a 32-bit platform using an Intel Core2 Duo 2.2 GHz processor, while the results presented in [65] have been obtained on a 64-bit platform and with a faster processor (AMD Athlon 64 X2 3.0 GHz).

The computational results are shown in Table 3.4. The name of each problem is prefixed by the number of clusters and suffixed by the number of nodes. Average times better than those obtained by the GK algorithm are in boldface. For each problem and for each CGS algorithm variant, we also report the p-values of the one-sided Wilcoxon rank sum tests for the null hypothesis (H_0), which states that, for the given problem, there is no difference between the running times of the algorithm variant under consideration and the running times of the GK algorithm, as well as for the alternative hypothesis (H_1), which states that the considered algorithm outperforms the GK algorithm for the given problem. Applying the Bonferroni correction for multiple comparisons, we obtain the adjusted α-level: $0.05/18 = 0.00278$. The p-values in boldface indicate the cases where the null hypothesis is rejected at this level of significance.

It can be seen that CGS-C-3 outperformed GK for 9 out of 18 instances and in 7 cases these results are significantly better. CGS-C-2 outperformed GK for 8 of the 18 instances and in cases the results are significantly better. The variants without use of confidence have poorer performance and for a few instances they need considerably more time to find the optimal solution.

For several pairs of algorithms, we use the one-sided Wilcoxon signed rank test to compute the p-values for the null hypothesis (H_0), which states that there is no difference between the running times of the first and the running times of the second algorithm, and the alternative hypothesis (H_1), which states that the running times of the first algorithm are better than the running times of the second algorithm. The p-values are given in Table 3.5, where significant p-values ($p < 0.05$) are in boldface.

It can be seen that GK outperforms our algorithms. However, in the case of CGS-C-3, the differences are not statistically significant. Similarly, CGS-C-3 outperforms CGS-C-2, but the numbers are not statistically significant. The fact that 3-opt lo-

Table 3.4. Times (in seconds) needed to find the optimal solutions, averaged over 25 trials.

Problem	Optimal	GK	CGS-C-3		CGS-C-2		CGS-3		CGS-2	
instance	cost	time	time	p-value	time	p-value	time	p-value	time	p-value
40d198	10 557	0.46	*0.36*	*0.0004*	*0.33*	*0.0012*	0.47	0.0034	*0.45*	0.0050
40kroA200	13 406	0.38	*0.33*	*0.0000*	*0.25*	*0.0000*	*0.37*	0.5711	*0.30*	*0.0001*
40kroB200	13 111	0.48	0.60	0.9460	*0.37*	*0.0008*	0.59	0.9689	0.60	0.6156
41gr202	23 301	0.71	*0.64*	0.0141	0.91	0.4674	1.35	1.0000	1.10	0.9101
45ts225	68 340	0.61	3.32	1.0000	4.06	0.9957	1.92	0.9999	2.67	1.0000
45tsp225	1 612	0.51	4.83	1.0000	3.25	0.9994	4.07	1.0000	2.28	0.9967
46pr226	64 007	0.28	*0.13*	*0.0000*	*0.07*	*0.0000*	*0.13*	*0.0000*	*0.09*	*0.0000*
46gr229	71 972	0.81	*0.36*	*0.0000*	*0.33*	*0.0000*	*0.39*	*0.0000*	*0.37*	*0.0000*
53gil262	1 013	0.83	1.22	0.1071	2.63	0.9999	1.63	1.0000	3.49	1.0000
53pr264	29 549	0.67	*0.57*	0.0070	*0.49*	*0.0005*	0.94	0.9482	1.08	0.9406
56a280	1 079	0.94	1.79	0.8215	3.71	0.9999	2.02	0.9998	4.46	1.0000
60pr299	22 615	1.10	3.54	0.9992	2.91	0.9992	3.23	1.0000	4.74	0.9999
64lin318	20 765	1.16	*0.85*	*0.0000*	2.68	0.9929	1.28	0.8946	3.81	1.0000
80rd400	6 361	2.57	10.30	0.9996	13.27	1.0000	87.96	1.0000	270.04	1.0000
84fl417	9 651	1.91	*1.10*	*0.0000*	*1.59*	*0.0001*	*1.51*	*0.0012*	2.27	0.0512
87gr431	101 946	6.01	8.16	0.8361	12.86	0.9916	477.38	1.0000	866.53	1.0000
88pr439	60 099	4.07	*1.56*	*0.0000*	*1.32*	*0.0000*	*3.68*	0.0104	10.71	0.9999
89pcb442	21 657	4.24	11.11	0.9980	13.53	1.0000	395.93	1.0000	1430.13	1.0000

Table 3.5. Performance comparison using the one-sided Wilcoxon signed rank test.

First algorithm	Second algorithm	p-value
GK	CGS-C-3	0.106 1
GK	CGS-C-2	*0.036 8*
GK	CGS-3	*0.006 9*
GK	CGS-2	*0.000 5*
CGS-C-3	CGS-C-2	0.070 8
CGS-C-3	CGS-3	*0.015 2*
CGS-C-2	CGS-2	*0.002 8*

cal search does not significantly improve the results obtained with 2-opt local search could be a consequence of the greater complexity of 3-opt. There are, however, significant differences between the running times of CGS variants with and without use of confidence. Due to the very poor results obtained in some cases by the algorithm variants that don't use confidence, these differences are not only statistically, but also practically significant, thus indicating the importance of the confidence component.

Figure 3.6 shows how the candidate list size affects the time needed by CGS-C-3 to find the optimal solution of the problem instance 64lin318. The results are averaged over 25 trials.

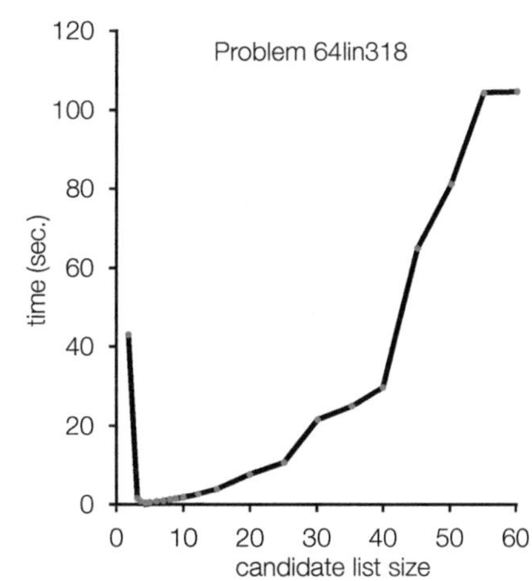

Figure 3.6. The influence of the candidate list size on the time needed to find the optimal solution for the problem 64lin318.

It can be seen that the size of the candidate list has a huge influence on the time needed by the algorithm to find the optimal solution. Therefore, the use of candidate lists is a key component contributing to the success of our algorithm.

The best results are obtained for candidate lists of size 4 or 5, but we should note that the algorithm is able to find the optimum even for candidate lists with only 2 elements. However, in this case, the time needed increases considerably. This is due to the fact that the probability to find the next cluster of the optimal tour in the candidate list of the current cluster is significantly smaller when using a candidate list with only 2 elements.

For the 64lin318 instance, only 44 of the 64 clusters are present in the candidate list of their precedent cluster when using candidate lists with 2 elements. In contrast, 59 of the 64 clusters are present when using candidate lists with 5 elements. The algorithm is able to find the optimal solution even for very small sized candidate lists, because during the construction phase a client may visit clusters that are not in the current candidate list, if all clusters in this candidate list have already been visited or when the consultant recommends it.

For candidate lists with a large number of elements, the algorithm performance in terms of running time worsens, due to the increase in the number of exchanges considered during the local search.

To conclude this section, we note that the obtained computational results show that there are no statistically significant differences between our algorithm variant with use of confidence and the memetic algorithm of Gutin and Karapetyan (GK) [65], which is currently the best published heuristic for the GTSP. The GK algorithm uses a sophisticated local improvement strategy, which combines many local search heuristics. One goal of our future research is to adopt a similar approach to the local improvement part of our algorithm, whilst still using candidate lists for each local search heuristic considered.

A dynamic version of the GTSP has been considered by Pintea et al. [138]. The same authors have proposed an ant colony system for solving this variant of the GTSP, in which one cluster is blocked at each iteration.

3.6 The drilling problem

In this section we present a new metaheuristic algorithm called the *Sensitive Robot Metaheuristic (SRM)*, which aims to solve the *drilling problem.*

The large drilling problem is a practical application of the generalized traveling salesman problem which involves a large graph and targets finding the minimal tour for drilling on a large scale printed circuit board. This problem plays an important role for an economical process for manufacturing printed circuit boards. A printed circuit board is used to mechanically support and electrically connect electronic components, using conductive pathways, tracks, or signal traces.

The process of manufacturing the printed circuit board is difficult and complex and is characterized by many layers that have to be linked through drilling small holes, using the shortest connecting route. These holes require precision and are done with use of an automated drilling machine controlled by computer programs. A layer is identified as a cluster of nodes and a feasible path must contain a node from each cluster. In order to minimize the drilling time, the optimal route connecting all layers has to be detected (see Figure 3.7).

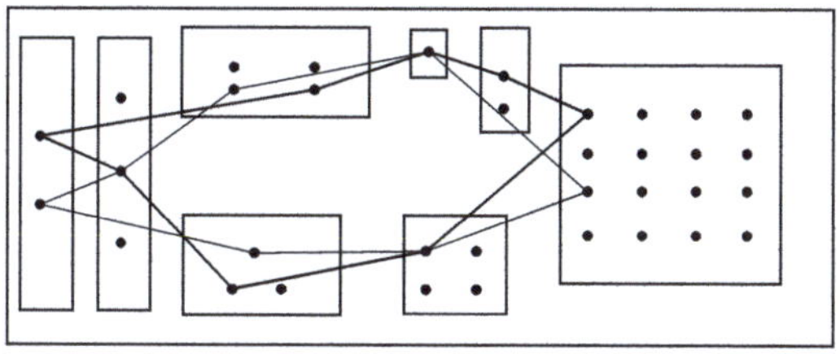

Figure 3.7. A schematic representation of the drilling problem on a Printed Circuit Board.

An example of printed circuit board is presented in Figure 3.8.

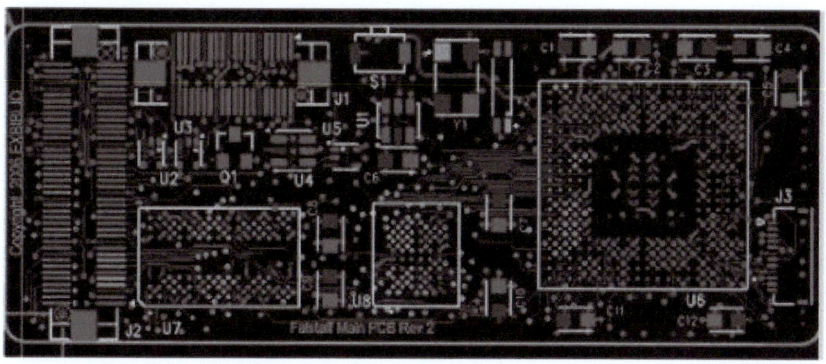

Figure 3.8. Printed Circuit Board.

In what follows we introduce the concept of the stigmergic autonomous robot and describe the proposed metaheuristic for solving the drilling problem.

3.6.1 Stigmergy and autonomous robots

The proposed metaheuristic combines the concepts of stigmergic communication and autonomous robotic search. Stigmergy occurs when an action of an insect is determined or influenced by the consequences of previous actions of another insect [12].

Stigmergy provides a general mechanism to relate individual and colony-level behaviors. Individual behavior modifies the environment, which in turn modifies the behavior of other individuals. Self-organization of social insects [15] can emerge, due to stigmergic interactions among individuals.

The behavior-based approach to design of intelligent systems has produced promising results in a wide variety of areas, including military applications, mining, space exploration, agriculture, factory automation, service industries, waste management, health care, and disaster intervention. Autonomous robots can accomplish real-world tasks without being told exactly how to do it.

Researchers try to make the coupling between perception and action as direct as possible. This aim remains the distinguishing characteristic of behavior-based robotics. The proposed *SRM* technique attempts to address this goal in an intelligent, stigmergic manner.

3.6.2 Sensitive robots

A stigmergic robot action is determined by the environmental modifications caused by prior actions of other robots. Quantitative stigmergy regards stimulus as a continuous variable. The value of such a variable modulates the intensity or probability of future actions. Qualitative stigmergy [32] involves discrete stimulus. In this case the action is not modulated. Instead, the agent switches to a different action, [32].

A qualitative stigmergic mechanism suits our aims better, since robot stigmergic communication does not rely on chemical deposition. The robot communication relies on local environmental modifications that can trigger specific actions. "Micro-rules" define action-stimuli pairs for a robot. The set of all micro-rules used by a homogeneous group of stigmergic robots defines their behavioral repertoire and determines the type of structure the robots will create [32].

Within the proposed model, the term *Sensitive robots* refers to artificial entities with a *Stigmergic Sensitivity Level (SSL)*, expressed by a real number in the unit interval [0, 1]. Extreme situations are:

- $SSL = 0$ indicates that the robot completely ignores stigmergic information (is a 'non-stigmergic' robot);
- $SSL = 1$ means that the robot has maximum stigmergic sensitivity.

Robots with small *SSL* values are highly independent and can be considered environment explorers. They have the potential to autonomously discover new promising regions of the search space. Therefore, search diversification can be sustained.

Robots with high *SSL* values are able to intensively exploit the promising search regions already identified. In this case the robot behavior emphasizes search intensification. The *SSL* value can increase or decrease according to the search space topology encoded in the robot's previous experience.

3.6.3 Sensitive robot metaheuristic for solving the drilling problem

The proposed *Sensitive Robot Metaheuristic (SRM)* can be implemented using two teams of sensitive robots. Robots of the first team have small *SSL* values. These sensitive-explorer robots are called *small SSL-robots (sSSL)* and can sustain search diversification. Robots of the second team have high *SSL* values. These sensitive-exploiter robots called *high SSL-robots (hSSL)* intensively exploit promising search regions already identified by the first team.

The *SRM* is applied for solving a robotic travel problem called the drilling problem. This problem can be viewed as an instance of the *Generalized Traveling Salesman Problem*.

To apply the *SRM* for solving the drilling problem under consideration, the robots are placed in the starting point and will search objects within a specific area. Assuming that each cluster has specific objects and the robots are capable to recognize this they will each time choose a different cluster. The stigmergic values guide the robots to the shorter path–a solution of the *Robotic Travel Problem*.

Solving the Drilling Problem with the SRM

Initially, the robots are randomly placed in the search space. In each iteration, a robot moves to a new node and the parameters controlling the algorithm are updated.

A robot chooses the next move with a probability based on the distance to the candidate node and the stigmergic intensity on the connecting edge and unit evaporation takes place each time, to stop the stigmergic intensity from increasing unbounded.

In order to prevent robots from visiting a cluster twice in the same tour, a *tabu list* is maintained. The stigmergic value of an edge is denoted by τ and the visibility value is η.

Let us consider ${J^k}_i$ as the unvisited successors of node i by robot k and $u \in {J^k}_i$. The *sSSL* robots choose the next node probabilistically. Let i be the current robot position (the current node).

Similar to the *ACS* technique [28], the probability of choosing u as the next node is given by:

$$
{p^k}_{iu}(t) = \frac{[\tau_{iu}(t)][\eta_{iu}(t)]^{\beta}}{\sum_{o \in {J^k}_i} [\tau_{io}(t)][\eta_{io}(t)]^{\beta}}, \tag{3.31}
$$

where β is a positive parameter, $\tau_{iu}(t)$ is the stigmergic intensity and $\eta_{iu}(t)$ is the inverse of the distance on edge (i, u) at time t.

The membership of robots to one of the two teams is modulated by a random variable, uniformly distributed in [0, 1]. Let q be a realization of this random variable and q_0 a constant, $0 \leq q_0 \leq 1$ (a parameter similar to the temperature in simulated annealing).

The *sSSL* robots are characterized by the inequality $q > q_0$, while for the *hSSL* robots $q \leq q_0$ holds.

An *hSSL-robot* uses the information supplied by the *sSSL* robots. *hSSL* robots choose the new node j in a deterministic manner, according to the following rule:

$$
j = \text{argmax}_{u \in J_i^k} \{\tau_{iu}(t)[\eta_{iu}(t)]^{\beta}\}, \tag{3.32}
$$

where the value of β determines the relative importance of stigmergy versus heuristic information.

The trail stigmergic intensity is updated using the local stigmergic correction rule:

$$\tau_{ij}(t+1) = q_0^2 \tau_{ij}(t) + (1-q_0)^2 \cdot \tau_0. \tag{3.33}$$

Only the elitist robot, which generates the best intermediate solution, is allowed to *globally* update the stigmergic value. The elitist robot can take advantage of global knowledge of the best tour found to date and reinforce this tour, in order to focus future searches more effectively.

The global update rule is:

$$\tau_{ij}(t+1) = q_0^2 \tau_{ij}(t) + (1-q_0)^2 \cdot \Delta\tau_{ij}(t), \tag{3.34}$$

where $\Delta\tau_{ij}(t)$ is the inverse value of the best tour length. In the update rules, q_0 is reinterpreted as the evaporation rate.

One run of the algorithm returns the shortest tour found.

Let n denote the number of nodes, e the number of edges and m the number of clusters in the input graph, r the number of robots and NC the number of cycles. The complexity of this algorithm leads to $O(m \cdot n \cdot r \cdot NC)$, [28]. For an exact algorithm, obtained by trying all $(m-1)!$ possible cluster sequences [153], the complexity is $O((m-1)!(ne + n \log n))$.

The description of the *Sensitive Robot Metaheuristic* for solving the drilling problem is depicted in the following algorithm:

Algorithm 3.11 (Sensitive Robot Algorithm).

Begin
  Set parameters, initialize stigmergic values of the trails;
  Loop
    Place robot k on a randomly chosen node
    from a randomly chosen cluster;
    Loop
      Each *robot* incrementally builds a solution
      based on autonomous search sensitivity;
      sSSL robots are characterized by the inequality
      $q > q_0$ while for *hSSL* robots $q \leq q_0$ holds;
      The *sSSL* robots probabilistically
      choose the next node using (3.24)
      A *hSSL-robot* uses the information supplied by
      the *sSSL* robots to find the new node j using (3.25);
      A local stigmergic updating rule (3.26);
    Until end_condition;
    A global updating rule is applied by the elitist robot (3.27);
  Until end_condition
End

3.6.4 Numerical experiments

The validation of the *SRM* concerns the minimization of the drilling operational time on printed circuit boards. The numerical experiments are based on four drilling problems, with Euclidean distances drawn from the *TSPLIB* [209] that provides optimum objective values for each problem.

As in all other metaheuristics, the choice of parameters is critical to the algorithm. Currently, no mathematical analysis has been developed that gives the optimum parameter in each situation.

Based on preliminary experiments, the values of the parameters in the *SRM* algorithm were chosen as $\beta = 5$, τ_0=0.01, $q_0 = 0.9$. The parameters were chosen based on [28, 137]. The total number of robots considered is 25. The sensitivity level q for the *hSSL* robots is considered to be distributed in the interval $(q_0, 1)$, while *sSSL* robots have a sensitivity level in the interval $(0, q_0)$.

The program is implemented in Java and was run on an AMD Athlon 2600+, 333MHz with 2 GB memory.

For all algorithms, the solutions represent the average of five consecutive runs for each problem. The termination criterion is given by a maximum of 200 trials and 100 tours.

The next table, Table 3.6, illustrates the *SRM* results after five consecutive runs of the algorithm. To divide the set of nodes into subsets, we used the procedure proposed in [43]. This procedure sets the number of clusters $nc = [n/5]$, identifies the nc farthest nodes from each other (called centers) and assigns each remaining node to its nearest center.

Table 3.6. Sensitive Robotic Metaheuristic results for five runs.

Drilling problem	Reported optimum	# optimum values	Mean value	Minimum value	Maximum value
32U159	22 664	5	*22 664*	*22 664*	*22 664*
40D198	10 557	5	*10 557*	*10 557*	*10 557*
84FL417	9 651	1	9 654.4	*9 651*	9 657
89PCB442	21 657	2	21 659.6	*21 657*	21 662

The table shows the values of the reported optimum values [209], the minimum, maximum and the mean values of the *SRM*, after five runs. The number of optimum values within the specified number of runs are also shown.

In order to evaluate the performance of the proposed algorithm, the *SRM* has been compared to the *Nearest Neighbor (NN)* [178], the composite heuristic (GI^3) [178], the Random Key Genetic Algorithm ($RKGA$) [188] and Reinforcing Ant Colony System (*RACS*) for *GTSP* [137], described in Section 3.4.

For each problem, the resulting averages of five successively runs have been examined. and the comparative results are shown in Table 3.7.

Table 3.7. *Sensitive Robotic Metaheuristic (SRM)* versus other algorithms: Nearest Neighbor (NN), composite heuristic GI^3, [178], Random Key Genetic Algorithm (RKGA), [188] and Ant Colony System (ACS) for *GTSP* (Mean values).

Drilling Problem	Reported Optimum	*NN*	GI^3	*RKGA*	*RACS*	*SRM*
32U159	22 664	26 869	23 254	*22 664*	22 729.2	*22 664*
40D198	10 557	12 038	10 620	*10 557*	10 575.2	*10 557*
84FL417	9 651	10 553	9 697	9 656	9 766.2	*9 654.4*
89PCB442	21 657	26 756	22 936	22 026	22 137.8	*21 659.6*

After analyzing the results reported in Table 3.7, we observe that our proposed SRM algorithm provides good solutions. For the instances $32U159$ and $40D198$, we obtained the optimal solutions and for the other two instances the percentage gap between the known optimal values and the values obtained using our SRM approach was at most 0.03%.

In the following, we conduct a statistical analysis. The Expected Utility Approach [56] technique was employed to determine the most accurate heuristic. Let x be the percentage deviation of the heuristic solution from the best known solution of a particular heuristic for a given problem:

$$x = \frac{\text{heuristic solution} - \text{best known solution}}{\text{best known solution}} \times 100.$$

The expected utility function may be: $\gamma - \beta(1 - \overline{b}t)^{-\overline{c}}$, where $\gamma = 500$, $\beta = 100$ and $t = 0.05$. $\overline{b}$ and $\overline{c}$ are the estimated parameters of the Gamma function. Because four problems were used for the testing, the following notations are used in Table 3.8:

$$\overline{x} = \frac{1}{4}\sum_{j=1}^{4} x_j, \quad s^2 = \frac{1}{4}\sum_{j=1}^{4}(x_j - \overline{x})^2, \quad \overline{b} = \frac{s^2}{\overline{x}}, \quad \overline{c} = \left(\frac{\overline{x}}{s}\right)^2.$$

The last column provides the rankings 1 to 5 of the entries. As indicated in the Table 3.8, the *SRM* ranks first for being the most accurate algorithm among the compared algorithms.

The results presented in Table 3.8 indicate that the novel SRM algorithm outperforms the other considered heuristics.

The new model may be improved in terms of execution time. Potential improvements involve the parameter values or employ an efficient combination with other algorithms. Another way to improve the algorithm is fully parallelizing the robots in the inner loop of the algorithm.

Table 3.8. Statistical analysis. Calculations for the expected utility function for the compared heuristics.

Heuristic	$\overline{x}$	s^2	$\overline{b}$	$\overline{c}$	$\gamma - \beta(1 - \overline{b}t)^{-\overline{c}}$	Rank
NN	16.5	31.25	1.8939	8.7122	262.0747	5
GI^3	2.3956	4.8206	2.0123	1.1905	386.5441	4
$RKGA$	0.4385	0.5558	1.2675	0.3459	397.7087	2
ACS	0.97	0.6783	0.6993	1.3871	394.9359	3
SRM	0.01	0.0001	0.01	1.0000	399.9499	1

3.7 Notes

The generalized traveling salesman problem (GTSP) is an extension of the well-known traveling salesman problem and has several applications in location problems, telecommunication problems, railway optimization, planning, postal routing, manufacture of microchips, logistics, computer file sequencing, scheduling with sequence dependent process times, etc.

Within this chapter, we have described the following aspects concerning the GTSP:

- an exact exponential time algorithm for the problem;
- integer programming formulations of the GTSP;
- two methods for solving the problem: a reinforcing ant colony system and a hybrid heuristic technique, obtained by combining the consultant-guided search technique with a local-global approach to the GTSP;
- a practical application of the GTSP: the drilling problem that consists of finding the minimal tour for drilling on a large-scale printed circuit board;
- a sensitive robot metaheuristic algorithm for solving the drilling problem, characterized by a stigmergic sensitivity level that facilitates exploration (by low-sensitive robots) as well as the exploitation (by high-sensitive robots) of the search space.

The material presented in this chapter is based on the results published by Pop [152], Pop et al. [161, 166] and Pintea et al. [137, 139, 141].

Chapter 4

The Railway Traveling Salesman Problem (RTSP)

In this chapter, we consider a problem of central interest in railway optimization called the *Railway Traveling Salesman Problem* (RTSP). This problem is a practical extension of the classical traveling salesman problem, which considers a railway network and train schedules, and was introduced by Hadjicharalambous et al. [67].

In addition, we present a variant of the RTSP called the dynamic RTSP, where we assume that the distances between cities are interpreted as travel times and are no longer fixed. In real life we are faced with this problem, when delays may be introduced, due to maintenance work, accidents, etc., which leads to possibly varying travel times.

4.1 Definition of the RTSP

We assume that we are given a set of stations, a timetable with trains connecting these stations, an initial station, a subset B of the stations and a starting time. A salesman wants to travel from the initial station, starting no earlier than the designated time, to every station in B and finally returns to the initial station, subject to the constraint that he/she spends the necessary amount of time in each station in B, to carry out his/her business. The goal is to find a set of train connections such that the overall time of the journey is minimized. This problem is called the *Railway Traveling Salesman Problem* and was introduced by Hadjicharalambous et al. [67].

The RTSP is related to the Generalized Traveling Salesman Problem (GTSP), in which a clustered graph is given and a round trip of minimal length, connecting exactly one node per cluster, is desired. By virtue of the TSP, the GTSP is an $\mathcal{NP}$-hard problem. It can be easily seen, that the TSP is also polynomially time reducible to the RTSP. For each pair of cities, to which there exists a connection, consider a train leaving from the first one to the second with a travel time equal to the cost of the corresponding connection in the TSP. This reduction puts the RTSP in the class of NP-hard problems.

Consider the so-called time-expanded digraph G, introduced by Schulz et al. [182], constructed from the timetable information. In that graph, there is a node for every time event (departure or arrival) at a station, and edges between nodes represent either elementary connections between the two events (i.e., served by a non-stop train), or waiting time at a station. The weight of an edge is the time difference between the time events associated with its endpoints. Now, roughly speaking, and considering each set of nodes belonging to a specific station as a node cluster, the RTSP reduces by finding a minimum-weight tour that starts at a specific node of a specific cluster, visits at least one node of each cluster in B, and ends at a node of the initial cluster.

Despite their superficial similarity, the RTSP differs from the GTSP in at least three aspects:

1. The GTSP is typically solved by transforming it to an instance of the TSP. The transformation is done by modifying the weights, such that inter-cluster edges are "penalized" by adding a large weight to them. Consequently the tour has to visit all nodes within a cluster before moving to the next one.

2. The RTSP starts from a specific node in a specific cluster, whereas the GTSP starts from any node within any cluster.

3. The GTSP requires visiting all clusters, whereas the RTSP requires visiting only a subset of them.

4.2 Preliminaries

In this section, we describe the input of an RTSP instance, provide some definitions and present the time-expanded graph.

In the following, we assume timetable information in a railway system, but the modeling and solution approaches that will be described can be applied to any other public transportation system, provided that it has the same characteristics.

A *timetable* consists of data concerning: *stations* (or bus stops, ports, etc), *trains* (or buses, ferries, etc), connecting stations, and *departure* and *arrival times* of trains at stations. More formally, we are given a set of trains Z, a set of stations S and a set of *elementary connections* C whose elements are 5-tuples of the form $(z, \sigma_1, \sigma_2, t_d, t_a)$. Such a tuple (elementary connection) is interpreted as train z leaving station σ_1 at time t_d and the *immediately nearest* stop of train z is σ_2 at time t_a. The *departure* and *arrival times* t_d and t_a are integers in the interval $T_{\text{day}} = [0, 1\,439]$, representing the time in minutes after midnight.

Given two time values t and t', $t \leq t'$, their *cycle-difference*(t, t') is the smallest nonnegative integer l such that $l \equiv t' - t$ (modulo 1440).

We are also given a starting station $\sigma_0 \in S$, a time value $t_0 \in T_{\text{day}}$, denoting the earliest possible departure time from station σ_0, and a set of stations $B \subseteq S - \sigma_0$, which represents the set of stations (cities) that the salesman must visit. A function $f_B : B \longrightarrow T_{\text{day}}$ is used to model the time that the salesman must spend in each city $b \in B$, i.e., the salesman must stay at station $b \in B$ for at least $f_B(b)$ minutes.

We naturally assume that the salesman does not travel continuously (i.e., through overnight connections) and if she/he arrives at some station late, she/he has to rest and spend the night. Moreover, the salesman's business for the next day may not require taking the first possible connection from that station. Consequently, we assume that the salesman never uses a train that leaves too late in the night or too early in the morning.

The formulations of the RTSP are based on the so-called *time-expanded* digraph, introduced by Schulz et al. [182]. Such a graph $G = (V, E)$ is constructed using the timetable information provided as follows:

- There is a node for every time *event* at a station (departure or arrival), and there are four types of edges.
- For every elementary connection $(z, \sigma_1, \sigma_2, t_d, t_a)$ in the timetable, there is a *train-edge* in the graph that connects a *departure node*, belonging to station σ_1 and associated with time t_d, with an *arrival node*, belonging to station σ_2 and associated with time t_a. In other words, the endpoints of the train-edges induce the set of nodes of the graph.
- For each station $\sigma \in S$, all departure nodes belonging to σ are ordered according to their time values. We denote with $v_1, v_2, \ldots, v_k$ the nodes of σ, in that order.
- There is a set of *stay-edges*, denoted by Stay(σ), (v_i, v_{i+1}), $1 \leq i \leq k-1$, and (v_k, v_1), connecting the time events within a station and representing a stay at that station;
- Additionally, for each arrival node in a station there is an arrival-edge to the immediately nearest (w.r.t. their time values) departure node of the same station.

The cost of an edge (u, v) is the *cycle-difference* (t_u, t_v), where t_u and t_v are the time values associated with u and v, respectively.

In the next figure, Figure 4.1, we present an example of two stations in the time-expanded graph, that illustrates our construction.

To formulate the RTSP, we introduce the following modifications to the time-expanded digraph:

- In the first place, we do not include any elementary connections that have departure times greater than the latest possible departure time, or smaller than the earliest.
- In the second place, we explicitly model the fact that the salesman has to wait for a time at least $f_B(b)$ at each station $b \in B$ by introducing a set of *busy-edges*, denoted by $Busy(b)$. We introduce a busy-edge from each arrival node in a station $b \in B$, to the first possible departure node of the same station that differs in time value by at least $f_B(b)$.
- In the third place, to model the fact that the salesman starts his journey at some station σ_0 and at time t_0, we introduce a *source node* s_0 in station σ_0 with time value t_0. Node s_0 is connected to the first departure node d_0 of σ_0 that has a time valuegreater than or equal to t_0, using an edge (called the *source edge*) with cost

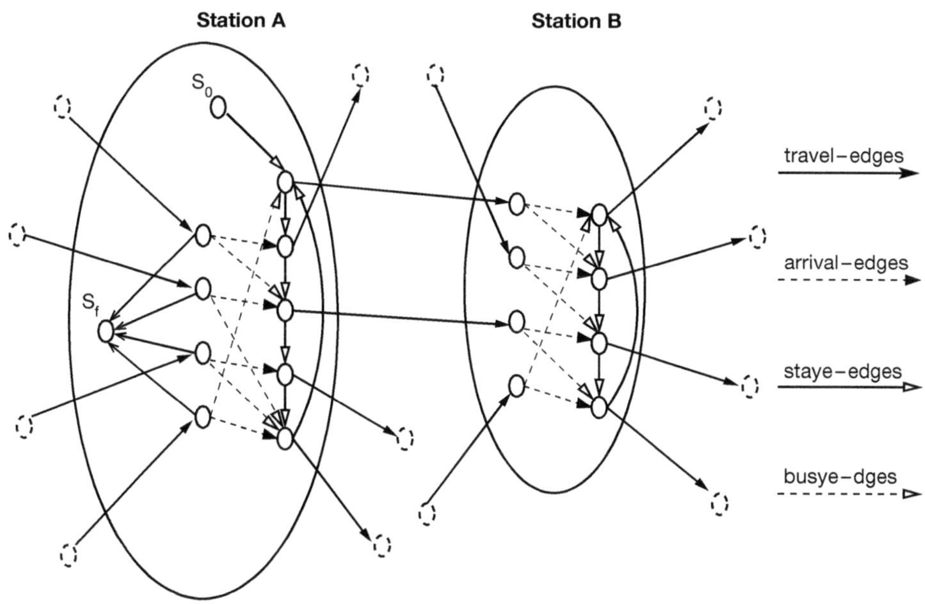

Figure 4.1. An example of two stations in the time-expanded graph, where A is the starting station

equal to the *cycle-difference* (t_0, t_{d_0}). In addition, we introduce a *sink node* s_f in the same station and we connect each arrival node of σ_0 with a zero-cost edge (called a *sink edge*) to s_f.

4.3 Methods for solving the RTSP

Four methods have been proposed for solving the RTSP, three of which are exact methods and one heuristic, which we will describe in the remainder of this chapter.

4.3.1 The size reduction method through shortest paths

Hadjicharalambous et al. [67] presented a preprocessing algorithm to reduce the size of the time-expanded graph. The idea is to consider that, for each station $\sigma \in B$, a new sink-node s_σ is added and all arrival-nodes in σ are connected to s_σ with zero cost edges. Then, for all departure-nodes in $d \in \sigma$, the shortest paths to sink-nodes in all other stations in B are computed. For each path, an edge between d and the last arrival-node of that path is added to G. The edge costs equal the costs of the corresponding shortest path. Then, all sink-nodes and their incoming edges are removed again, as well as all stations not in B, along with their nodes. Arrival nodes, which are not used by any of the shortest paths, and all arrival edges are also removed.

After the size reduction procedure, G consists only of stations $\sigma \in B$, their arrival and departure-nodes, the edges connecting them, and the newly added shortest path edges.

An integer linear programming formulation

The objective of the RTSP is to find a tour from node s_0 to node s_f that passes through each station $b \in B$, with minimum total cost.

In order to model the RTSP problem as an Integer Linear Program (ILP), we introduce for each edge (u, v) a variable $x_{(u,v)} \in \mathbb{N}_0$ that indicates the number of times the salesman uses the edge (u, v). If $c(u, v)$ denotes the cost of edge (u, v), then the ILP becomes:

$$\min \sum_{(u,v)\in E} c(u, v)x_{(u,v)}, \tag{4.1}$$

$$s.t. \quad \sum_{(v,u)\in E} x_{(v,u)} - \sum_{(u,v)\in E} x_{(u,v)} = 0, \qquad \forall u \in V \setminus \{s_0, s_f\} \tag{4.2}$$

$$x_{(s_0,d_0)} = \sum_{(v,s_f)\in E} x_{(v,s_f)} = 1 \tag{4.3}$$

$$\sum_{e\in Busy(b)} x_e \geq 1, \qquad \forall b \in B \tag{4.4}$$

$$x_e \in \mathbb{N}_0, \qquad \forall e \in E. \tag{4.5}$$

The constraints (4.2) and (4.3) are the flow conservation constraints that form a path from node s_0 to node s_f , whereas the constraints (4.4) ensure that the salesman spends the required time at each station $b \in B$.

The problem with the formulation so far is that it permits feasible solutions that contain cycles disjoint from the rest of the path (subtours). To deal with this problem, we have to add more constraints which will force the solution to not use any subtours.

Suppose that the selected stations are numbered from 0 to $|B| - 1$. For each such station i, we introduce a new node f_i called the *sink node* of that station. Moreover, we create a copy $\bar{d}^i_k$ for the k-th departure node d^i_k of station i that has one or more incoming busy edges. We connect this new node to the original node with a zero cost edge. All busy edges now point to $\bar{d}^i_k$ instead of to d^i_k, whereas all other edges remain unchanged. We also add an edge $(\bar{d}^i_k, f_i)$, for all k, such that $\bar{d}^i_k \in V$. An example is given in Figure 4.2.

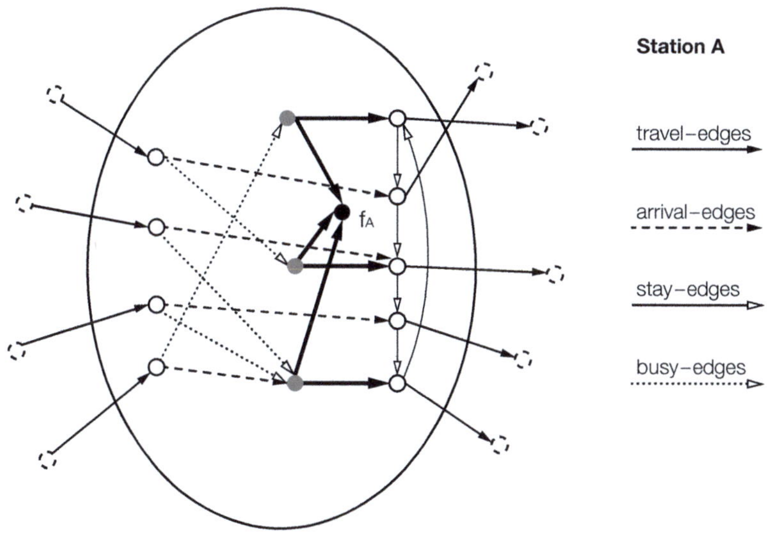

Figure 4.2. Station A from Figure 4.1, after introducing the new nodes and edges used for the subtour elimination constraints. For simplicity, nodes s_0 and s_f are not shown. The gray nodes are copies of the departure nodes with incoming busy edges, whereas the black node is the sink node of A. The thick black edges are the edges that have been newly introduced to the graph.

We now introduce, for each edge $e \in E$, a new set of variables y_e^i, $0 \leq i \leq |B|$ and the following constraints:

$$\sum_{(v,u)\in E} y^i_{(v,u)} - \sum_{(u,v)\in E} y^i_{(u,v)} = 0, \qquad \forall u \in V \setminus \{s_0, f_i\} \tag{4.6}$$

$$y^i_{(s_0,d_0)} = \sum_{(v,f_i)\in E} y^i_{(v,f_i)} = 1 \tag{4.7}$$

$$y^i_e \leq x_e, \qquad \forall e \in E \setminus \{(u, f_i) \in E\}. \tag{4.8}$$

The above constraints form a multicommodity flow problem, introducing a commodity for each selected station and asking for one unit of flow of each commodity $i \in [0, |B| - 1]$ to be routed from the source node to the corresponding selected station's sink node f_i. An additional condition imposed on the y-variables is that the flow y_e^i of commodity i on any edge $e \in E$ cannot have a value greater than x_e. This means that, if the x-flow is zero on some edge e, the y-flows will all be zero on that edge, too.

The constraints (4.6)–(4.8) force the x variables to be assigned appropriately, so that, for each selected station, there exists a path from so to some departure node in that station, using only edges for which the corresponding x variables are non-zero.

Since now the only way to reach f_i is through the busy edges of station i, the flow modeled by the x variables is forced to use some busy edge for each selected station, which makes the constraints (4.4) redundant.

The reason that the variables $x_{(u,v)}$, $(u, v) \in E$, are not restricted to be 0-1 variables is the fact that the salesman is allowed to pass through the same station σ more than one time, regardless of whether σ belongs to S, or to B, or is the starting station σ_0.

Knowing the values of the variables $x_{(u,v)}$, we can easily retrieve the path, using the Flow Decomposition Theorem (see e.g., [1, Chap. 3]). Let n (respectively m) be the number of nodes (respectively edges) of the time-expanded graph G. We decompose the flows into at most m simple cycles and one simple path (from s_0 to s_f) in time $O(nm)$. We can then construct the minimum cost tour, as a linked list of edges, as follows. Initially, we set the tour equal to the path, and then we iteratively expand it, by merging it with a cycle that has a common edge. This can be done in a time linearly related to the length of the final tour, which is at most $O(nm)$.

Size reduction through shortest paths

The time-expanded graph can be rather large. In this section, we present a method that reduces the size of the graph, which in turn may be beneficial in the case where the salesman wishes to visit a relatively small number of stations in comparison with the total number of stations. We reduce the size of the problem by transforming the time-expanded graph G to a new graph G_{sh}, called the *reduced time-expanded graph*. The graph G_{sh} can be constructed by precomputing the shortest paths among the stations that belong to B, as follows.

Let σ_0 again denote the starting station. For each station σ in $B \cup \{\sigma_0\}$ we introduce a sink-node s_σ and connect each arrival node of σ to s_σ with a zero cost edge. Then, for each departure node d of $\sigma \in B \cup \sigma_0$, we compute a shortest path to the sink-nodes of every other station in $B \cup \sigma_0$. If such a shortest path does not pass through some other station in $B \cup \sigma_0$, then we insert a *shortest path edge* from d to the last arrival node of that path. The cost of this edge is equal to the cost of the corresponding shortest path.

We can now transform G in the following way. We first remove the sink-nodes (and their incoming edges) that were previously introduced. Next, we delete from G all nodes and edges that belong to stations not in $B \cup \sigma_0$. The remaining arrival nodes (that belong to some station $\sigma \in B \cup \sigma_0$), that were not used by any of the shortest paths previously computed, are also deleted, together with the remaining travel-edges. This procedure results in the reduced time-expanded graph G_{sh}.

Implementation details and experiments

In order to speed-up the computations (by trying to reduce the number of variables used by the integer programs) when not using the shortest path reduction, we try to eliminate the unnecessary nodes of the graph. To be more precise, we remove all arrival nodes of all stations $\sigma \notin B$. These nodes have a single incoming edge (the one from the corresponding departure node), and a single out-going edge (to some departure node of the same station), whereas there are no busy edges starting from them. Therefore, we can eliminate these nodes (as well as their adjacent edges), and introduce a new virtual edge connecting the departure node that was the source of the incoming edge to the departure node that was the target of the outgoing edge, with a cost of the sum of the costs of the two eliminated edges. In this way, we can reduce the number of edges and nodes in the graph and, consequently, the number of variables and constraints in the integer program.

The construction of the graphs is based both on synthetic as well as on real-world data. For the synthetic case, we have considered grid graphs (with nodes representing stations). Each node has connections (in both directions) with all of its neighboring nodes, i.e., the stations located immediately next to it in its row or column in the grid.

Note that the number of connections is equal to half the number of nodes, since two nodes form one connection, and the number of edges is twice the number of nodes. The number of edges does not include the incoming edges of s_f and the outgoing edge of s_0 or any other artificial nodes or edges.

The connections among the stations were placed at random among neighboring stations, such that there is at least one connection in every pair of neighboring stations (for both directions) and the average number of elementary connections for each station is 10. The time-differences between the departure and arrival times of each elementary connection are independent uniform random variables, chosen in the interval [20, 60] (representing minutes), while the departure times are random variables in the time interval between the earliest and the latest possible departure time. We have created graphs with a number of stations varying from 20 to 80.

The real-world data represent parts of the railroad network of the Netherlands. The first data set, called *nd_ic*, contains the Intercity train connections among the larger cities in the Netherlands, which only stop at the main train stations, and are thus considered faster than normal trains. These trains operate at least every half an hour, while the number of stations is equal to 21. The second real-world data set, *nd_loc*, contains the schedules of trains that connect cities in only one region, including some main stations, whereas trains stop at each intermediate station between two main ones. In this case, the total number of stations is 23.

The characteristics of all graphs we used, for both real and synthetic data, are shown in Table 4.1.

For each data set, several instances of each problem were created, varying the number $|B|$ of selected stations, i.e., the set of stations that the salesman must visit. For both graphs based on real and synthetic data, we have used two values for $|B|$, namely 5 and 10. Note that $|B|$ does not contain the starting station. Because of that, when $|B|$ is set to the total number of stations, the actual value that will be used is $|B| - 1$, since one station must be the starting one.

Table 4.1. Graph parameters for the time expanded graph in each data set.

Data Set	No. of Stations	No. of Nodes	No. of Edges	No. of Connections	Conn/ Stations
nd_ic	21	684	1 368	342	16.29
nd_loc	23	778	1 556	389	16.91
synthetic 1	20	400	800	200	10
synthetic 2	30	600	1 200	300	10
synthetic 3	40	800	1 600	400	10
synthetic 4	50	1 000	2 000	500	10
synthetic 5	60	1 200	2 400	600	10
synthetic 6	70	1 400	2 800	700	10
synthetic 7	80	1 600	3 200	800	10

For each combination of data set and value of B, we have randomly selected the stations that belong to B, independently from each other. The selection of stations has been repeated many times, and the mean values among all corresponding instances were computed. For each instance we have created, the corresponding integer program was given as input to GLPSOL v4.6 (GNU Linear Programming Kit LP/MIP Solver, Version 4.6). The time needed by GLPSOL to find the optimum solution for each case has been measured.

We have tested both the original version (with our modifications) of the time-expended graph, as well as the reduced version, based on precomputed shortest paths. The time for this precomputation also has been measured.

Computational results and conclusions

The results of the experiments are reported in Tables 4.2 and 4.3 (a graphical comparison is illustrated in Figure 4.3).

The measured values are the average values for 50 different sets of selected stations. The standard deviation of the running times, provided by GLPSOL for instances of the same parameters, was large, showing a large dependence on the graph structure.

It can be easily seen that, when the original graphs were used, the value of $|B|$ greatly impacts the running time. The larger the $|B|$, the larger the running time.

The size reduction approach, based on precomputed shortest paths, results in much smaller graphs than the original one. Table 4.2 shows the characteristics of the reduced graphs that have resulted from the use of the size reduction technique.

Additionally, it is clear from Table 4.3 that the size reduction approach results in a great speedup, w.r.t. the time achieved, if we take into account the original time-expanded graph to find an optimal (or close to optimal) solution. Moreover, the smaller the value of $|B|$, the larger the speedup.

Table 4.2. Graph parameters for the reduced graphs for all data sets (average values).

(a) Real graphs

		$\lvert\mathcal{B}\rvert = 5$		$\lvert\mathcal{B}\rvert = 10$	
Data Set	$\lvert S\rvert$	Nodes	Edges	Nodes	Edges
nd_loc	23	209.4	415.2	343.9	698.1
nd_ic	21	181.6	385	303.6	708.7

(b) Synthetic graphs

		$\lvert\mathcal{B}\rvert = 5$		$\lvert\mathcal{B}\rvert = 10$	
Data Set	$\lvert S\rvert$	Nodes	Edges	Nodes	Edges
1	20	95.75	209.35	197.7	420.3
2	30	91.3	204.7	181.1	430.4
3	40	95.7	213.0	188.3	437.8
4	50	92.7	205.2	184.6	440.3
5	60	94.8	216.2	181.9	456.8
6	70	93.0	215.4	184.0	491.7
7	80	91.2	207.2	174.6	457.7

Table 4.3. Results for real and synthetic data sets. Time is measured in seconds.

(a) Running times for real data sets

Reduced graphs

$\lvert B\rvert$	5	10
nd_ic	319.0	9 111.9
nd_loc	29.1	6 942.6

(b) Running times for the synthetic data sets

Original graphs

$\lvert S\rvert$	20	30	40
$\lvert B\rvert$=5	13.12	32.24	72.06
$\lvert B\rvert$=10	781.12	1 287.00	16 239.80

Reduced graphs

$\lvert S\rvert$	20	30	40	50	60	70	80
$\lvert B\rvert$=5	1.12	1.12	1.50	0.80	1.45	1.30	1.00
$\lvert B\rvert$=10	214.76	369.59	244.18	181.85	257.96	431.80	233.26

(c) Shortest path computation times for synthetic data sets

$\lvert S\rvert$	20	30	40	50	60	70	80
$\lvert B\rvert$=5	0.13	0.18	0.24	0.29	0.35	0.41	0.46
$\lvert B\rvert$=10	0.28	0.38	0.48	0.59	0.69	0.79	0.88

(a)

(b)

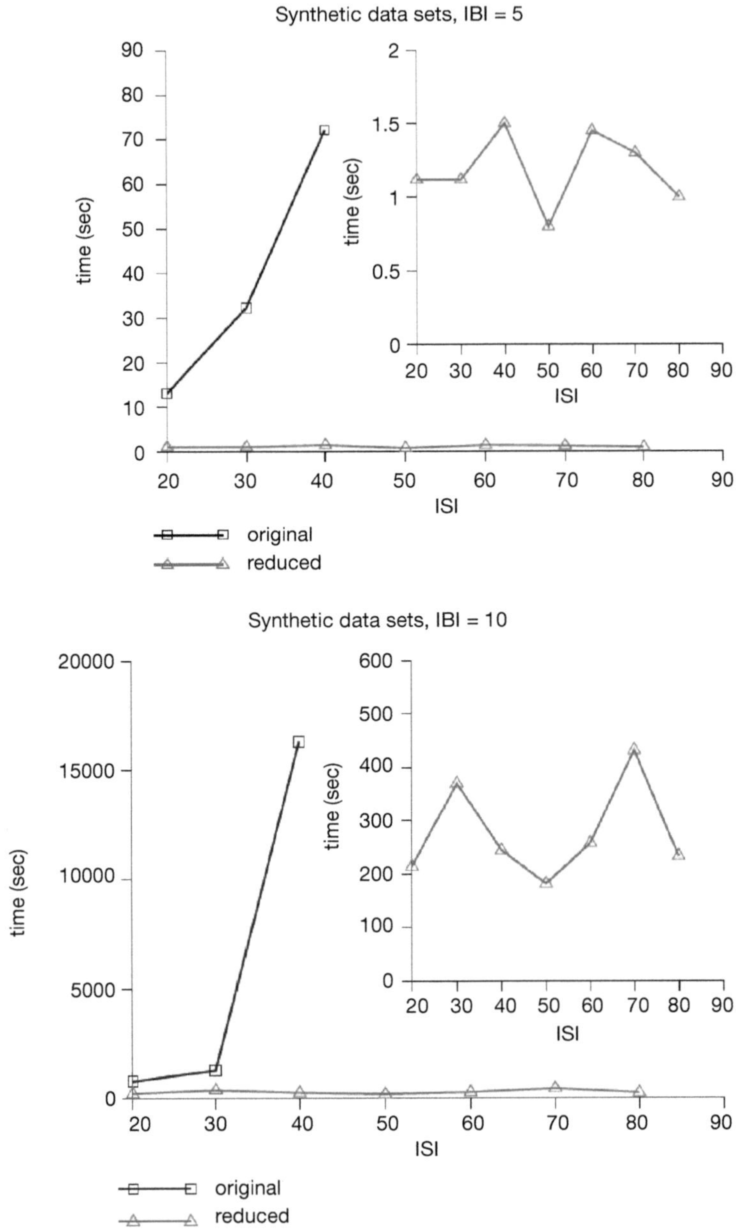

Figure 4.3. Graphical comparison of running times for synthetic data sets.

For a fixed size of B the running time seems to grow rapidly with increasing S, if the shortest path technique is not being used, in contrast with the case where it is used. In this case, when we consider synthetic data, the running times fluctuate. Nevertheless, it is quite clear that the time depends mainly on the size of B, rather than on the size of S.

4.3.2 A cutting plane approach for the RTSP

On the basis of the time-expanded graph, we can see that the RTSP reduces to finding a Hamiltonian tour H of minimum cost in the subgraph of G, induced by S, with $S \subseteq V$, such that S contains exactly two nodes from every cluster. This leads to a variant of the generalized traveling salesman problem which we will denote by 2-GTSP. Clearly, a solution to 2-GTSP is a solution to the RTSP.

In the following we present an efficient cutting plane approach for solving the RTSP (see [157]).

Definition of the 2-GTSP

The *2-generalized traveling salesman problem* (2-GTSP) aims to find a minimum-cost tour H, spanning a subset of nodes, such that H contains exactly two nodes from each cluster $V_i, i \in \{1, \dots, m\}$. The problem involves two related decisions:

- Choosing a node subset $S \subseteq V$, such that $|S \cap V_k| = 2$, for all $k = 1, \dots, m$.
- Finding a minimum cost Hamiltonian cycle in the subgraph of G, induced by S.

We will call such a cycle a *2-generalized Hamiltonian tour*. An example of a 2-generalized Hamiltonian tour, for a graph with nodes partitioned into 6 clusters, is presented in the next figure.

Integer programming formulations of the 2-GTSP

In order to formulate the 2-GTSP as an integer program, we introduce the following binary variables:

$$x_e = \begin{cases} 1 & \text{if edge } \; e = \{i, j\} \in E \; \text{ is included in the selected subgraph} \\ 0 & \text{otherwise} \end{cases},$$

$$z_i = \begin{cases} 1 & \text{if node } i \text{ is included in the selected subgraph} \\ 0 & \text{otherwise} \end{cases}.$$

For $F \subseteq E$ and $S \subseteq V$, let $E(S) = \{e = \{i, j\} \in E \mid i, j \in S\}$, $x(F) = \sum_{e \in F} x_e$ and $z(S) = \sum_{i \in S} z_i$. Also, let $x(V_k, V_k) = \sum_{i,j \in V_k, i<j} x_{ij}$, where $k \in \{1, \dots, m\}$.

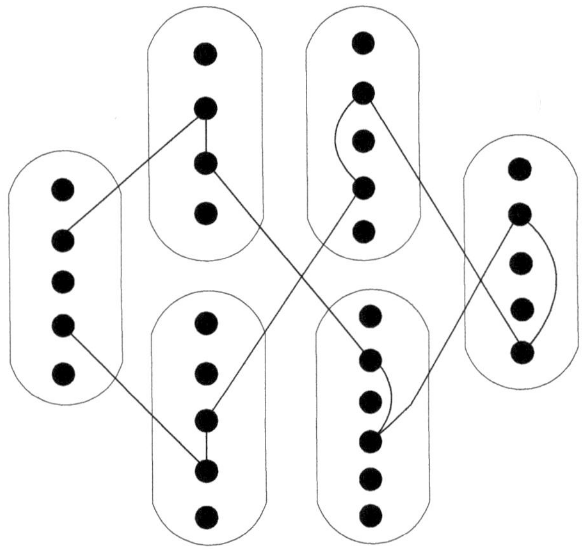

Figure 4.4. Example of a 2-generalized Hamiltonian tour.

A feasible solution to the 2-GTSP may be seen as a cycle free subgraph with $2m - 1$ edges, connecting all clusters, such that exactly two nodes from each cluster are selected .

Therefore, the 2-GTSP can be formulated as the following 0-1 integer programming problem:

$$
\begin{aligned}
&\min \sum_{e \in E} c_e x_e, && \\
&s.t.\quad z(V_k) = 2, && \forall\, k \in \{1, \dots, m\}, && (4.9)\\
&x(\delta(i)) = 2z_i, && \forall\, i \in V \setminus V_1, && (4.10)\\
&x(E) = 2m - 1, && && (4.11)\\
&x(V_k, V_k) = 1, && \forall\, k \in \{2, \dots, m\}, && (4.12)\\
&x(E(S)) \le 2r - 1, && \forall\, S = \cup_{i=1}^{r} V_{l_i}, 2 \le r \le m - 1, && (4.13)\\
&x_e \in \{0, 1\}, && \forall\, e \in E, &&\\
&z_i \in \{0, 1\}, && \forall\, i \in V. &&
\end{aligned}
$$

In the above formulation, constraint (4.9) guarantees that we select exactly two nodes from every cluster. The constraints (4.10) require the number of edges incident with a node i to be either 2 (if node i is visited) or 0 (otherwise). Constraint (4.11) guarantees that the selected subgraph has $2m - 1$ edges, constraints (4.12) guarantee that we select one edge from each cluster (except the starting cluster). Finally, the constraints (4.13) eliminate all cycles connecting at most $m - 1$ clusters.

Replacing the subtour elimination constraints (4.13) with connectivity constraints, we obtain the so-called *generalized cut-set formulation*:

$$\begin{aligned}
\min \sum_{e \in E} c_e x_e, & \\
s.t. \quad (1), (3) - (5) \quad \text{and} & \\
x(\delta(S)) \geq 2(z_i + z_j - 1), \quad & \forall\, S \subset V, \\
& \text{with} \quad 2 \leq |S| \leq m - 1 \\
& \text{and} \quad \forall\, i \in S, \quad j \in V \setminus S, \qquad (4.14) \\
x_e, z_i \in \{0, 1\}, \quad & \forall\, e \in E, \quad i \in V.
\end{aligned}$$

where, as usual for $S \subseteq V$, we denoted the cut-set by $\delta(S)$.

In the above formulation, the constraints (4.14) are the connectivity constraints, which state that each cut separating two visited nodes (i and j) must be crossed at least twice.

In the addition to the constraints that appear in the previous formulations, we consider the following constraints specific to the railway traveling salesman problem:

$$t_d z_d - t_a z_a \geq t_k, \quad \forall\, a, d \in V_k, \quad 2 \leq k \leq m. \tag{4.15}$$

The above constraints state that the difference between departure and arrival times must be at least a specified period t_h (depending on the city). This means that the traveling salesman has to stay in each city for some time, in order to finish business. If the difference is too small, the salesman may fail to achieve this. If, on the other hand, the difference is too large, the waiting time at the station will be inconvenient.

The disadvantage of the described integer programming formulations is their exponential number of constraints (we have to choose subsets of V, constraints (4.13) and (4.14)). These constraints can be omitted and can then be generated, as needed, by a *separation algorithm*: one can start without constraints (4.13), solve the corresponding relaxation, and then generate subtour inequalities that are violated by the current solution. The separation algorithm for subtour constraints is based on network flow techniques. For further details see [43].

Solution procedure and computational results

We used the following *cutting plane algorithm*, in order to solve the 2-GTSP:

1. Let the integer programming (IP) formulation consist of the constraints (4.9)–(4.12), (4.15) and the integrality constraints.
2. Solve the IP and assume that the optimal solution consists of r subtours: $S_1, \dots, S_r$.
3. If $r = 1$, then STOP. The solution to the 2-GTSP is optimal. Otherwise, add the corresponding constraints that eliminate the cycles $S_1, \dots, S_r$ to the IP formulation and go to Step 2.

The algorithm was implemented in C and for each instance we have created the corresponding integer program, which we solved with CPLEX 6.5.

Test data for our algorithm are real networks from the Dutch railroad company *Nederlandse Spoorwegen*. The first three data sets contain the Intercity train connections among the larger cities in the Netherlands, stopping only at the main train stations, and thus are considered faster than the normal trains. These trains operate at least every half hour. The second real-world data set, contains the schedules of the Intercity trains and regional trains. The regional trains connect the cities in only one region, including some main stations and stop at each intermediate station between two main ones.

Some characteristics of the graphs used for the real-world data sets are displayed in the next table together with the computational results obtained using the cutting plane algorithm.

The first four columns in the table give the name and size of the problem as the number of stations, the number of nodes and the number of edges. The next two columns describe the cutting plane procedure and contain the lower bounds, obtained as a percentage of the optimal value of the RTSP (LB/OPT), and the computational times (CPU) for solving the RTSP to optimality.

Table 4.4. Computational results for solving the RTSP.

Pb. name	*No. stations*	*No. nodes*	*No. edges*	*LB*/OPT	*Sol. time*
NS1 (IC)	5	394	4 240	100	14.08 s
NS2 (IC)	7	674	9 754	100	64.57 s
NS3 (IC)	9	926	16 271	100	206.52 s
NS4 (IC+IR)	12	1 470	23 850	100	39.30 min
NS5 (IC+IR)	12	1 586	27 383	100	72.28 min
NS6 (IC+IR)	15	1 722	28 200	100	1.05 h
NS7 (IC+IR)	15	1 946	34 450	100	5.45 h
NS8 (IC+IR)	18	2 182	38 650	100	4.52 h

4.3.3 Solving the RTSP via a transformation into the classical TSP

As we have already seen, the RTSP is related to the GTSP. Due to the fact that there are several efficient algorithms for solving the GTSP, such as genetic algorithms [184, 188], branch-and-cut [43], the algorithms described in Section 3.5, or approaches that exploit the transformation of the GTSP into the TSP [8, 26],it seems a reasonable idea to transform the RTSP into TSP. Such an approach was described by Hu and Raidl [78].

Transformation of the RTSP to the asymmetric TSP

Applying ideas similar to those from Behzad et al. [8], who presented an efficient transformation of the GTSP into the classical TSP, Hu et al. proposed a scheme to transform the reduced RTSP into an asymmetric TSP, defined on a graph $G' = (V', E')$. The transformation works as follows: for each station $\sigma \in B$ in the original reduced graph, the following procedure is applied to obtain a corresponding new station $\sigma' \in V'$.

Let $a'_1, \dots, a'_k \in \sigma'$ be exact copies of arrival-nodes $a_1, \dots, a_k \in \sigma$, ordered according to their times. We connect them to a cycle with zero cost edges $(a'_1, a'_2), \dots, (a'_{k-1}, a'_k), (a'_k, a'_1)$. The same procedure is applied on the departure-nodes $d'_1, \dots, d'_l$, which get connected with zero cost edges $(d'_1, d'_2), \dots, (d'_{l-1}, d'_l)$, (d'_l, d'_1). Let pred(v) denote the cyclic predecessor of node v, according to these cycles. Every arrival-node a'_i, $i = 1, \dots, k$ is connected with every departure-node d'_j, $j = 1, \dots, l$. For the initial station, $c(a'_i, d'_j)$ is set to $t_{d'_j} - t_{d'_1}$, which does not depend on $t(a'_i)$.

The reason for choosing such a value is that the salesman has to end his tour at the initial station, but not at the initial time. So it only makes a difference when he starts the journey, thus

$$c(a'_i, d'_j) = t_{d'_j} - t_{d'_1},$$

where $t(d'_1)$ is the earliest possible departure time from the initial station. For all other stations in B, the costs are set to

$$c(a'_i, d'_j) = [1\,440 + t_{d'_j} - t_{\mathrm{pred}(a'_i)} - f_B(\sigma)] \mod 1\,440 + f_B(\sigma).$$

Note that, since not all original stay-edges are present, their costs are included in these new edges. For example, in Figure 4.5, going from a_1 to d_4 means taking the busy-edge (a_1, d_3) and the stay-edge (d_3, d_4). The costs are $t_{d_4} - t_{a_1}$, which equal the costs of the new edge (a'_3, d'_4).

Finally, the outgoing edges from d'_j, $j = 1, \dots, l$ were also adapted. For each original travel-edge (d_j, v) an edge $(\mathrm{pred}(d'_j), v)$ was introduced with cost

$$(\mathrm{pred}(d'_j), v) = c(d_j, v) + M,$$

where M must be a sufficiently large number, e.g. $\sum_{(u,v)\in E} c(u, v)$, so that no more than one travel-edge will be included in the optimal solution.

This transformation is illustrated in Figure 4.5 and additionally shows an example of how a route is adapted to the new graph. We may notice that the resulting graph is directed and, therefore, the TSP is asymmetric.

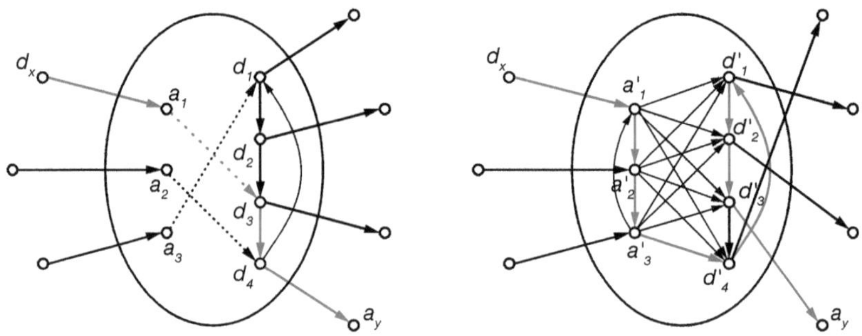

Figure 4.5. Transforming the original station σ into the station σ' in the antisymmetric TSP. The gray path marks the travel route from d_x to a_y in the original station σ and in the transformed station σ'.

Transformation of the RTSP to the symmetric TSP

Since the previous transformation results in an asymmetric TSP, Hu et al. [78] proposed an alternative procedure for generating a symmetric TSP, defined on the graph $G' = (V', E')$. Their approach follows the same basic ideas as the previous transformation, with the observation that the number of nodes must be doubled during the process.

Let $a'_1, \dots, a'_k \in \sigma'$ again be exact copies of the arrival-nodes $a_1, \dots, a_k \in \sigma$, ordered according to their arrival times. These nodes are duplicated once more, thus producing the nodes denoted by $a''_1, \dots, a''_k$. Then, these nodes are connected to a cycle by edges $(a'_2, a''_1), (a'_3, a''_2), \dots, (a'_k, a''_{k-1}), (a'_1, a''_k)$ with zero costs and the edges (a'_i, a''_i), $i = 1, \dots, k$ with high costs M. High costs on edges (a'_i, a''_i) cause as few of them as possible to be present in the solution.

For example, in Figure 4.6, when a travel-edge leads to a_1 there are two possible routes to go through $\{a'_1, a'_2, a'_3\} \cup \{a''_1, a''_2, a''_3\}$, one ending in a''_3 and one ending at a'_1. The latter one is cheaper, since the edge (a'_1, a''_1) with high costs is ignored.

The same procedure is applied to the nodes $d'_1, \dots, d'_l$ and their duplicates $d''_1, \dots, d''_l$. Then, all nodes $a''_i, i = 1, \dots, k$ are connected to nodes $d'_j, j = 1, \dots, l$ and the cost of the edges (a''_i, d'_l) are set to

$$c(a''_i, d'_j) = [1440 + t_{d'_j} - t_{a''_i} - f_B(\sigma)] \mod 1440 + f_B(\sigma).$$

Similar to the previous transformation, the costs of the original stay-edges are implicitly included in these edges.

Finally, for each original travel-edge (d_j, v) a new edge (d''_j, v) is added, with the cost given by

$$c(d''_j, v) = c(d_j, v) + M'.$$

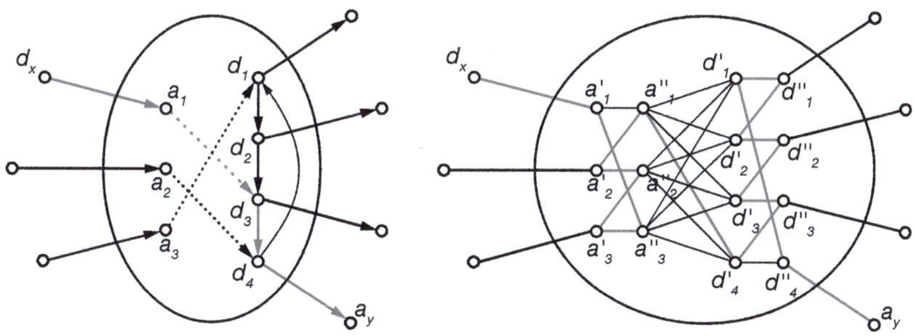

Figure 4.6. Transforming the original station σ into the station σ' in the symmetric TSP. The Gray path marks the travel route from d_x to a_y in the original station σ and in the transformed station σ'.

Since M was already used by edges (a'_i, a''_i) and (d'_j, d''_j) inside σ', an even larger value for M' was chosen, to ensure that only one travel-edge is included in the solution on G', namely $M' = M \cdot |B|$.

Computational results

In order to test the proposed TSP transformation schemes, the authors used instances based on two railway timetables, containing real-world data of train schedules in the Netherlands. The instance *nd_loc* contains 23 cities and represent trains in a local region and the instance *nd_ic* contains 21 larger cities in the Netherlands, which are connected by Intercity trains.

The requirements for the traveling salesman were generated at random. They were created by letting the salesman visit either 5 or 10 cities, and staying times were chosen between 10 and 240 minutes at each location. Since [46, 92] also used randomly generated data for the salesman with the same number of cities to be visited, the results should be at least roughly comparable. However, the staying times were not mentioned.

The time-expanded digraph, based on both timetables, contains more than 2000 nodes and around 4000 edges (depending on the traveling person). After reducing the graph, the transformation procedure for generating a symmetric TSP instance was carried out, and it was solved with the branch-and-cut (B&C) algorithm from the Concorde library.

The experiments were performed on an Intel Core 2 Quad 2.4 GHz PC with 4 GB memory and B&C uses ILOG CPLEX 9.0 as solver.

Tables 4.5 and 4.6 show the performance of the Concorde branch-and-cut (B&C) algorithm, on the transformed instances. For each instance, 30 runs were performed, since B&C uses random seeds to decide its branching order, thus resulting in variable running times. Listed are the instance's indices, the number of cities the salesman

Table 4.5. Computational results for the instance *nd_loc*

Index	$\|B\|$	*No. nodes*	*Time*	*std. deviation*
1	5	348	1.14 s	0.35
2	5	514	1.02 s	0.12
3	5	514	0.98 s	0.07
4	5	688	7.14 s	3.42
5	5	852	2.65 s	0.45
6	5	762	2.41 s	0.34
7	5	1 312	56.81 s	22.89
8	5	782	4.03 s	1.20
9	5	854	2.84 s	0.73
10	5	456	0.84 s	0.11
11	10	1 786	15.48 s	6.57
12	10	1 562	8.67 s	4.04
13	10	1 960	36.29 s	13.39
14	10	1 260	5.17 s	1.07
15	10	1 410	7.89 s	3.03
16	10	1 478	219.67 s	267.30
17	10	2 276	52.95 s	13.69
18	10	1 748	24.64 s	13.40
19	10	2 252	20.77 s	4.88
20	10	1 342	6.20 s	2.98

must visit, the number of nodes in the TSP after graph reduction and transformation have been applied, the average running times needed for B&C to find and prove the optimal solutions, and the standard deviations the of running times.

For instances 12, 14, and 18 in Table 4.6, all 30 runs were terminated after 10 000 s when B&C still could not prove optimality. However, in all these cases, the remaining gap is less than 0.1%.

In analyzing the results, we observe large differences in running times between the instances, especially in the case of the *nd_ic* instance. Among instances with the same size of B, trips can contain a quite different number of nodes. This is due to the random selection of stations in B. Looking at the data from the railway timetables, some contain only a few arrival and departure-nodes, whereas others have a full lineup. This imbalance appears to a greater extent for timetable *nd_ic*, in which some stations contain as few as 2 nodes and others are as large as 254 nodes. This might be the reason why these instances seems to be harder to solve for Concorde B&C.

Table 4.6. Computational results for the instance *nd_ic*.

Index	$\|B\|$	*No. nodes*	*Time*	*std. deviation*
1	5	422	3.35 s	1.35
2	5	542	5.50 s	1.93
3	5	712	6.19 s	1.36
4	5	364	0.52 s	0.04
5	5	1 178	30.26 s	9.38
6	5	450	0.77 s	0.06
7	5	516	3.47 s	1.09
8	5	1 140	10.16 s	3.57
9	5	324	29.68 s	16.17
10	5	568	8.50 s	3.29
11	10	1 786	1 907.47 s	990.37
12	10	1 562	10 000.00 s	0.00
13	10	1 960	45.68 s	27.39
14	10	1 260	10 000.00 s	0.00
15	10	1 410	71.29 s	48.41
16	10	1 478	1 471.14 s	781.79
17	10	2 276	239.52 s	72.94
18	10	1 748	10 000.00 s	0.00
19	10	2 252	1 071.12 s	377.63
20	10	1 342	346.39 s	134.51

4.3.4 An ant-based heuristic for solving the RTSP

Due to the complexity of the RTSP, the problem cannot be solved by means of exact methods, within polynomially bounded computation times. Nevertheless, sub-optimal solutions are sometimes easy to find. Consequently, it seems reasonable to focus on metaheuristic algorithms, that can find near optimal solutions within a reasonable running time. Such an approach, based on ant colony optimization, was described by Pop et al. [156].

The ant colony algorithm for solving the RTSP

In [137], Pintea et al. proposed a new algorithm, based on the Ant Colony System, called the Reinforcing Ant Colony System (RACS), in order to solve the Generalized Traveling Salesman Problem.

For solving the RTSP, the RACS algorithm must be changed, following the characteristics of the railway problem. The salesman starts from a specific station (cluster), has to visit a given number of stations (cities), spends a minimum amount of time in each station, and must finally return to the starting station.

In the following, we describe the new algorithm called the Railway Ant Colony (RAC) algorithm for the Railway Traveling Salesman Problem.

In the RAC algorithm m, ants individually construct candidate solutions in an incremental fashion. The choice of the nest node is based on two main components, pheromone trails and a heuristic value called visibility.

The algorithm works similar to the case of the reinforcing ant colony system, described in Section 3.5.1 for solving the GTSP. We considered the same probability function that favors the selection of an edge with a high pheromone value, τ, and high visibility value, η and the same correction rules to update the trail intensity locally respectively globally.

At the outset, the ants are in their nests and will start to search for food within a specific area. An ant goes through a given number of clusters, B exactly once, and then returns to the nest. An ant stays in each cluster for a period of time, in order to find desirable food and after that, starting from another node, continues the tour. The choice of the next node is probabilistic. The more pheromone there is on a certain road (edge) the bigger the chance that this specific road will be taken. Therefore, the pheromone trails will guide the ants to the shorter path, which is a solution of the RTSP.

When each ant has completed a tour, the amounts of pheromone are updated. On every road (edge) some fraction of pheromone evaporates whilst en route. New pheromones are deposited in those that have been visited by at least one ant. All roads on the best tour so far get an extra amount of pheromone. This is called an elitist ant approach - the best ants reinforce their associated tour even more.

The process continues until a certain condition, such as a certain number of iterations, amount of CPU time or solution quality, has been achieved.

The Railways Ant Colony (RAC) algorithm can be stated as follow:

Algorithm 4.1 (*Railways Reinforcing System* algorithm for the *RTSP*).

For every edge (i, j) do $\tau_{ij}(0) = \tau_0$
  for $k = 1$ to m do
    place ant k on a *specified* chosen node from a *specified* cluster
  end for
  let T^+ be the shortest tour found and L^+ its length
  for $iter = 1$ to Nb_{iter}
    for $k = 1$ to m do
      build tour $T^k(t)$ by applying $|B| - 1$ *times*
        choose the next *arrival* node j from an unvisited cluster,

$$j = \begin{cases} \text{argmax}_{u \in J_i^k} [\tau_{iu}(t)]^\alpha [\eta_{iu}(t)]^\beta & \text{if } q \le q_0 \\ J & \text{if } q > q_0 \end{cases}$$

```
      where $J \in J_i^k$ is chosen with probability:
```

$$p^k{}_{iu}(t) = \frac{[\tau_{iu}(t)]^\alpha[\eta_{iu}(t)]^\beta}{\sum_{o\in J^k{}_i}[\tau_{io}(t)]^\alpha[\eta_{io}(t)]^\beta}$$

```
        and where i is the current starting node
        update pheromone trails by applying the local rule:
```

$$\tau_{ij}(t+1) = (1-\rho)\cdot\tau_{ij}(t) + \rho\cdot\tau_0$$

```
      end for
    for k = 1 to m do
      compute the length L^k(t) of the tour T^k(t)
    if an improved tour is found, then update T^k(t) and L^k(t)
    for every edge (i, j) ∈ T^+ do
      update pheromone trails by applying the global rule:
```

$$\tau_{ij}(t+1) = (1-\rho)\cdot\tau_{ij}(t) + \rho\cdot\Delta\tau_{ij}(t)$$

```
      where Δτ_ij(t) = 1/L^+
     end for
  end for
  Print the shortest tour T^+ and its length L^+
```

where by Nb_{iter} we denoted the number of iterations.

To evaluate the performance of the proposed algorithm, we compared our results with the results obtained by Hadjicharalambous et al. in [67]. The computational results are reported in the next section, where we also compare them with those obtained in the case of the Dynamic Railway Traveling Salesman Problem.

4.4 Dynamic Railway Traveling Salesman Problem

After an initial emphasis on static problems, part of our focus now shifts towards dynamic variants of combinatorial optimization problems. Recently, research has been done on *ACO* for dynamic problems, see [61, 62, 63, 108].

The *Dynamic Railway Traveling Salesman Problem*, denoted *DRTSP*, and introduced by Pop et al., is a variation on the *RTSP* in the sense that it is assumed that the distances between cities, determined as the travel times, are no longer fixed. In real life, This situation is encountered, when delays due to maintenance work, accidents, etc., may occur. Therefore, the travel times may vary.

In the definition given below, we verbally define what we mean when we talk about a static RTSP.

The Static Railway Traveling Salesman Problem

- All information relevant to the planning of the tour is assumed to be known by the salesman, before the process begins.
- Information relevant to planning of the tour does not change after the tour has been constructed.

The information that is assumed to be relevant includes all attributes of the customers in different cities, such as the geographical location of the customers and times spent within each of the visited cities in order carry out his/her business. Furthermore, system information as for example travel times of the trains between the cities must be known by the salesman.

The dynamic counterpart of the static RTSP, as defined in the previous definition, may then be formulated as:

The Dynamic Railway Traveling Salesman Problem

- Not all information relevant to the planning of the tour is assumed to be known by the salesman before the process begins.
- Information relevant to planning of the tour may change after the tour has been constructed.

We mention that several other dynamic variants for combinatorial optimization problems may be considered, such as variants resulting from insertion or deletion of cities, see [61, 63, 108].

4.4.1 Ant colony approach to the Dynamic RTSP

The combination of positive and negative reinforcement, as mentioned in the case of the static RTSP, works well for static problems. In the beginning, there is relatively much exploration. After a while all non-promising connections will be slowly excluded from the search space, because they do not get any positive reinforcement and the associated pheromones have evaporated over time.

In the dynamic case, solutions that are bad before a change in the environment occurs, may be good afterward. Now, if the ant system has converged to a state where those solutions are ignored, very promising connections will be lost and the result will be a suboptimal solution. We have considered two ways to overcome this effect:

- we use a technique called *shaking* in order to smooth all pheromone levels in a certain way. If the amount of pheromones on an edge becomes much higher than all the other edges, this edge will be always be chosen. This way, in the static case, it can be ensured that a good connection will always be followed, but it prevents ants from taking another connection if the good connection is blocked. The formula used in shaking is a new local update rule:

$$\tau_{ij}(t+1) = \tau_0 \cdot \left(1 + \log\left(\frac{\tau_{ij}(t)}{\tau_0}\right)\right). \tag{4.16}$$

This formula will cause pheromone values close to τ_0 to move a little towards τ_0 and causes higher values to move relatively closer to τ_0;

- the heuristic information is typically problem-dependent and static throughout each single run of the algorithm. But, by limiting the influence of the pheromone trails, one can easily avoid the relative differences between the pheromone trails from becoming too extreme during the run of the algorithm. To achieve this goal, our algorithm uses a minimum limit (e.g. τ_0) as a lower bound on the amount of pheromone on every edge in the dynamic case, as in [193]. This prevents the chance of an ant choosing a particular edge to approach zero beyond a certain point.

$$\textit{if} \quad \tau_{ij}(t) < \tau_0 \quad \textit{then} \quad \tau_{ij}(t+1) = \tau_0. \tag{4.17}$$

In the dynamic *RTSP*, at each iteration of the *RAC* algorithm, one station is blocked, which means that no trains are going in or out of that station.

The dynamic *Railways Ant Colony* algorithm can be stated as follows:

Algorithm 4.2 (*Dynamic Railways Ant Colony* algorithm for the *Dynamic RTSP*).

For every edge (i, j) do $\tau_{ij}(0) = \tau_0$
for $k = 1$ to m do
    place ant k on a *specified* chosen node from a *specified* cluster
let T^+ be the shortest tour found and L^+ its length
for $i = 1$ to N_{iter} do
    choose a cluster at random and set the cluster to visited
    for $k = 1$ to m do
        build tour $T^k(t)$ by applying $B - 1$ *times*
            choose the next *arrival* node j from an unvisited cluster (1)(2)
            update pheromone trails by applying the local rule (5)
        end for
    for $k = 1$ to m do
        compute the length $L^k(t)$ of the tour $T^k(t)$
    if an improved tour is found, then update $T^k(t)$ and $L^k(t)$
    for every edge $(i, j) \in T^+$ do
        update pheromone trails by applying the global rule (4)

if the amount of pheromone is lower than the lower-bounded value, then
apply the pheromone correction phase (6)
end for
end while
Print the shortest tour T^+ and its length L^+

4.4.2 Implementation details and computational results

The initial value of all pheromone trails, $\tau_0 = 0.1$. As in all other ant systems, the parameters chosen for the algorithm are critical. Currently, no mathematical analysis has been developed to give optimal parameter values in each situation. In our algorithm, for both static and dynamic RTSP, the values of the parameters were chosen as $\alpha = 1$, $\beta = 2$, $\rho = 0.001$, $q_0 = 0.5$. The number of ants considered is $m = 10$ for $B = 5$ and $m = 100$ for $B = 10$. Each ant makes 100 tours in order to find a good one.

The experiments were conducted on synthetic as well on real-world data. We used the same real-world data as in [67] representing part of the railroad network of the Netherlands.

For the synthetic case we consider the grid graphs constructed in [67], where each node of the graph has connections with all of its neighboring nodes, i.e., the stations located immediately next to it in its row or column in the grid, with the difference that in our case the number of stations varies between 20 to 200. The connections among the stations were randomly placed among neighboring stations, such that there is at least one connection for every pair of neighboring stations (for both directions) and the average number of elementary connections for each station is 10. The time differences between the departure and the arrival times of each elementary connection are independent, uniform, random variables, chosen in the interval [20, 60] (representing minutes), whereas the departure times are random variables from the time interval between the earliest and the latest possible departure time.

The characteristics of all new graphs that were used for the synthetic data, corresponding to a number of stations between 100 and 200, are shown in Table 4.7.

Table 4.7. Graph parameters for the time expanded graph in each data set

Data Set	Number of Stations	Number of Nodes	Number of Edges	Number of Connections	Conn/ Stations
syn. 8	100	2 000	4 000	1 000	10
syn. 9	125	2 500	5 000	1 250	10
syn. 10	150	3 000	6 000	1 500	10
syn. 11	175	3 500	7 000	1 750	10
syn. 12	200	4 000	8 000	2 000	10

For each data set, several problem instances were created, where the number $|B|$ of selected stations, i.e., stations that the salesman has to visit, has varied. For both graphs based on real and synthetic data, we used two values for $|B|$, namely 5 and 10. Note that $|B|$ does not contain the starting station. The stations that belong to $|B|$ were selected at random and independent of each other.

For each combination of data set and value of B, we have selected the stations that belong to B at random and independent of each other.

The station selections have been repeated many times, and the mean values over all corresponding instances have been computed.

Each of the instances we created was solved with the *RAC* algorithm for both the static and dynamic variant of the RTSP. The algorithm was written in Java. The solutions we obtained have been compared with the optimal solutions obtained Hadjicharalambous et al. using GLPSOL v.4.6 for solving the corresponding integer linear programs.

In Table 4.8, we compare the computational results for solving the *RTSP* using the *RAC* algorithm with the computational results of Hadjicharalambous et al. from [67], in the case of original graphs and reduced graphs. In addition, we present the results obtained in the case of the dynamic *RTSP*.

The first column in Table 4.8 contains the data sets used in our experiments. The following columns contain the number of selected stations (clusters) $|B|$ that must be included a tour, together with the corresponding running times. The first two rows, called nd_ic and nd_loc, are real data representing parts of of the rail network of the

Table 4.8. Computational results for the static and dynamic *RTSP* using the *RAC* algorithm.

Data set	$\|B\| = 5$				$\|B\| = 10$			
	RAC	*CPU* [67]	*CPU red.*	*DRTSP*	*RAC*	*CPU* [67]	*CPU red.*	*DRTSP*
nd_ic	16.60	–	29.1	22.36	374.28	–	6 942.6	375.04
nd_loc	18.68	–	319.0	22.53	677.36	–	9 111.9	390.71
syn. 1	5.43	13.12	1.12	4.33	78.04	781.12	214.76	49.1
syn. 2	5.51	32.24	1.12	5.46	99.95	1 287.00	369.59	50.4
syn. 3	8.38	72.06	1.50	6.67	119.54	16 239.80	214.18	79.98
syn. 4	6.47	–	0.80	7.87	132.66	–	181.85	105.9
syn. 5	8.11	–	1.45	9.14	132.66	–	257.96	160.1
syn. 6	8.74	–	1.30	10.32	189.80	–	431.80	168.02
syn. 7	9.95	–	1.00	11.48	196.76	–	233.26	229.82
syn. 8	34.04	–	–	32.27	499.90	–	–	501.82
syn. 9	40.56	–	–	41.08	613.42	–	–	611.37
syn. 10	48.20	–	–	45.63	746.09	–	–	720.29
syn. 11	54.48	–	–	53.20	845.28	–	–	860.45
syn. 12	61.02	–	–	61.87	986.46	–	–	991.77

Netherlands. The last rows represent the synthetic data constructed on grid graphs. In the table the sign '–' means that the corresponding information was not provided in [67].

For both graphs (based on real and synthetic data), we used two values for $|B|$, namely 5 and 10. For each combination of data set and value of B, we provided the running times obtained with the *RAC* algorithm, in comparison with the running times obtained by Hadjicharalambous et al. from [67], for the original and reduced graphs. The computational values are the result of an average of 50 successive runs of the *RAC* algorithm. The termination criterion is given by the total number of iterations, $N_{\text{iter}} = 2\,500$.

In the next figure, we present an example of *Railway Ant Colony (RAC)* average time values for real data, the *nd_ic* data file, starting from different clusters.

As we can see from Table 4.8 the *Railway Ant Colony* algorithm performs very well compared to the approach considered by Hadjicharalambous [67] using CPLEX for solving an integer programming formulation of the problem.

The *RAC* algorithm for the *Railway Traveling Salesman Problem* can be improved by using more appropriate values for the parameters. Also, an efficient combination of the *RAC* with other algorithms can potentially improve the results.

4.5 Notes

The Railway Traveling Salesman Problem (RTSP) is a practical extension of the classical traveling salesman problem that considers railway network and train schedules and is related to the Generalized Traveling Salesman Problem. It was introduced by Hadjicharalambous [67].

In this chapter, we have presented the following results concerning the RTSP:

- A model of the problem, based on the time-expanded digraph G, constructed from the timetable information introduced by Schulz et al. [182].
- Four methods for solving the RTSP: the size reduction method through shortest paths, a cutting plane approach, a method based on transformation of the RTSP into the classical TSP, and an ant-based heuristic.
- A variant of the RTSP called the dynamic railway traveling salesman problem, in which we assume that the distances between cities seen as travel times are no longer fixed.
- An ant colony approach for solving the dynamic railway traveling salesman problem.

The material presented in this chapter is mainly based on the results published by Hadjicharalambous [67], Hu et al. [78] and Pop et al. [155, 156, 157].

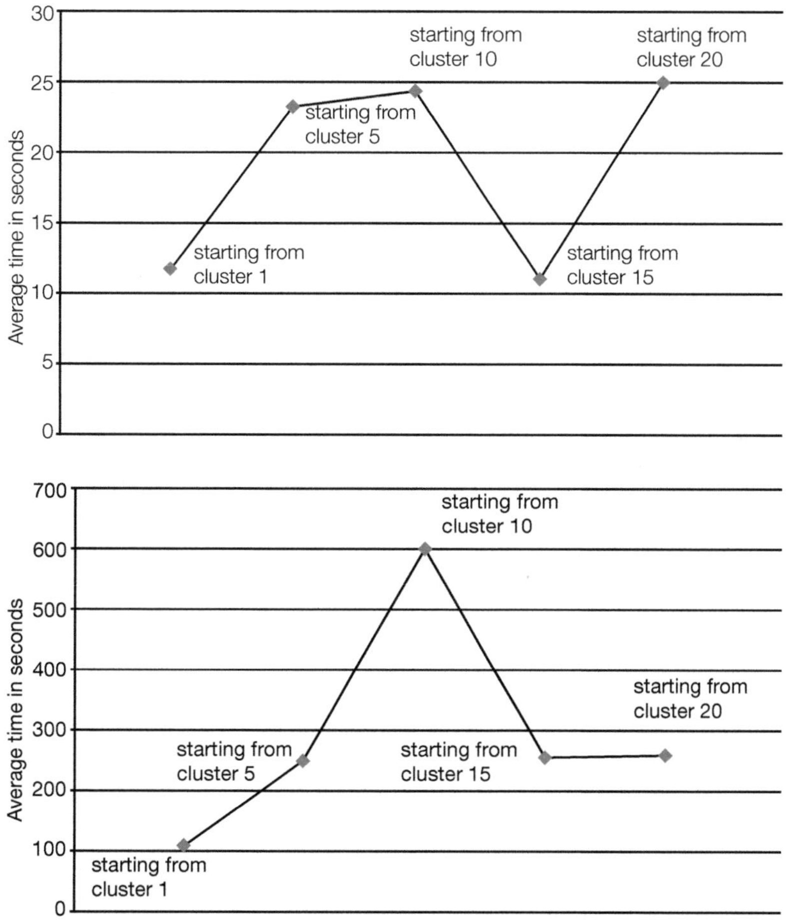

Figure 4.7. Railway Ant Colony (RAC) average time values for the *nd_ic* data file, starting from different clusters. The first set is for $B = 5$ and the second one is for $B = 10$.

Chapter 5

The Generalized Vehicle Routing Problem (GVRP)

Problems associated with determining optimal routes for vehicles from one or several depots to a set of locations/customers are known as Vehicle Routing Problems (VRPs). They have many practical applications in the field of distribution and logistics. A wide body of literature on the problem exists (for an extensive bibliography, see Laporte and Osman [104], Laporte [107] and the book edited by Ball et al. [6]).

Given a set of vehicles, a set of locations containing the depot location as well as the distance between each pair of locations, the VRP consists of finding the minimum cost tour for each vehicle, such that all locations are visited, and each vehicle returns to the depot. Because of the simplicity of the VRP, variations of the VRP, built on the basic VRP with extra features, proved most attractive to many researchers

- The Capacitated VRP [201], in which each vehicle has finite capacity and each location has a finite demand.
- The VRP with Time Windows [24, 190], in which there is a specified temporal window of opportunity in which to visit each location.
- The VRP with Multiple Depots [19], which generalizes the idea of a depot, in such a way that there are several depots from which each customer can be served.
- The Multi-Commodity VRP [180] in which each location has an associated demand for different commodities and each vehicle has a set of compartments in which only one commodity can be loaded. The problem then becomes that of deciding which commodities to place in which compartments, in order to minimize distance traveled.
- The Generalized Vehicle Routing Problem (GVRP) [48] is the problem of designing optimal delivery or collection routes, from a given depot to a number of predefined, mutually exclusive and exhaustive node-sets (clusters), subject to capacity restrictions. The GVRP can be viewed as a particular type of location-routing problem (see, e.g. Laporte [101], Nagy and Salhi [129]) for which several algorithms, mostly heuristics, exist.

In this section, we are concerned with the GVRP introduced by Ghiani and Improta [48]. They proposed solution procedure by transforming the GVRP into a capacitated arc routing problem, for which an exact algorithm and several approximate procedures are reported in literature.

5.1 Definition and complexity of the GVRP

Let $G = (V, A)$ be a directed graph, with $V = \{0, 1, 2, \ldots, n\}$ the set of vertices and $A = \{(i, j) \mid i, j \in V, i \neq j\}$ the set of arcs. A nonnegative cost c_{ij} is associated with each arc $(i, j) \in A$. The set of vertices (nodes) is partitioned into $m+1$ mutually exclusive nonempty subsets, called clusters, $V_0, V_1, \ldots, V_m$. The cluster V_0 has only one vertex 0, which represents the depot, and the remaining n nodes, belonging to the remaining m clusters, represent geographically dispersed customers. Each cluster V_k, $k \in \{1, \ldots, m\}$ has a nonnegative demand denoted by q_k and $q_0 = 0$. Each customer $i \in V \setminus \{0\}$ has a certain demand, denoted by d_i, and the total demand of each cluster can be satisfied via any of its nodes. There are K identical vehicles, each with a capacity Q.

The GVRP consists of finding the minimum total cost tours, starting and ending at the depot, such that each cluster is visited exactly once, the entering and leaving nodes of each cluster are the same and the sum of all the demands of any tour (route) does not exceed the capacity Q of the vehicle.

An illustrative scheme of the GVRP is shown in the next figure together with one feasible tour.

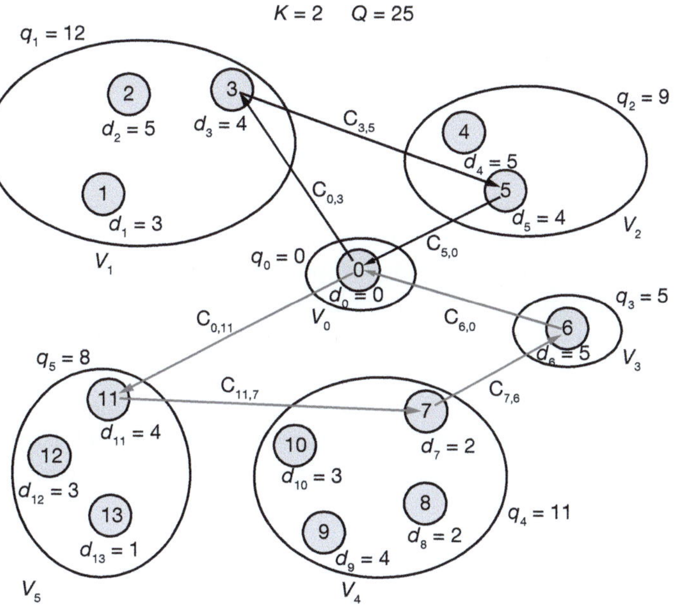

Figure 5.1. An example of a feasible solution of the GVRP.

The GVRP reduces to the classical Vehicle Routing Problem (VRP), when all the clusters are singletons, and to the Generalized Traveling Salesman Problem (GTSP), when $K = 1$ and $Q = \infty$.

The GVRP is NP-hard because it includes the generalized traveling salesman problem as a special case for $K = 1$ and $Q = \infty$.

Several real-world situations can be modeled as a GVRP. The post-box collection problem, described in Laporte et al. [102], becomes an asymmetric GVRP if more than one vehicle is required. Furthermore, the GVRP is able to model the distribution by sea of goods to a number of customers situated in an archipelago as in Philippines, New Zealand, Indonesia, Italy, Greece and Croatia. In this application, a number of potential harbors are selected for every island and a fleet of ships is required to visit exactly one harbor for every island.

Several applications of the GTSP (Laporte et al. [105]) may be extended naturally to the GVRP. In addition, several other situations can be modeled as a GVRP, including

- The Traveling Salesman Problem (TSP) with profits (Feillet et al. [36]).
- A number of Vehicle Routing Problem (VRP) extensions: the VRP with selective backhauls, the covering VRP, the periodic VRP, the capacitated general windy routing problem, etc.
- The design of tandem configurations for automated guided vehicles (Baldacci et al. [5]).

5.2 An efficient transformation of the GVRP into a capacitated arc routing problem

As we have seen in Section 3.2, transforming generalized network design problems into classical combinatorial optimization problems for which there exist efficient solution methods, seems to be a very promising approach.

In the case of the GVRP, Ghiani and Improta [48] showed that the problem can be transformed into a capacitated arc routing problem (CARP) and Baldacci et al. [5] proved that the reverse transformation is also valid.

Given an undirected graph $G = (V, E)$, a depot located at a vertex $v_0 \in V$ where a given fleet of vehicles are based, a subset $R \subseteq E$ of edges called *required* if they are serviced by a vehicle, a nonnegative cost c_e associated with each edge $e \in E$, and a positive demand d_e attached to each edge $e \in R$ then the CARP consists of finding a set of shortest tours such that:

- each route starts and ends at the depot;
- each required edge is traversed at least once and is serviced by exactly one vehicle;
- the total demand serviced by any vehicle does not exceed vehicle capacity Q.

CARP was introduced by Golden and Wong [55] and the problem finds many interesting applications including street garbage collection, postal delivery, winter gritting, routing of electric meter readers, etc. The same authors proved that the problem is strongly $\mathcal{NP}$-*hard*. However, there exist efficient polynomial algorithms for solving special classes of the CARP for special structures of graph, demand or costs, see Assad et al. [4]. Due to the complexity of the CARP, several approximation procedures (see Assad and Golden [3], Eiselt et al. [33, 34])and heuristic algorithms (see Chapleau et al. [17], Hertz et al. [74], etc.) to solve the problem have been developed.

In what it follows, we will describe the efficient transformation of the GVRP into the CARP, introduced by Ghiani and Improta [48].

For the sake of simplicity, we assume that the GVRP is *symmetric* (i.e. $c_{ij} = c_{ji}$ for all customers $i, j \in V$). It is easy to show that the assumption mentioned above is not restrictive and every GVRP can be transformed into an instance of our problem.

Assad and Golden [3] and Eiselt [33] proved that, in the case of every feasible solution of the CARP, every vertex is entered by an even number of copies of edges. For the sake of simplicity, the parity condition is formulated using the word "edge" instead of "copy of an edge".

Let us denote by G_R the subgraph induced by edges in R, by N_h the set of nodes of the h-th connected component of G_R, $h = 1, \dots, m$ and by N_R the set of nodes v_i such that an edge (v_i, v_j) exists in R, i.e. $V_R = \cup_{h \in \{1,\dots,m\}} N_h$.

In this case, the transformation of the GVRP into a CARP is defined as follows:

1. A loop of "very expensive" required edges is created with the vertices $v^{(h)_1}, \dots, v^{(h)_{r(h)}}$ of the cluster V_h for $h = 1, \dots, m$, where $r(h)$ denotes the number of customers belonging to cluster V_h. The following edges $(v_i^{(h)}, v_{i \bmod (r(h))+1}^{(h)})$ with $i = 1, \dots, r(h)$ and cost equal to a large positive number M, are inserted into R. In particular, if $r(h) = 1$, a required loop $v^{(h)_1}, v^{(h)_1}$ is introduced. These loops create m connected components of required edges. The remaining edges are non-required.

2. The non-required intra-cluster edges are removed.

3. The cost of each inter-cluster edge is increased by $\frac{M}{2}$ if an endpoint coincides with the depot, or by M otherwise.

4. Strictly positive demands, having sum equal to d_h are assigned to the edges of the loop corresponding to V_h, $h = 1, \dots, m$.

In the next figure, we illustrate the described transformation procedure and show the corresponding CARP solution on the transformed graph belonging to the feasible solution of the GVRP shown in Figure 5.1.

The following results hold:

Proposition 5.1. *(Ghiani and Improrota [48]) The optimal solution to the CARP on the transformed graph corresponds to the GVRP optimal solution on the original graph.*

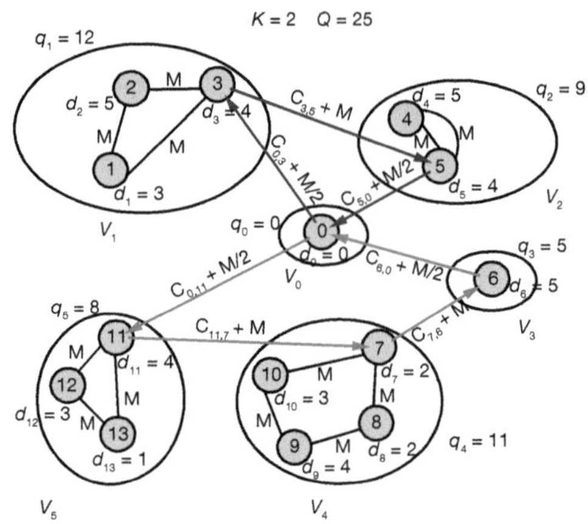

Figure 5.2. The corresponding CARP solution on the transformed graph.

Proposition 5.2. *(Ghiani and Improrota [48]) If M is set equal to $2 \cdot c_{\mathrm{MAX}}$, then the optimal CARP solution corresponds to the solution of the GVRP.*

Here, we denoted by c_{MAX} the largest intra-cluster edge cost in the GVRP graph.

5.3 Integer linear programming formulations of the GVRP

In this section, we first propose a general semi-closed formulation for the GVRP and then we specify its explicit forms and present four formulations of the problem (see as well [9, 94, 163]).

5.3.1 A general formulation

We define the related sets, associated with the GVRP, decision variables and parameters as follows

Sets:

$V = \{0, 1, 2, \dots, n\}$ the set of nodes corresponding to customers, where 0 represents the origin (depot).

$V_0, V_1, \dots, V_m$ a partition of V into mutually exclusive and exhaustive non-empty sub-sets, each of which represents a cluster of customers with $V_0 = \{0\}$ the origin (home city, depot).

$C = \{0, 1, 2, \dots, m\}$ the set of clusters.

$A = \{(i, j) \mid i, j \in V, i \neq j\}$ the set of arcs.

Decision Variables:
We define the following binary variables:

$$x_{ij} = \begin{cases} 1 & \text{if the arc } (i,j) \in A \text{ is included in the tour of a vehicle} \\ & i \in V_p,\ j \in V_r,\ p,r \in C \\ 0 & \text{otherwise} \end{cases},$$

$$w_{pr} = \begin{cases} 1 & \text{if there is a path from cluster } V_p \text{ to cluster } V_r,\ p,r \in C \\ 0 & \text{otherwise} \end{cases}.$$

Parameters:
c_{ij} is the cost of traveling from node i to node j, $i \neq j$, $i \in V_p$, $j \in V_r$, $p \neq r$, $p,r \in C$;

d_i is the demand of customer i, $i = 1, 2, \ldots, n$;

q_r is the demand of the cluster V_r, $q_r = \sum_{i \in V_r} d_i$, $r \in C$;

$K = \{1, \ldots, s\}$ is the set of the uniform vehicles (tours);

Q is the capacity of each vehicle.

Throughout this paper, we assume that $m \geq s$.

Constraints to the problem, are given below, under the corresponding header:

Cluster Degree Constraints
For each cluster, excluding V_0, there can be only a single outgoing arc to any other node belonging to other clusters. This condition is implied by the following constraints:

$$\sum_{i \in V_p} \sum_{j \in V \setminus V_p} x_{ij} = 1, \qquad p \neq 0, \quad p \in C \tag{5.1}$$

There can be only a single incoming arc to a cluster from any other node belonging to other clusters, excluding V_0. This condition is implied by the following constraints:

$$\sum_{i \in V \setminus V_p} \sum_{j \in V_p} x_{ij} = 1, \qquad p \neq 0, \quad p \in C \tag{5.2}$$

There should be s outgoing arcs from and at most s incoming arcs to the home city (origin), which is implied by the following constraints:

$$\sum_{i=1}^{n} x_{0i} = s \tag{5.3}$$

$$\sum_{i=1}^{n} x_{i0} = s \tag{5.4}$$

Cluster Connectivity Constraints
The incoming and outgoing nodes should be the same for each cluster, which is satisfied by:

$$\sum_{i \in V \setminus V_p} x_{ij} = \sum_{i \in V \setminus V_p} x_{ji}, \quad j \in V_p, \quad p \in C \tag{5.5}$$

Flows from cluster V_p to the cluster V_r are defined by w_{pr}. Thus, w_{pr} must be equal to the sum of x_{ij}'s from V_p to V_r. Hence, we write:

$$w_{pr} = \sum_{i \in V_p} \sum_{j \in V_r} x_{ij} \tag{5.6}$$

With the above definitions and constraints, a general integer programming formulation for the GVRP may be given as:

$$\begin{aligned} \text{minimize} \quad & \sum_{i=1}^{n} \sum_{i=1}^{n} x_{ij} \\ \text{subject to} \quad & (5.1)\text{– }(5.6) \\ & \text{Capacity bounding constraints} \qquad (5.7) \\ & \text{Subtour elimination constraints} \qquad (5.8) \\ & x_{ij} \in \{0, 1\}, \quad \forall (i, j) \in A. \end{aligned}$$

Note that w_{pr} will automatically be 0 or 1 due to the degree constraints given by (5.1), (5.2), (5.3) and (5.4). Therefore we do not need to impose integrality constraints on w_{pr}.

The constraints given in (5.1)–(5.5) are similar to those given by Noon and Bean [133] for the GTSP. The constraints given in (5.6) have been introduced by Kara and Bektaş [92].

Model formulations for the GVRP may differ from each other with respect to the "capacity bounding" and/or "subtour elimination" constraints given in (5.7) and (5.8) above.

In the case of an integer programming formulation, if the number of constraints of the formulation grows exponentially, with respect to the number of nodes of the graph, it is called the *exponential size formulation.* If the number of constraints grows polynomially, it is called the *polynomial size formulation.* In addition, if a formulation has additional auxiliary decision variables, it may be classified with respect to the new decision variables as, *node based formulation* if the additional variables are relative to the nodes of the graph, and *flow based formulation* if the new variables are relative to the arcs of the graph.

In the following subsections, we propose integer programming formulations for the GVRP, by defining new auxiliary variables.

5.3.2 A node based formulation

In addition to the decision variables defined before, let us define the following auxiliary variables:

Let u_p be the load of a vehicle just after leaving the cluster V_p (collection case) or the delivered amount of goods from a vehicle just after leaving the cluster V_p (delivery case), $p \neq 0$, $p \in C$.

Capacity Bounding Constraints

The following proposition gives capacity bounding constraints for a node based formulation.

Proposition 5.3. *The following inequalities are valid capacity bounding constraints for the GVRP:*

$$u_p - \sum_{r \in C, r \neq p} q_r w_{pr} \geq q_p \qquad p \neq 0, \quad p \in C \tag{5.9}$$

$$u_p + (Q - q_p) w_{0p} \leq Q \qquad p \neq 0, \quad p \in C \tag{5.10}$$

$$u_p + \sum_{r \in C, r \neq p} q_r w_{pr} \leq Q \qquad p \neq 0, \quad p \in C \tag{5.11}$$

where $q_0 = 0$.

Proof. Any cluster V_p may be on the first, last or an intermediate position in a vehicle tour, i.e., either $w_{0p} = 1$ and/or $w_{p0} = 1$, or there exist clusters V_s and V_t, such that $w_{sp} = w_{pt} = 1$.

If $w_{0p} = 1$ (or $w_{0p} = w_{p0} = 1$), the inequalities (5.10), (5.11) and (5.12) imply that $u_p = q_p$ and if $w_{p0} = 1$, the bounding constraints imply $q_p = u_p = Q$. If a vehicle comes from cluster V_s, goes to cluster V_p and then travels from V_p to the cluster V_t, which means that $w_{rp} = 0$ and $w_{pr} = 0$ for all $r \neq s$ and $r \neq t$, then, from the bounding constraint we obtain $q_p + q_s \leq u_p \leq Q - q_t$. □

Subtour elimination constraints

Formation of any subtour between clusters excluding V_0 is not allowed by the following constraint:

$$u_p - u_r + Qw_{pr} + (Q - q_p - q_r)w_{rp} \leq Q - q_r \quad p \neq r \neq 0, p, r \in C \quad (5.12)$$

where $w_{pr} = 0$, whenever $q_p + q_r > Q$ with $p \neq r$ and $p, r \in C$.

These constraints are adapted from the subtour elimination constrains of the capacitated VRP (see Desroshers and Laporte [23] and Kara et al. [93]).

Note that the constraints given in (5.10) guarantee that $u_p \geq 0$ for all $p \in C$. Therefore, we do not need nonnegativity constraints.

The first integer linear programming formulation of the GVRP is given by:

$$(NB) \quad \text{minimize} \quad \sum_{i=1}^{n} \sum_{i=1}^{n} x_{ij}$$

$$\text{subject to} \quad (5.1)\text{–}(5.6) \text{ and } (5.9)\text{–}(5.12),$$

$$x_{ij} \in \{0,1\}, \quad \forall (i,j) \in A.$$

where, $x_{ij} = 0$ whenever $i, j \in V_r$, $r \in K$ and $w_{pr} = 0$ whenever $q_p + q_r > Q$.

In the above formulation, the total number of binary variables is $n(n+1)$, i.e. $O(n^2)$. The variables w_{pr} automatically take on binary values. The respective numbers of constraints implied by (5.1), (5.2), (5.3), (5.4), (5.5), (5.6), (5.9), (5.10), (5.11) and (5.12) are $m, m, 1, 1, (n+1), m(m+1), m, m, m$ and $m(m-1)$, respectively, where n is the number of customers, m is the number of clusters and $m \leq n$. Hence the total number of constraints is $2m^2 + n + 5m + 3$, i.e. $O(m^2) = O(n^2)$.

(NB), the node based formulation of the GVRP, is structurally similar to the Kara-Bektaş [92] formulation, but it contains completely different bounding constraints. We will show below, that the node based formulation produces a stronger lower bound than the formulation described by Kara and Bektaş [92].

We denote the feasible set of the linear programming relaxation of the node based formulation of the GVRP by P_{NB}, where we replace the integrality constraints $x_{ij} \in \{0,1\}$ by $0 \leq x_{ij} \leq 1$. If we denote the feasible set of the linear programming relaxation of the Kara-Bektaş formulation by $P_{\text{K-B}}$, then the following result holds:

Proposition 5.4. $P_{\text{NB}} \subseteq P_{\text{K-B}}$.

Proof. The degree constraints, subtour elimination constraints and route continuity constraints of the two formulations are the same. They differ from each other with respect to their bounding constraints. As we discussed in the proof of the previous proposition, any cluster V_p may be on the first, or last or an intermediate position of a vehicle tour, i.e., either $w_{0p} = 1$ or $w_{p0} = 1$ or there exist some clusters V_s and V_t such that $w_{sp} = w_{pt} = 1$.

If $w_{0p} = 1$ or $w_{p0} = 1$, both formulations produce the same u_p values. If a vehicle travels from cluster V_s to V_p and then goes from V_p to cluster V_t, which means that $w_{rp} = 0$ and $w_{pr} = 0$ for all $r \neq s$ and $r \neq t$, we get, from the bounding constraints given in (5.10), (5.11) and (5.12)

$$q_p + q_s \leq u_p \leq Q - q_t.$$

For the same case, the formulation described by Kara and Bektaş [92] implies

$$q_p + \overline{q}_p \leq u_p \leq Q - \overline{q}_p,$$

where $\overline{q}_p = \min\{q_r \mid r \in C, r \neq p\}$.

Since $q_s \geq \overline{q}_p$ and $q_t \geq \overline{q}_p$, our formulation produces narrower bounds for auxiliary variables. Hence, every feasible solution of the proposed node based model is also a feasible solution to Kara-Bektaş model. □

5.3.3 Flow based formulations

In addition to the defined decision variables, let us define the following auxiliary variables:

y_{pr} is the amount of goods picked up (or delivered in the case of delivery) on the route of a vehicle, just after leaving the cluster V_p, if the vehicle goes from cluster V_p to cluster V_r, and zero otherwise.

Proposition 5.5. *The following inequalities are valid capacity bounding and subtour elimination constraints for the GVRP:*

$$y_{rp} \leq (Q - q_p)w_{rp} \qquad r \neq p, r, p \in C \tag{5.13}$$

$$y_{rp} \geq q_r w_{rp} \qquad r \neq p, r, p \in C \tag{5.14}$$

$$\sum_{p=1}^{m} y_{p0} = \sum_{p=1}^{m} q_p \tag{5.15}$$

$$\sum_{p=1}^{m} y_{rp} - \sum_{p=1}^{m} y_{pr} = q_r \qquad r \neq 0, r, p \in C \tag{5.16}$$

where $y_{0p} = 0$ for all $p \in C$ and $q_0 = 0$.

Proof. When any arc (r, s) is on the tour of a vehicle, i.e., $w_{rs} = 1$, the constraints given in (5.13) and (5.14) imply that

$$q_r \leq y_{rs} \leq Q - q_s.$$

If the arc (r, s) is the final arc of a vehicle tour, i.e., $w_{r0} = 1$, then $q_r \leq y_{rs} \leq Q$. These constraints also guarantee that $y_{rp} = 0$ if there is no flow from cluster V_r to cluster V_p. Therefore, the constraints (5.13) and (5.14) produce upper and lower bounds for the flow variables. □

The constraints given in (5.15) guarantee that the sum of the in flow to the origin is equal to the total demand.

Finally, the constraints in (5.16) are the classical conservation of flow equation, which also eliminates illegal tours, these constraints correspond to the subtour elimination constraints of the flow based formulation.

The second integer linear programming formulation of the GVRP is given by:

$$\begin{aligned}(FF)\quad & \text{minimize} && \sum_{i=1}^{n}\sum_{i=1}^{n} x_{ij}\\ & \text{subject to} && (5.1)\text{–}(5.6) \text{ and } (5.13)\text{–}(5.16),\\ & && x_{ij} \in \{0,1\}, \quad \forall (i,j) \in A\end{aligned}$$

where $x_{ij} = 0$ whenever $i, j \in V_r, r \in C$ and $w_{pr} = 0$ whenever $q_p + q_r > Q$. In addition, $y_{0p} = 0$ for all $p \in C$ and $q_0 = 0$. Note that the constraints given in (5.14) guarantee that $y_{pr} \geq 0$, for all $p \neq r,\ p, r \in C$.

In this formulation, as we observed in the previous formulation, the total number of binary variables is n(n+1), i.e. $O(n^2)$. The variables w_{pr} take on binary values automatically. The respective numbers of constraints implied by (5.1)–(5.6) and (5.13)–(5.16) are $m, m, 1, 1, (n+1), m(m+1), m(m+1), m(m+1), 1$ and m, respectively, where n is the number of customers, m is the number of clusters and $m \leq n$. Hence, the total number of constraints is $3m^2 + n + 6m + 4$, i.e., $O(m^2) \leq O(n^2)$.

An equivalent flow formulation, called the compact single-commodity formulation, was provided by Bektaş et al. [9]:

$$\begin{aligned}(F_1)\quad & \text{minimize} && \textstyle\sum_{(i,j)\in A} c_{ij} x_{ij} && &&\\ & \text{subject to} && x(\delta^+(V_k)) = 1 && \forall\, k \in C \setminus \{0\}, && (5.17)\\ & && x(\delta^-(V_k)) = 1 && \forall\, k \in C \setminus \{0\}, && (5.18)\\ & && x(\delta^+(V_0)) = s, && && (5.19)\\ & && x(\delta^+\{i\}) = x(\delta^-\{i\}) && \forall\, i \in V, && (5.20)\\ & && \sum_{p\in C\setminus\{r\}} y_{rp} = \sum_{p\in C\setminus\{r\}} y_{pr} + q_r && \forall\, r \in C \setminus \{0\}, && (5.21)\\ & && 0 \leq y_{pr} \leq Q x(V_p, V_r) && \forall\, p, r \in C,\ p \neq r, && (5.22)\\ & && x_{ij} \in \{0,1\} && \forall\, (i,j) \in A. &&\end{aligned}$$

In this formulation, constraints (5.17) and (5.18), corresponding to the restriction that each cluster is visited exactly once, are the same as constraints (5.1)–(5.2). Constraints (5.19), that ensure that exactly s vehicles depart from the depot, are the same as constraints (5.3) and (5.4). Constraints (5.20), which model the continuity of the route are the same as constraints (5.5), but written using different notations.

By taking into account the demands on clusters at the endpoints of the arc $(i, j) \in A$, the constraints (5.22) can be strengthen as follows:

$$q_p \leq y_{pr} \leq (Q - q_r)x(V_p, V_r) \quad \forall\ p, r \in C. \tag{5.23}$$

In model F_1 we will use constraints (5.23) instead of the constraints (5.22).

Taking into account the above observations and based on the fact that constraints (5.21) that model the increasing flow as the vehicle traverses the route are the same as constraints (5.16) and constraints (5.23) are the same as the constraints (5.13) and (5.14), we can conclude that our flow model and the compact single-commodity formulation described by Bektaş et al. [9] are equivalent.

A weaker formulation, described by Bektaş et al. [9], was obtained by considering flow variables for every pair of vertices, as opposed to every pair of clusters. Defining the continuous variables f_{ij}, with $\forall (i, j) \in A$ indicating the amount that the vehicle carries from vertex i to vertex j, the single-commodity formulation of the GVRP is:

$$
\begin{aligned}
(F_2) \quad & \text{minimize} \quad \sum_{(i,j)\in A} c_{ij}x_{ij} \\
& \text{subject to} \quad (5.17)\text{–}(5.20), && \qquad (5.24)\\
& f(\delta^+\{i\}) - f(\delta^-\{i\}) = \frac{1}{2}q_{(\alpha(i))}(x(\delta^-\{i\}) - x(\delta^+\{i\})) \\
& && \forall\ i \in V \setminus \{0\}, \qquad (5.25)\\
& 0 \leq f_{ij} \leq Qx_{ij} && \forall\ (i, j) \in A, \qquad (5.26)\\
& x_{ij} \in \{0, 1\} && \forall\ (i, j) \in A,
\end{aligned}
$$

where for each $i \in V$ by $\alpha(i)$ we denoted the index of the cluster that vertex i belongs to.

As in the case of the previous formulation, taking into consideration the demands on clusters at the endpoints of the arc (i, j), constraints (5.26) can be strengthened as follows:

$$q_{(\alpha(i))} \leq f_{ij} \leq (Q - q_{(\alpha(i))})x_{ij} \quad \forall\ (i, j) \in A. \tag{5.27}$$

The following proposition compares the feasible sets of the linear programming relaxations of the formulations FF, F_1 and F_2:

Proposition 5.6. $P_{FF} = P_{F_1} \subseteq P_{F_2}$.

Proof. The proof $P_{F_1} \subseteq P_{F_2}$ can be found in Bektaş et al. [9]. □

5.4 A numerical example

In this section, we optimally solve the numerical example described by Ghiani and Improta [48] using our novel integer programming formulations of the GVRP. The example was derived from a VRP instance, namely test problem 7, introduced by Araque et al. [2] and contains 50 vertices, 25 clusters and 4 vehicles. This instance is a randomly generated problem and was generated by randomly placing the customers on a square area. The vertex coordinates of the depot are $(50, 50)$. Those of the customers are:

1.	$(10,42)$	2.	$(23,6)$	3.	$(8,46)$	4.	$(51,29)$	5.	$(64,24)$
6.	$(28,6)$	7.	$(30,69)$	8.	$(87,3)$	9.	$(13,56)$	10.	$(80,87)$
11.	$(95,41)$	12.	$(92,21)$	13.	$(86,10)$	14.	$(39,45)$	15.	$(76,44)$
16.	$(30,34)$	17.	$(47,3)$	18.	$(84,70)$	19.	$(65,5)$	20.	$(98,18)$
21.	$(3,13)$	22.	$(52,98)$	23.	$(73,17)$	24.	$(48,82)$	25.	$(28,32)$
26.	$(16,86)$	27.	$(50,56)$	28.	$(53,72)$	29.	$(75,89)$	30.	$(41,38)$
31.	$(11,28)$	32.	$(76,57)$	33.	$(86,18)$	34.	$(34,19)$	35.	$(70,25)$
36.	$(79,50)$	37.	$(25,13)$	38.	$(55,94)$	39.	$(4,33)$	40.	$(8,66)$
41.	$(51,70)$	42.	$(3,22)$	43.	$(93,17)$	44.	$(96,45)$	45.	$(71,85)$
46.	$(39,32)$	47.	$(37,71)$	48.	$(79,45)$	49.	$(96,66)$	50.	$(69,92)$

The distances between customers are the Euclidean distances and have been rounded to obtain integer values. The set of vertices is partitioned into 25 clusters as follows:

$V_0 = \{0\}$	$V_1 = \{22, 38\}$	$V_2 = \{26\}$	$V_3 = \{24\}$
$V_4 = \{10, 29, 45, 50\}$	$V_5 = \{40\}$	$V_6 = \{7, 47\}$	$V_7 = \{28, 41\}$
$V_8 = \{18, 49\}$	$V_9 = \{1, 3, 9\}$	$V_{10} = \{14\}$	$V_{11} = \{32\}$
$V_{12} = \{21, 42, 39, 31\}$	$V_{13} = \{16, 25\}$	$V_{14} = \{30, 46\}$	$V_{15} = \{4\}$
$V_{16} = \{15, 36, 48\}$	$V_{17} = \{11, 44\}$	$V_{18} = \{2, 6, 37\}$	$V_{19} = \{34\}$
$V_{20} = \{5, 23, 35\}$	$V_{21} = \{12, 20, 33, 43\}$	$V_{22} = \{17\}$	$V_{23} = \{8, 13, 19\}$
$V_{24} = \{27\}$			

In Figure 5.3, we represent the vertices (customers) and their partitioning into clusters.

Each customer has a unit demand. The demand of a cluster is given by the cardinality of that cluster. The capacity of each vehicle is equal to 15.

The solution reported by Ghiani and Improta [48] was obtained by transforming the GVRP into a Capacitated Arc Routing Problem (CARP), which was then solved using the heuristic proposed by Hertz et al. [74], to yield the objective value 532.73. The same instance was solved to optimality by Kara and Bektaş [92], using their proposed integer programming formulation, by CPLEX 6.0 on a Pentium 1100 MHz PC with 1 GB RAM in 17 600.85 CPU seconds.

Using our proposed formulations for the GVRP, we solved the same instance to optimality using CPLEX 12.2, getting the same value as reported by Kara and Bektaş [92]. The required computational times were 1210.01 CPU seconds in the case of the node based formulation, and 1002.87 CPU seconds in the case of the flow based formulation. The computations were performed on an Intel Core 2 Duo 2.00 GHz with 2 GB RAM.

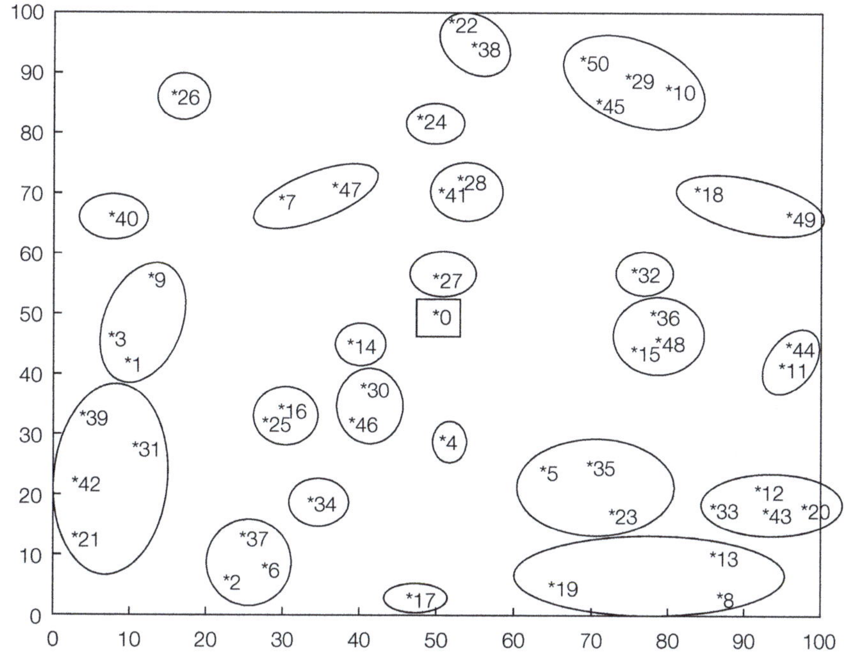

Figure 5.3. Representation of the 51 vertices and their partitioning into 25 clusters.

In the next figure we point out the optimal solution obtained for the instance described by Ghiani and Improta [48].

These results prove that we are able to cope with the considered instance and shows the superiority of our novel compact integer programming based models, compared to the existing ones.

5.5 Special cases of the proposed formulations

In this section, we show that one can obtain integer linear programming formulations for some routing problems as special cases of the proposed formulations for the GVRP.

5.5.1 The Generalized multiple Traveling Salesman Problem

In the GVRP, let the total demand of each cluster be equal to 1. There is no restriction on the capacity of the vehicles. We will call this version of the GVRP the Generalized Multiple Traveling Salesman Problem (GmTSP).

For GmTSP, the meaning of the auxiliary variables u_i's and y_{rp}'s and parameters of the model will be as follows:

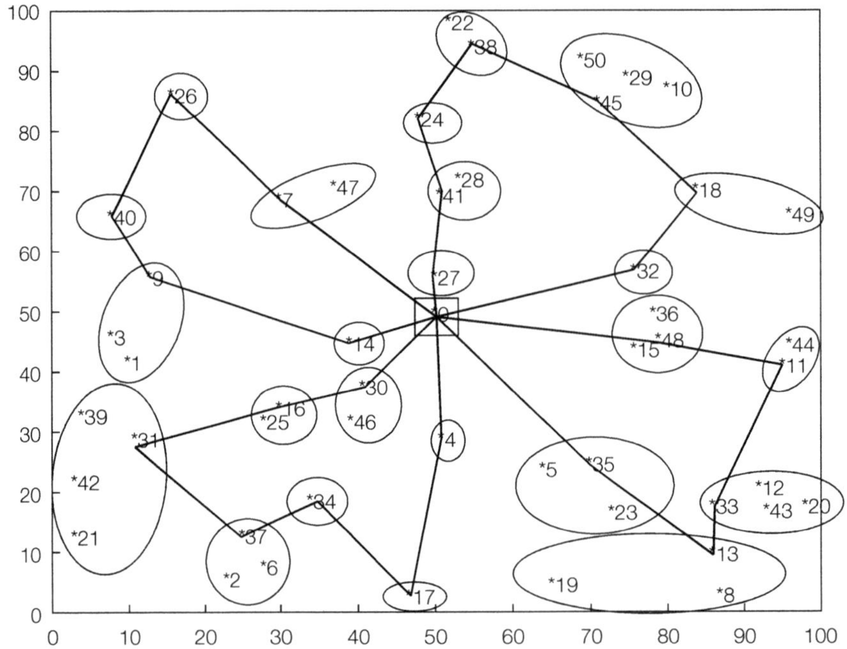

Figure 5.4. The optimal solution of the GVRP instance described by Ghiani and Improta.

u_p is the rank order of cluster V_p on the tour of a vehicle (visit number), $p \in C$;

y_{pr} is is the total number of arcs traveled on the route of a vehicle just after leaving the cluster V_p, if a vehicle goes from cluster V_p to cluster V_r, and zero otherwise.

$q_p = 1$, for all $p \in C$, with $q_0 = 0$ and $m - s + 1$ the maximum number of clusters that a vehicle can visit.

Substituting these conditions into constraints (5.9), (5.10) and (5.11) of the node based formulation (NB), we obtain a polynomial size, node based formulation for the GmTSP, and substituting those in (5.13), (5.14), (5.15) and (5.16) of the flow based formulation (FF), we obtain a polynomial size, flow based formulation for the GmTSP, as follows:

$$\text{minimize} \quad \sum_{i=1}^{n}\sum_{i=1}^{n} c_{ij}x_{ij}$$

subject to (5.1)–(5.6),

$$u_p - \sum_{r=1, r\neq p}^{m} w_{rp} \geq 1, \qquad \forall p \in C, \quad p \neq 0, \tag{5.28}$$

$$u_p + (m-s)w_{0p} \le m-s+1, \quad \forall\, p \in C, \quad p \ne 0, \tag{5.29}$$

$$u_p - u_r + (m-s+1)w_{pr} + (m-s-1)w_{rp} \le m-s \quad \forall\, p,r \in C, \quad p \ne r \ne 0, \tag{5.30}$$

$$x_{ij} \in \{0,1\}, \quad \forall (i,j) \in A,$$

where $m-s-1 \ge 0$ and $x_{ij} = 0$ whenever $i, j \in V_r, r \in C$.

$$\text{minimize} \quad \sum_{i=1}^{n}\sum_{i=1}^{n} c_{ij}x_{ij}$$

subject to (5.1)–(5.6),

$$y_{rp} \le (m-s)w_{rp}, \quad \forall\, r,p \in C,\ r \ne p \tag{5.31}$$

$$y_{rp} \ge w_{rp}, \quad \forall\, r,p \in C,\ r \ne p \tag{5.32}$$

$$\sum_{p=1}^{m} y_{p0} = m \tag{5.33}$$

$$\sum_{p=1}^{m} y_{rp} - \sum_{p=1}^{m} y_{pr} = 1, \quad \forall\, r \in C \tag{5.34}$$

$$x_{ij} \in \{0,1\}, \quad \forall\, (i,j) \in A,$$

where $k \ge m$ and $x_{ij} = 0$ whenever $i, j \in V_r, r \in C$. In addition, $y_{0p} = 0$ for all $p \in C$ and $q_0 = 0$.

This new variant of the TSP, called the generalized multiple traveling salesman problem (GmTSP), and the above integer programming formulations were introduced by Kara and Pop [94].

5.5.2 The Generalized Traveling Salesman Problem

The GmTSP reduces to the Generalized Traveling Salesman Problem (GTSP) when $m = 1$, i.e., when there is a single traveler.

If we substitute $s = 1$ in the previous two formulations, we obtain two polynomial size, integer linear programming models for the GTSP as follows:

$$\text{minimize} \quad \sum_{i=1}^{n}\sum_{i=1}^{n} c_{ij}x_{ij}$$

subject to (5.1), (5.2), (5.5), (5.6), (5.28),

$$\sum_{i=1}^{n} x_{i0} = 1, \tag{5.35}$$

$$\sum_{i=1}^{n} x_{0i} = 1, \tag{5.36}$$

$$u_p + (m-1)w_{0p} \le m, \quad p \ne 0, \quad p \in C, \tag{5.37}$$

$$u_p - u_r + mw_{pr} + (m-2)w_{rp} \le m-1,$$
$$p \ne r \ne 0, \quad p, r \in C, \tag{5.38}$$

$$x_{ij} \in \{0,1\}, \quad \forall (i,j) \in A,$$

where $x_{ij} = 0$ whenever $i, j \in V_r, r \in C$ and $m \ge 2$.

$$\text{minimize} \quad \sum_{i=1}^{n}\sum_{i=1}^{n} c_{ij}x_{ij}$$

$$\text{subject to} \quad (5.1), (5.2), (5.5), (5.6), (5.28), (5.35), (5.36)$$

$$y_{rp} \le (m-1)w_{rp}, \quad \forall\ p, r \in C,\ p \ne r, \tag{5.39}$$

$$x_{ij} \in \{0,1\}, \quad \forall\ (i,j) \in A,$$

where $x_{ij} = 0$ whenever $i, j \in V_r, r \in C$.

In most previous studies, the GTSP has been introduced without a fixed depot. The above formulation can be easily adapted to the non-fixed depot case by selecting any cluster as the starting and ending cluster (depot), denoted as V_0 ($|V_0| > 1$), and then rewriting the constraints (5.1) and (5.2), including the cluster V_0, and omitting the constraints (5.35) and (5.36).

5.5.3 The Clustered Generalized Vehicle Routing Problem

Consider the GTSP with the restriction that the traveler must visit all nodes of each cluster consecutively. According to Laporte and Palekar [106], this variant of the GTSP is defined as the Cumulative Traveling Salesman Problem by Chisman [18] and we will denote it by CTSP. Several applications of the CTSP are outlined by Laporte and Palekar [106]. Recently, Ding et al. [27] proposed a two level genetic algorithm for CTSP.

As far as we are aware, there is no such extension for GVRP.

In the following, we define a similar extension of the GVRP, which we call the Clustered Generalized Vehicle Routing Problem, denoted by CGVRP, where all nodes of each cluster must be visited consecutively on a route of a vehicle. Thus, for the case of the CGVRP, we are interested in finding a collection of s tours of minimum cost, starting and ending at the depot, such that each node of the entire graph is visited exactly once, by performing a Hamiltonian path within each cluster. Each cluster should be visited by exactly one vehicle at any of its nodes, and the load of each vehicle does not exceed its capacity Q.

A clustered GVRP and a feasible solution of the problem are presented in Figure 5.5.

The proposed formulations of the GVRP can be easily adapted to the CGVRP by adding additional constraints to each formulation, as we will show below. There is no

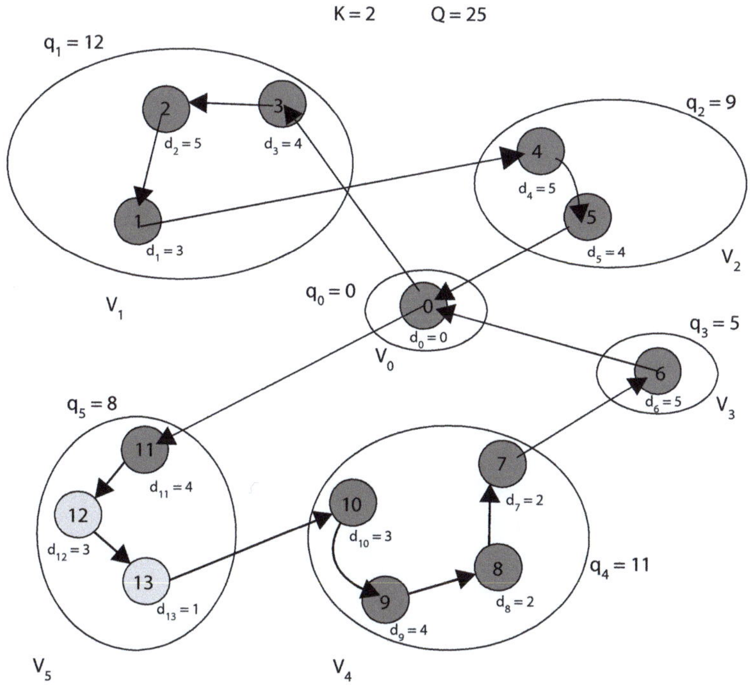

Figure 5.5. An example of a feasible solution of the CGVRP.

need to define new parameters and new decision variables for CGVRP. It is enough to redefine the cost parameters c_{ij} for all $i \neq j, i, j \in V$.

Node degree constraints for CGVRP

In the case of the CGVRP, each node should be included in the tour of a vehicle. Therefore, in addition to the cluster degree constraints given in (5.1) and (5.2), we have to write the degree constraints for all nodes. Therefore, we omit the connectivity constraints given in (5.5) and add the following node degree constraints to both formulations:

$$\sum_{i=0}^{n} x_{ij} = 1, \quad j = 1, \ldots, n \tag{5.40}$$

$$\sum_{j=0}^{n} x_{ij} = 1, \quad i = 1, \ldots, n \tag{5.41}$$

In the following, we present the node based and flow based formulations for the case of the clustered generalized vehicle routing problem.

A node based formulation for CGVRP

We define the new variable v_i to be the rank order of node i on the tour of a vehicle, $i \in \{1, \ldots, n\}$. The CGVRP can then be modeled as an integer program in the following way

$$\begin{aligned}
\text{minimize} \quad & \sum_{i=1}^{n}\sum_{i=1}^{n} c_{ij}x_{ij} \\
\text{subject to} \quad & (5.1)\text{–}(5.4), (5.6), (5.9)\text{–}(5.12), (5.40), (5.41) \\
& v_i - v_j + |V_p|x_{ij} + (|V_p| - 2)x_{ji} \le |V_p| - 1, \\
& \qquad\qquad \forall\, p \in C, i, j \in V_p \qquad (5.42) \\
& v_i \ge 1, \qquad \forall\, i \in V_p, p \in C \qquad (5.43) \\
& x_{ij} \in \{0, 1\}, \qquad \forall\, (i, j) \in A,
\end{aligned}$$

where, as in the other node based formulations, $x_{ij} = 0$, whenever $q_i + q_j > Q$ for all $i \ne j$, and $c_{ii} = 0$ for all $i \in V$.

The constraints presented in (5.42) were introduced by Desrochers and Laporte [23] for the case of the traveling salesman problem, in order to eliminate the subtours. These constraints avoid the formation of any subtours in each of the clusters. The constraints (5.43) initialize the variables v_i's for each cluster.

A flow based formulation for CGVRP

We introduce a new flow variable t_{ij} as the number of arcs visited within a given cluster on the route of a vehicle after just leaving the i-th node, if the vehicle goes from node i to node j, or zero otherwise.

$$\begin{aligned}
\text{minimize} \quad & \sum_{i=1}^{n}\sum_{i=1}^{n} c_{ij}x_{ij} \\
\text{subject to} \quad & (5.1)\text{–}(5.4), (5.13)\text{–}(5.16), (5.40), (5.41) \\
& \sum_{j=0}^{n} t_{ij} - \sum_{j=0}^{n} t_{ji} = 1, \qquad \forall\, i \in V_p, p \in C \qquad (5.44) \\
& t_{ij} \le (|V_k| - 1)x_{ij}, \qquad \forall\, i, j \in V_k, k \in C \qquad (5.45) \\
& x_{ij} \in \{0, 1\}, \qquad \forall\, (i, j) \in A,
\end{aligned}$$

where $m \ge s$.

In the above formulation, the constraints presented in (5.44) are the standard conservation of flow equations, increasing the t_{ij}'s of the consecutive arc of any path by one after each node. Hence, they are the subtour elimination constraints of the formulation. The constraints given in (5.45) are the bounding constraints of the flow variables. They guarantee that, if the arc (i, j) is not on any route, i.e. $x_{ij} = 0$, the corresponding flow variables $t_{ij} = 0$.

5.6 Solving the Generalized Vehicle Routing Problem

Despite the importance of the GVRP in the real world, due to the many interesting applications, not much much research has been done on the problem until recently. Ghiani and Improta [48] proposed a solution procedure that transforms the GVRP into a capacitated arc routing problem for which an exact algorithm and several approximate procedures are reported in literature.

Recently, some specific algorithms have been proposed for solving the GVRP, which are an ant colony based algorithm, described by Pop et al. [158], a genetic algorithm, proposed by Pop et al. [162], a branch-and-cut algorithm and an adaptive large neighborhood search, proposed by Bektaş et al. [9], an incremental tabu search heuristic described by Moccia et al. [126], and an improved genetic algorithm proposed by Pop et al. [168].

In this section, we will describe two algorithmic methods for solving the generalized vehicle routing problem:

- A hybrid algorithm based on a genetic algorithm applied to the global graph.
- a novel memetic algorithm, that combines a genetic algorithm (GA) with a powerful local search (LS) procedure.

5.6.1 An improved hybrid algorithm for Solving the GVRP

In this section we describe a hybrid heuristic algorithm that combines a genetic algorithm (GA) with a local-global approach to the problem and with a powerful local search (LS) procedure. Some of the advantages of the proposed heuristic are that we considerably reduce the search space by using the GA with respect to the global graph, and the power and diversity of the local search heuristic.

The local-global approach to the GVRP

In the case of the GVRP, as in the case of the other generalized network design problems, the local-global approach aims at distinguishing between global connections (connections between clusters) and local connections (connections between nodes belonging to different clusters).

We denote by G' the graph obtained from G after replacing all nodes of a cluster V_i with a supernode representing V_i, $\forall i \in \{1, \dots, m\}$, the cluster V_0 (depot) already consists of one vertex. We will call the graph G' the *global graph*. For convenience, we identify V_i with the supernode that represents it and we suppose that the edges of the graph G' are defined between each pair of graph vertices $V_0, V_1, \dots, V_m$.

Given a solution in the global graph, i.e. a collection of r global routes of form $(V_0, V_{k_1}, \dots, V_{k_p})$ in which the clusters are visited, we want to find the best feasible route R^* (w.r.t. cost minimization), i.e. a collection of r generalized routes visiting the clusters according to the given sequence.

This can be done in polynomial time, by solving the following r shortest path problems.

For each global route $(V_0, V_{k_1}, \dots, V_{k_p})$, the best generalized route that visits the clusters according to the given sequence can be determined in polynomial time by constructing a layered network (LN) with $p+2$ layers, corresponding to the clusters $V_0, V_{k_1}, \dots, V_{k_p}$. In addition, we duplicate the cluster V_0. The layered network contains all nodes of clusters $V_0, V_{k_1}, \dots, V_{k_p}$ plus an extra node $0' \in V_0$ and the arcs are defined as an arc $(0, i)$ for each $i \in V_{k_1}$ with cost c_{0i}, an arc (i, j) for each $i \in V_{k_l}$ and $j \in V_{k_{l+1}}$ $(l = 1, \dots, p-1)$ having a cost c_{ij} and an arc $(i, 0')$ for each $i \in V_{k_p}$ with cost $c_{i0'}$.

In the next figure, we present the constructed layered network. The bold lines represent a route visiting the clusters according to the given sequence $(V_0, V_{k_1}, \dots, V_{k_p})$.

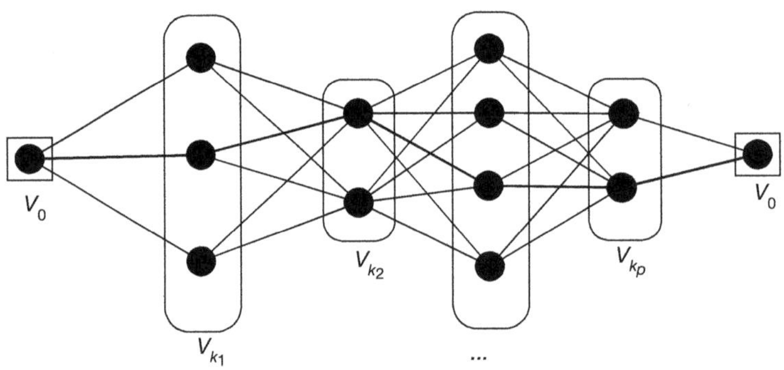

Figure 5.6. Example showing a route visiting the clusters $V_0, V_{k_1}, \dots, V_{k_p}$ in the constructed layered network LN.

We consider paths from 0 to $0'$, visiting exactly one node from each cluster $V_{k_1}, \dots,$ V_{k_p}, hence giving a feasible generalized route.

Conversely, every generalized route that visits the clusters according to the sequence $(V_0, V_{k_1}, \dots, V_{k_p})$, corresponds to a path in the layered network from $0 \in V_0$ to $0' \in V_0$.

Therefore, it follows that the best (w.r.t. cost minimization) collection of routes R^* can be found by determining the r shortest paths from $0 \in V_0$ to the corresponding $0' \in V_0$ with the property that it visits exactly one node from each of the clusters $(V_{k_1}, \dots, V_{k_p})$.

The proposed computational model to approach the problem is a genetic algorithm, applied with respect to the global graph, which substantially reduces the size of the solution space.

Genetic algorithms are predisposed to optimizing general combinatorial problems and therefore are serious candidates for solving the GVRP. However, it is known that

GA are not well suited for fine-tuning structures which are close to optimal solutions. Therefore, we use local search optimization in order to refine the solutions explored by GA. The general scheme of our hybrid heuristic is:

Algorithm 5.7.

Step 1. *Initialization.* Construct the first generation of global solutions, i.e. collection of global routes.

Step 2. *Creation of next generation.* Use genetic operators (crossover and mutation) to produce next generation of global solutions. The parents are selected from the previous generation.

Step 3. *Producing the corresponding best GVRP generation.* Using the procedure described in section 5.2, for each global solution, determine its best corresponding GVRP solution.

Step 4. *Improving next generation.* Use a local search procedure to replace each of the current generation solutions by the local optimum. Eliminate duplicate solutions.

Step 5. *Termination condition.* Repeat Steps 3, 4 and 5 until a stopping condition is reached.

The genetic algorithm

Genetic encoding

In our algorithm, we used the following genetic representation of the solution domain: a chromosome is represented as a list of clusters

$$(V_0, V_{l_1}^{(1)}, V_{l_2}^{(1)}, \dots, V_{l_p}^{(1)}, V_0, \dots, V_0, V_{l_1}^{(r)}, V_{l_2}^{(r)}, \dots, V_{l_t}^{(r)}) \tag{5.46}$$

representing a collection of r global routes $V_0 - V_{l_1}^{(1)} - V_{l_2}^{(1)} - \dots - V_{l_p}^{(1)} - V_0, \dots,$ $V_0 - V_{l_1}^{(r)} - V_{l_2}^{(r)} - \dots - V_{l_t}^{(r)} - V_0$, where $p, t \in N$ with $1 \le p, t \le k$.

For the feasible solution of the GVRP illustrated in Figure 5.7, the corresponding chromosome is: (1 2 0 3 4 5 0 6) and represents the collection of 3 global routes, passing through the clusters in the following order:

$$(V_1\ V_2\ V_0\ V_3\ V_4\ V_5\ V_0\ V_6). \tag{5.47}$$

The values $\{1, \dots, 6\}$ represent the clusters, while the depot, denoted by $\{0\}$, is the route splitter. Route 1 starts at the depot then visits the clusters V_1, V_2 and returns to the depot. Route 2 starts at the depot and visits the clusters $V_3 - V_4 - V_5$. Finally, in route 3 only the cluster V_6 is visited.

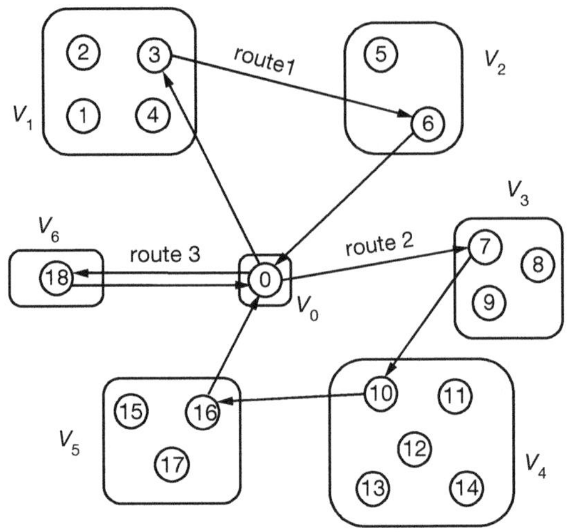

Figure 5.7. An example of a feasible solution of the GVRP.

Several collections of generalized routes correspond to the given collection of global routes, the one represented in figure 1 consisting of the following nodes: $3 \in V_1$, $6 \in V_2$, $7 \in V_3$, $10 \in V_4$, $16 \in V_5$ and $18 \in V_6$. The best w.r.t. cost minimization collection of generalized routes can be determined using the approach described at the beginning of this section.

We can see that our described representation has variable length and allows empty routes by simply placing two route splitters together, without clients between them. Some routes in the chromosome may cause the vehicle to exceed its capacity. When this happens, in order to guarantee that the interpretation is always a valid candidate solution, we modify any route that exceeds the vehicle capacity by splitting it in several ones.

The fitness value

The fitness function is defined over the genetic representation and measures the quality of the represented solution. In our case, the fitness value of a feasible solution, i.e. a collection of global routes, is given by the cost of the best corresponding collection of generalized routes (w.r.t. cost minimization), obtained by constructing the layered network and determined by solving a given number of shortest path problems.

Initial population

The construction of the initial population is of great importance to the performance of GA, since it contains most of the material of which the final best solution is made. Experiments were carried out with a randomly generated initial population and with an initial population of structured solutions. In order to generate the population of structured solutions, we used a Monte Carlo based method. However, from carrying out the experiments we learned that the Monte Carlo method of generating the initial population has not brought any improvements. Randomly generating the initial population has the advantage that it is representative for any area of the search space.

Genetic operators

Crossover

Two parents are selected from the population by the binary tournament method. Offspring are produced from two parent solutions, using a 2-point order crossover procedure that creates offspring which preserves the order and position of symbols in a subsequence of one parent, while preserving the relative order of the remaining symbols from the other parent. The procedure is implemented by selecting two random cut points which define the boundaries for a series of copying operations.

The recombination of two collections of global routes requires some further explanation. First, the symbols between the cut points are copied from the first parent to the offspring. Then, starting just after the second cut-point, the symbols are copied from the second parent to the offspring, omitting any symbols that were copied from the first parent. When the end of the second parent sequence is reached, this process continues with the first symbol of the second parent until all symbols have been copied to the offspring. The second offspring is produced by swapping the parents and then using the same procedure.

Next, we present the application of the proposed 2-point order crossover in the case of a problem consisting of 8 clusters and the depot. We assume two well-structured parents chosen randomly, with cutting points between nodes 3 and 4, respectively 6 and 7:

$$\begin{array}{lllllllllllll} P_1 = & 6 & 8 & 1 & | & 0 & 2 & 7 & | & 0 & 5 & 4 & 3 \\ P_2 = & 8 & 2 & 1 & | & 6 & 0 & 4 & | & 3 & 5 & 7 & \end{array}$$

Note that the length of each individual differs according to the number of routes.

The sequences between the two cutting-points are copied to the two offspring

$$\begin{array}{lllllllllllll} O_1 = & x & x & x & | & 0 & 2 & 7 & | & x & x & x & x \\ O_2 = & x & x & x & | & 6 & 0 & 4 & | & x & x & x & \end{array}.$$

The nodes of the parent P_1 are copied to the offspring O_2 if O_2 does not contain already the clusters of P_1 and therefore the offspring O_2 is

$$\begin{array}{lllllllllllll} O_2 = & 8 & 1 & 2 & | & 6 & 0 & 4 & | & 7 & 0 & 5 & 3 \end{array}.$$

Then the nodes of parent P_2 are copied to the offspring O_1 in the same manner. The nodes of the clusters not present in O_1 are copied into the remaining positions

$$\begin{array}{lllllllllllll} O_1 = & 8 & 1 & 6 & | & 0 & 2 & 7 & | & 0 & 4 & 3 & 5 \end{array}.$$

Mutation

In our genetic algorithm, we use the following random mutation operator, called the inter-route mutation operator, which is a swap operator. It picks two random locations in the solution vector and swaps their values.

Let the parent solution be (6 8 1 | 0 2 7 | 0 5 4 3). The inter-route mutation operator picks two random clusters, for example V_8 and V_5 and swaps their values obtaining the new chromosome: (6 5 1 | 0 2 7 | 0 8 4 3).

Selection

Selection is the stage of a genetic algorithm in which individuals are chosen from a population for later breeding (crossover or mutation). The selection process is deterministic.

In our algorithm we investigated and used the properties of (μ, λ), selection, where μ parent produce λ $(\lambda > \mu)$ and only the offspring undergo selection. In other words, the lifetime of every individual is limited to only one generation. The limited life span allows us to forget the inappropriate internal parameter settings. This may lead to short periods of recession, but it avoids long stagnation phases due to unadapted strategy parameters.

Genetic parameters

The genetic parameters are very important for the success of the algorithm, equally important as the other aspects, such as the representation of the individuals, the initial population and the genetic operators. The most important parameters are the population size μ, set to 10 times the number of clusters, the intermediate population size λ, chosen ten times the size of the population, $\lambda = 10 \cdot \mu$ and the mutation probability, set at 5%. The number of epochs used in our algorithm was set to 2000.

Local search improvement

We here present five routing plan improvements, used in our local search procedure. Two of them operate within the routes (the intra-route 2-opt and the 3-opt operators) and the rest operate between routes and are called relocate, exchange and cross operators. These last operators are extensions of the operators described by van Breedam [202] in the case of the classical VRP.

1. *2-opt.* The two-opt procedure is used for improving single vehicle routes, and consists of eliminating two arcs and reconnecting the two resulting paths in a different way to obtain a new route. In Figure 5.8, the two-opt operator is shown and as we can see there are different ways to reconnect the paths that yield a different tour. Among all pairs of edges whose 2-opt exchange decreases the length, we choose the pair that gives the shortest tour. This procedure is then iterated until no such pair of edges is found.

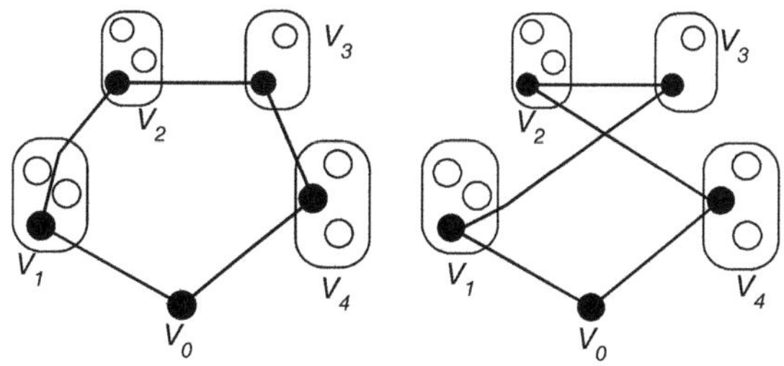

Figure 5.8. An example of a possible 2-opt exchange.

2. *3-opt.* The three-opt procedure extends the 2-opt and involves deleting 3 arcs in a route and reconnecting the network in all other possible ways, and then evaluating each reconnection method to find the optimum one.

3. *Exchange operator:* two strings of at most k vertices are exchanged between two routes. A new solution is obtained from the current one by selecting k vertices from one route and exchanging them with k vertices belonging to another route. The exchange is acceptable only if it produces a feasible solution. This is pictured in Figure 5.9.

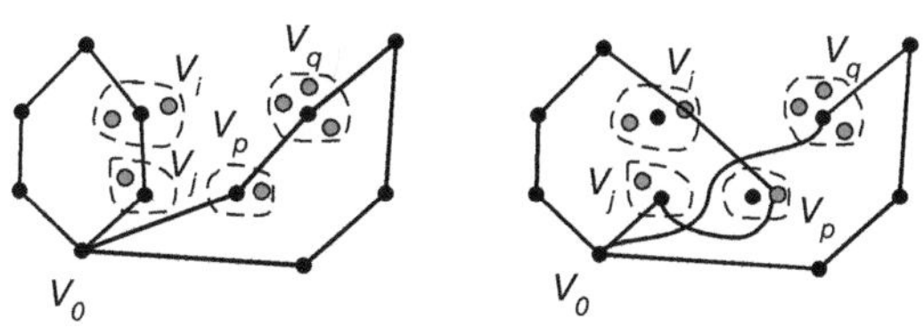

Figure 5.9. An example of a possible string exchange for $k = 1$.

4. *Relocate operator.* The relocate operator simply moves a customer visit from one route to another. It is illustrated in the next figure.

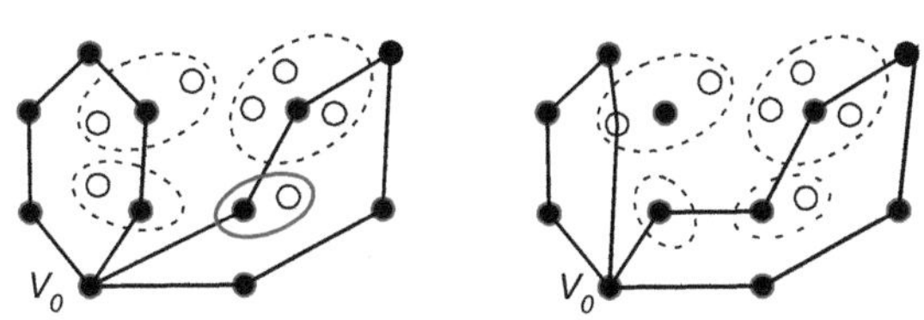

Figure 5.10. An example of a possible relocation.

5. *Cross operator.* This operator includes all feasible solutions obtained by the replacement of two arcs by two others. Given two different routes r_1 and r_2, an arc is deleted from each of them, dividing the routes into two components that are combined such that a new route is made of the first component of r_1, the second of r_2 and another new route is made of the first component of r_2 and the second component of r_1.

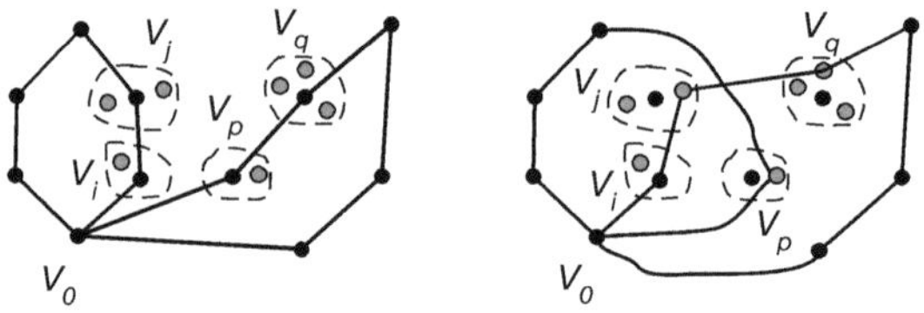

Figure 5.11. An example of a possible string cross.

5.6.2 An efficient memetic algorithm for solving the GVRP

Memetic algorithms have been introduced by Mascato [127] to denote a family of metaheuristic algorithms that emphasize the use of a population-based approach with separate individual learning or local improvement procedures for problem search. Therefore a memetic algorithm is a genetic algorithm (GA), hybridized with a local search procedure, to intensify the search space.

Genetic algorithms are not well suited for fine-tuning structures close to optimal solutions. Therefore, incorporating local improvement operators into the recombination step of a GA is essential, in order to obtain a competitive GA.

Our effective heuristic algorithm for solving the GVRP is a memetic algorithm, which combines the power of a genetic algorithm with that of local search. The general scheme of our heuristic is

Memetic algorithms have been recognized as a powerful algorithmic paradigm for evolutionary computing, having been successfully applied to solve combinatorial optimization problems such as the VRP (Vehicle Routing Problem) [119, 131], the CARP (Capacitated Arc Routing Problem) [172], the generalized traveling salesman problem [13], etc.

The genetic algorithm

Solution encoding

We used a natural, compact and efficient encoding of solutions of GVRP similar to that described by Pop et al. [162]. Specifically, 0 represents the depot and each customer is tagged with a non-duplicated natural number from 1 to n. We represent a chromosome

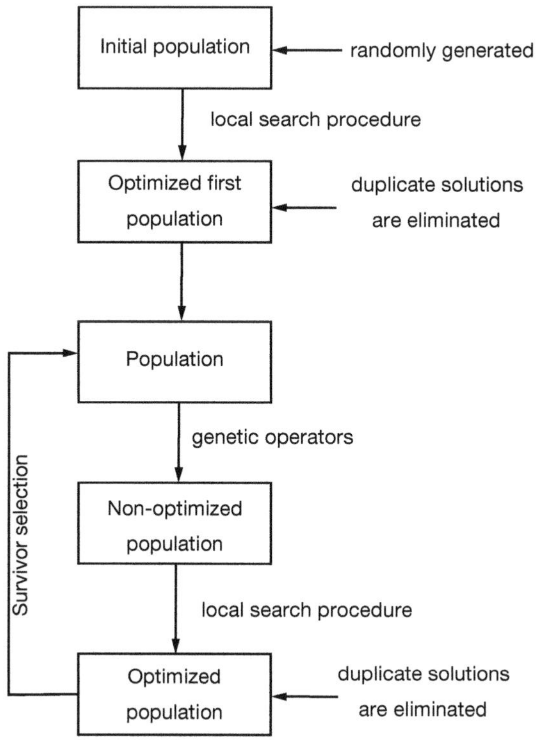

Figure 5.12. Generic form of our memetic algorithm.

by a variable length array, so that the gene values correspond to the nodes selected to form the collection of routes which are delimited by 0 representing the depot.

The corresponding chromosome representation of the feasible solution of the GVRP presented in Figure 5.7 is:

$$(3 \quad 6 \quad 0 \quad 7 \quad 10 \quad 16 \quad 0 \quad 18)$$

where the values $\{1, \ldots, 18\}$ represent the customers, and the depot denoted by 0 is the route splitter. Route 1 begins at the depot, then visits customers 3 and 6, belonging to the clusters V_1, respectively V_2, and returns to the depot. Route 2 starts at the depot and visits the customers 7–10–16 belonging to the clusters $V_3 - V_4 - V_5$. Finally, in route 3 only customer 18 from the cluster V_6 is visited.

The described representation allows empty routes by simply placing two route splitters together without clients between them.

Suppose we have given a representation of a solution of the GVRP. Then we first have to check whether it satisfies the constraints of the problem. If some routes in the chromosome may cause the vehicle to exceed its capacity, we modify the route that exceeds the vehicle capacity by splitting it in several ones. Assuming that the original route is composed of the following ordered set of nodes $\{i_1, \dots, i_k, i_{k+1}, \dots, i_p\}$, and that the vehicle capacity is exceeded at node i_{k+1}, then it will be divided in two parts $\{i_1, \dots, i_k\}$ and $\{i_{k+1}, \dots, i_p\}$. If necessary, further divisions can be made in the second part.

Initial population

The first step in any genetic algorithm is to generate a set of possible solutions as an initial generation or population to the problem. Although it seems simple, the convergence, performance and the ability of the GA to provide the best possible solutions are critically affected by the initial generation.

In the case of the GVRP, we carried out experiments with the initial population generated randomly and with an initial population of structured solutions.

The initial randomly generated population has the advantage that it is representative from any area of the search space.

In order to generate the population of structured solutions, we used a Monte Carlo based method. Based on computational experiments, we observed that this method assures an initial population with a average fitness which is 20% better than a population generated entirely at random, but the major drawback was that such a population lacks the diversity needed to obtain near-optimal solutions.

Analyzing the advantages and disadvantages of these ways of generating the initial population, and based on computational experiments we decided to choose the initial population randomly.

The fitness value

In order for genetic algorithms to work effectively, it is necessary to be able to evaluate how "good" a potential solution is, relative to other potential solutions. The "fitness function" is responsible for performing this evaluation and returning a positive number, or "fitness value", that reflects how optimal the solution is. The higher the number, the better the solution. The fitness values are then used in a process of natural selection to choose which potential solutions will continue on to the next generation, and which will die out.

In our case, the fitness value of a feasible solution, i.e. collection of routes, is given by the sum of the costs of the arcs selected in the routes. The aim is to find the minimum cost collection of routes.

Crossover operator

The crossover operator requires some strategy to select two parents from the previous generation. In our algorithm, we used an elitist approach: the parents are chosen randomly from among the best 40% of all solutions in the previous generation.

We randomly select two solutions from the best 40% of the all the solutions of the previous generation, assign the cheapest (w.r.t. cost minimization) to parent 1, and repeat the procedure to select parent 2.

Offspring are produced from two parents, using the 2-point order standard crossover procedure, which creates offspring that preserves the order and position of symbols in a subsequence of one parent, while preserving the relative order of the remaining symbols from the other parent. It is implemented by selecting two random cut points which define the boundaries for a series of copying operations. First, the symbols between the cut points are copied from the first parent to the offspring. Then, starting just after the second cut point, the symbols are copied from the second parent to the offspring, omitting any symbols that were copied from the first parent. When the end of the second parent sequence is reached, this process continues with the first symbol of the second parent, until all symbols have been copied to the offspring. The second offspring is produced by swapping the parents and then using the same procedure.

Next, we present the application of the proposed 2-point order crossover in the case of the problem presented in Figure 5.7. We assume two well-structured parents, chosen at random, with the cutting points between nodes 2 and 3, respectively 5 and 6:

$$\begin{array}{rcccccccccc} P_1 = & 18 & 0 & | & 3 & 6 & 0 & | & 7 & 10 & 16 \\ P_2 = & 5 & 2 & | & 18 & 0 & 15 & | & 13 & 8 & \end{array}$$

Note that the length of each individual differs according to the number of routes. According to Figure 5.7, the cluster representation of the previously selected parents is:

$$\begin{array}{rcccccccccc} C_1 = & 6 & 0 & | & 1 & 2 & 0 & | & 3 & 4 & 5 \\ C_2 = & 2 & 1 & | & 6 & 0 & 5 & | & 4 & 3 & \end{array}$$

The sequences between the two cutting-points are copied to the two offspring:

$$\begin{array}{rcccccccccc} O_1 = & x & x & | & 3 & 6 & 0 & | & x & x & x \\ O_2 = & x & x & | & 18 & 0 & 15 & | & x & x & x \end{array}$$

The nodes of parent P_1 are copied to the offspring O_2 if O_2 does not already contain nodes in the same clusters as the nodes of P_1. The sequence of nodes of P_1 is 18–0–3–6–0–7–10–16, and the clusters are 6–0-1–2-0–3-4–5. Note that cluster 6 is already represented in O_2 by node 18 and cluster 5 by node 16. The depot (0) is kept anyway, since it is a splitter. Therefore, the remaining sequence of nodes in P_1 is 0–3-6–0-7–10. Hence, the offspring O_2 is

$$\begin{array}{rcccccccccc} O_2 = & 0 & 3 & | & 18 & 0 & 15 & | & 6 & 0 & 7 & 10\,. \end{array}$$

Following this, the nodes of the parent P_2 are copied to the offspring O_1 in the same manner. The nodes of the clusters not present in O_1 are copied to the remaining positions:

$$\begin{array}{rcccccccccc} O_1 = & 18 & 0 & | & 3 & 6 & 0 & | & 15 & 13 & 8 \end{array}$$

Mutation operators

The mutation operator is a genetic operator that alters one or more gene values in a chromosome from its initial state. This can result in entirely new gene values being added to the gene pool. With these new gene values, the genetic algorithm may be able to arrive at better solution than was previously possible. Mutation is an important part of the genetic search as it helps to prevent the population from stagnating at any local optima.

In our GA, we use two random mutation operators. The first one (intra-route mutation) randomly selects a cluster to be modified and replaces its current node by another one randomly selected from the same cluster. The second one (inter-route mutation) is a swap operator, it picks two random locations in the solution vector and swaps their values. The new chromosome is accepted immediately if it results in a feasible GVRP. If not, the route that exceeds the vehicle capacity is decomposed into several ones as described before.

In the following mutation example, let the parent solution be as in Figure 5.7:

$$P = \quad 18 \quad 0 \quad | \quad 3 \quad 6 \quad 0 \quad | \quad 7 \quad 10 \quad 16$$

with the corresponding cluster representation

$$C = \quad 6 \quad 0 \quad | \quad 1 \quad 2 \quad 0 \quad | \quad 3 \quad 4 \quad 5$$

in this case, the intra-route mutation operator randomly selects a cluster, for example V_4, and replaces its current node 10 with another random node from V_4, for example 13 obtaining the new chromosome

$$(18, 0, 3, 6, 0, 7, 13, 16) \tag{5.48}$$

and the inter-route mutation picks two random locations, for example 3 and 10 and swaps their values obtaining the new chromosome

$$(18, 0, 10, 6, 0, 7, 3, 16). \tag{5.49}$$

As in the case of the crossover operator, we used an elitist strategy, where the parent is randomly selected from among 60% of all solutions in the previous generation.

The choice of which of the operators described above should be used to create an offspring is a probabilistic one. The probabilities for each operator to be applied are called operator rates. Typically, crossover is applied with highest probability, the crossover rate being 90% or higher. In contrast, the mutation rate is much smaller, typically being in the region of 10%.

Selection

The selection process is deterministic. In our algorithm, we use the $(\mu + \lambda)$ selection, where μ parents produce λ offspring. The new population of $(\mu + \lambda)$ is reduced again to μ individuals by a selection based of the "survival of the fittest" principle. In other words, parents survive until they are suppressed by better offspring. It might be possible for very well adapted individuals to survive forever.

Genetic parameters

The genetic parameters are very important for the success of a GA and are of equal importance as the other aspects, such as the representation of the individuals, the initial population and the genetic operators. The most important parameters are:

- The population size μ has been set to 2 times the number of the clusters. This turned out to be the best number of individuals in a generation.
- The intermediate population size λ was chosen as twice the size of the population: $\lambda = 2 \cdot \mu$.
- The mutation probability was set at 10%.

Termination criteria

The loop of chromosome generation is terminated when certain conditions are met. When the termination criteria are met, the elite chromosome is returned as the best solution found so far. In our algorithm, we used a termination condition based on the number of generations, namely 10^4 generations, which seems necessary to sufficiently explore the solution space.

Local improvement procedure

For some combinatorial optimization problems, classical GAs are not aggressive enough. One possibility to obtain more competitive heuristics is to combine the GAs with local search procedures.

For each solution belonging to the current generation we use a local improvement procedure that runs several local search heuristics sequentially. Once an improvement move is found, it is immediately executed.

In our algorithm, we used five local search heuristics. Two of them, the 2-opt intra-route and the cluster route optimization that uses the shortest path algorithm to find the best node from each cluster when the order of clusters is given, operate within the routes and the rest operate between routes and are called neighborhood cross, neighborhood exchange and neighborhood relocation. These local search heuristics have been described in the previous subsection.

Our improvement procedure applies all described local search heuristics cyclically.

5.6.3 Computational experiments

In this section, we present extensive computational results in order to assess the effectiveness of our proposed algorithms for solving the GVRP.

We conducted our experiments on a set of instances generated through an adaptation of the existing instances in the CVRP-library, available at `http://branchandcut.org/VRP/data/`. The naming of the generated instances follows the general convention of the CVRP instances available online, and follows the general format $X - nY -$

$kZ - C\Omega - V\Phi$, where X corresponds to the type of the instance, Y refers to the number of vertices, Z corresponds to the number of vehicles in the original CVRP instance, Ω is the number of clusters and Φ is the number of vehicles in the GVRP instance. These instances were used by Bektaş et al. [9] and Moccia et al. [126] in their computational experiments.

Originally, the set of nodes in these problems are not divided into clusters. The CLUSTERING procedure proposed by Fischetti et al. [43] divides data into node-sets. This procedure sets the number of clusters $s = [\frac{n}{\theta}]$, identifies the s nodes farthest from each other and assigns each remaining node to its nearest center. Fischetti et al. [43] used $\theta = 5$, but according to Bektaş et al. [9], the created GVRP instances were too easy to solve.

For our instances, we considered, as in [9, 126], two clustering procedures. One with $\theta = 2$ and the other one with $\theta = 3$. However, the proposed solution approaches are able to handle any cluster structure.

The testing machine was an Intel Dual-Core 1.6 GHz and 1 GB RAM. The operating system was Windows XP Professional. The algorithms were developed in Java, with JDK 1.6.

In Tables 5.1 and 5.2, we summarize the results of all methods on all 148 small to medium test instances, with $\theta = 2$ and $\theta = 3$. The first column in the tables gives the name of the instances, the second column provides the values of the best lower bounds in the branch-and-cut tree [11]. The next four columns contain the values of the best solutions, obtained using the adaptive large neighborhood search (ALNS) [9], the incremental tabu search (ITS) [126], and our proposed methods: the hybrid algorithm and the memetic algorithm.

If we analyze the computational results reported in Table 5.1 (Parts 1 and 2) and Table 5.2 (Parts 1 and 2), we can observe that our proposed memetic algorithm performs slightly better compared to all other approaches for solving both small and medium size instances. We were able to improve the best known solutions for 6 out of 148 instances. In one instance, *P-n55-k15-C28-V8*, the solution returned by our algorithm exhibits a gap of 0.01 %, with the best known solution provided by the ALNS. The rest of the instances for which the optimal solution is known, we solved optimally with our algorithm. The proposed hybrid algorithm is comparable with the other existing approaches for solving the GVRP, in terms of the quality of the solutions .

We next present the results of the computational experiments on the large-scale instances generated using $\theta = 2$ and $\theta = 3$ in Tables 5.3 and 5.4, respectively.

Analyzing Tables 5.3 and 5.4, we see that, overall, in the case of large-size instances, our memetic algorithm performs better, in comparison to ALNS and ITS heuristics, while the hybrid algorithm provides the same solution as ALNS and ITS heuristics in 7 out of 10 instances. In the other 3 instances, the solution returned by our algorithm exhibits a gap to the best known solution provided of at most 0.0089%.

Table 5.1. Computational results on small and medium instances with $\theta = 2$ (Part 1).

Instance	LB [9]	*ALNS* [9]	*ITS* [126]	*Hybrid alg.*	*Memetic alg.*
A-n32-k5-C16-V2	519	519	519	519	519
A-n33-k5-C17-V3	451	451	451	451	451
A-n33-k6-C17-V3	465	465	465	465	465
A-n34-k5-C17-V3	489	489	489	489	489
A-n36-k5-C18-V2	505	505	505	505	505
A-n37-k5-C19-V3	432	432	432	432	432
A-n37-k6-C19-V3	584	584	584	584	584
A-n38-k5-C19-V3	476	476	476	476	476
A-n39-k5-C20-V3	557	557	557	557	557
A-n39-k6-C20-V3	544	544	544	544	544
A-n44-k6-C22-V3	608	608	608	608	608
A-n45-k6-C23-V4	613	613	613	613	613
A-n45-k7-C23-V4	674	674	674	674	674
A-n46-k7-C23-V4	593	593	593	593	593
A-n48-k7-C24-V4	667	667	667	667	667
A-n53-k7-C27-V4	603	603	603	603	603
A-n54-k7-C27-V4	690	690	690	690	690
A-n55-k9-C28-V5	699	699	699	699	699
A-n60-k9-C30-V5	769	769	769	769	769
A-n61-k9-C31-V5	638	638	638	638	638
A-n62-k8-C31-V4	740	740	740	740	740
A-n63-k10-C32-V5	801	801	801	801	801
A-n63-k9-C32-V5	900.3	912	912	912	908
A-n64-k9-C32-V5	763	763	763	763	763
A-n65-k9-C33-V5	682	682	682	682	682
A-n69-k9-C35-V5	680	680	680	680	680
A-n80-k10-C40-V5	957.4	997	997	1 004	997
B-n31-k5-C16-V3	441	441	441	441	441
B-n34-k5-C17-V3	472	472	472	472	472
B-n35-k5-C18-V3	626	626	626	626	626
B-n38-k6-C19-V3	451	451	451	451	451
B-n39-k5-C20-V3	357	357	357	357	357
B-n41-k6-C21-V3	481	481	481	481	481
B-n43-k6-C22-V3	483	483	483	483	483
B-n44-k7-C22-V4	540	540	540	540	540
B-n45-k5-C23-V3	497	497	497	497	497
B-n45-k6-C23-V4	478	478	478	478	478
B-n50-k7-C25-V4	449	449	449	449	449
B-n50-k8-C25-V5	916	916	916	916	916

Table 5.1. Computational results on small and medium instances with $\theta = 2$ (Part 2).

Instance	LB [9]	*ALNS* [9]	*ITS* [126]	*Hybrid alg.*	*Memetic alg.*
B-n51-k7-C26-V4	651	651	651	651	651
B-n52-k7-C26-V4	450	450	450	450	450
B-n56-k7-C28-V4	486	486	486	486	486
B-n57-k7-C29-V4	751	751	751	751	751
B-n57-k9-C29-V5	942	942	942	942	942
B-n63-k10-C32-V5	816	816	816	816	816
B-n64-k9-C32-V5	509	509	509	509	509
B-n66-k9-C33-V5	808	808	808	808	808
B-n67-k10-C34-V5	673	673	673	673	673
B-n68-k9-C34-V5	704	704	704	704	704
B-n78-k10-C39-V5	803	803	803	803	803
P-n16-k8-C8-V5	239	239	239	239	239
P-n19-k2-C10-V2	147	147	147	147	147
P-n20-k2-C10-V2	154	154	154	154	154
P-n21-k2-C11-V2	160	160	160	160	160
P-n22-k2-C11-V2	162	162	162	162	162
P-n22-k8-C11-V5	314	314	314	314	314
P-n23-k8-C12-V5	312	312	312	312	312
P-n40-k5-C20-V3	294	294	294	294	294
P-n45-k5-C23-V3	337	337	337	337	337
P-n50-k10-C25-V5	410	410	410	410	410
P-n50-k7-C25-V4	353	353	353	353	353
P-n50-k8-C25-V4	378.4	392	421	392	392
P-n51-k10-C26-V6	427	427	427	427	427
P-n55-k10-C28-V5	415	415	415	415	415
P-n55-k15-C28-V8	545.3	555	565	560	558
P-n55-k7-C28-V4	361	361	361	361	361
P-n55-k8-C28-V4	361	361	361	361	361
P-n60-k10-C30-V5	433	443	443	443	443
P-n60-k15-C30-V8	553.9	565	565	565	565
P-n65-k10-C33-V5	487	487	487	487	487
P-n70-k10-C35-V5	485	485	485	485	485
P-n76-k4-C38-V2	383	383	383	383	383
P-n76-k5-C38-V3	405	405	405	405	405
P-n101-k4-C51-V2	455	455	455	455	455

Table 5.2. Computational results on small and medium instances with $\theta = 3$ (Part 1).

Instance	LB [9]	*ALNS* [9]	*ITS* [126]	*Hybrid alg*	*Memetic alg.*
A-n32-k5-C11-V2	386	386	386	386	386
A-n33-k5-C11-V2	315	318	315	315	315
A-n33-k6-C11-V2	370	370	370	370	370
A-n34-k5-C12-V2	419	419	419	419	419
A-n36-k5-C12-V2	396	396	396	396	396
A-n37-k5-C13-V2	347	347	347	347	347
A-n37-k6-C13-V2	431	431	431	431	431
A-n38-k5-C13-V2	367	367	367	367	367
A-n39-k5-C13-V2	364	364	364	364	364
A-n39-k6-C13-V2	403	403	403	403	403
A-n44-k6-C15-V2	503	503	503	503	503
A-n45-k6-C15-V3	474	474	474	474	474
A-n45-k7-C15-V3	475	475	475	475	475
A-n46-k7-C16-V3	462	462	462	462	462
A-n48-k7-C16-V3	451	451	451	451	451
A-n53-k7-C18-V3	440	440	440	440	440
A-n54-k7-C18-V3	482	482	482	482	482
A-n55-k9-C19-V3	473	473	473	473	473
A-n60-k9-C20-V3	595	595	595	595	595
A-n61-k9-C21-V4	473	473	473	473	473
A-n62-k8-C21-V3	596	596	596	596	596
A-n63-k10-C21-V4	593	593	593	593	593
A-n63-k9-C21-V3	625.6	642	643	643	642
A-n64-k9-C22-V3	536	536	536	536	536
A-n65-k9-C22-V3	500	500	500	500	500
A-n69-k9-C23-V3	520	520	520	520	520
A-n80-k10-C27-V4	679.4	710	710	710	710
B-n31-k5-C11-V2	356	356	356	356	356
B-n34-k5-C12-V2	369	369	369	369	369
B-n35-k5-C12-V2	501	501	501	501	501
B-n38-k6-C13-V2	370	370	370	370	370
B-n39-k5-C13-V2	280	280	280	280	280
B-n41-k6-C14-V2	407	407	407	407	407
B-n43-k6-C15-V2	343	343	343	343	343
B-n44-k7-C15-V3	395	395	395	395	395
B-n45-k5-C15-V2	410	422	410	410	410
B-n45-k6-C15-V2	336	336	336	336	336
B-n50-k7-C17-V3	393	393	393	393	393
B-n50-k8-C17-V3	598	598	598	598	598

Table 5.2. Computational results on small and medium instances with $\theta = 3$ (Part 2).

Instance	LB [9]	*ALNS* [9]	*ITS* [126]	*Hybrid alg.*	*Memetic alg.*
B-n51-k7-C17-V3	511	511	511	511	511
B-n52-k7-C18-V3	359	359	359	359	359
B-n56-k7-C19-V3	356	356	356	356	356
B-n57-k7-C19-V3	558	558	558	558	558
B-n57-k9-C19-V3	681	681	681	681	681
B-n63-k10-C21-V3	599	599	599	599	599
B-n64-k9-C22-V4	452	452	452	452	452
B-n66-k9-C22-V3	609	609	609	609	609
B-n67-k10-C23-V4	558	558	558	558	558
B-n68-k9-C23-V3	523	523	523	523	523
B-n78-k10-C26-V4	606	606	606	606	606
P-n16-k8-C6-V4	170	170	170	170	170
P-n19-k2-C7-V1	111	111	111	111	111
P-n20-k2-C7-V1	117	117	117	117	117
P-n21-k2-C7-V1	117	117	117	117	117
P-n22-k2-C8-V1	111	111	111	111	111
P-n22-k8-C8-V4	249	249	249	249	249
P-n23-k8-C8-V3	174	174	174	174	174
P-n40-k5-C14-V2	213	213	213	213	213
P-n45-k5-C15-V2	238	238	238	238	238
P-n50-k10-C17-V4	292	292	292	292	292
P-n50-k7-C17-V3	261	261	261	261	261
P-n50-k8-C17-V3	262	262	262	262	262
P-n51-k10-C17-V4	309	309	309	309	309
P-n55-k10-C19-V4	301	301	301	301	301
P-n55-k15-C19-V6	378	378	378	378	378
P-n55-k7-C19-V3	271	271	271	271	271
P-n55-k8-C19-V3	274	274	274	274	274
P-n60-k10-C20-V4	325	325	325	325	325
P-n60-k15-C20-V5	379.2	382	382	382	382
P-n65-k10-C22-V4	372	372	372	372	372
P-n70-k10-C24-V4	385	385	385	385	385
P-n76-k4-C26-V2	309	320	309	309	309
P-n76-k5-C26-V2	309	309	309	309	309
P-n101-k4-C34-V2	370	374	370	370	370

Table 5.3. Computational results for large-scale instances with $\theta = 2$.

Instance	LB [9]	*ALNS* [9]	*ITS* [126]	*Hybrid alg.*	*Memetic alg.*
M-n101-k10-C51-V5	542	542	542	542	542
M-n121-k7-C61-V4	707.7	719	720	720	720
M-n151-k12-C76-V6	629.9	659	659	659	659
M-n200-k16-C100-V8	744.9	791	805	791	791
G-n262-k25-C131-V12	2 863.5	3 249	3 319	3 278	3 176

Table 5.4. Computational results for large-scale instances with $\theta = 3$.

Instance	LB [9]	*ALNS* [9]	*ITS* [126]	*Hybrid alg.*	*Memetic alg.*
M-n101-k10-C34-V4	458	458	458	458	458
M-n121-k7-C41-V3	527	527	527	527	527
M-n151-k12-C51-V4	465	483	483	483	483
M-n200-k16-C67-V6	563.13	605	605	609	605
G-n262-k25-C88-V9	2 102	2 476	2 463	2 484	2 476

With regard to the computational times, it is difficult to make a fair comparison between algorithms, because they have been evaluated on different computers and are implemented in different languages. The running times of our described approaches are proportional to the number of generations. From the computational experiments, it seems that 10^4 generations are enough to explore the solution space of the GVRP. Our proposed heuristic algorithms seem to be slower than ALNS and comparable with ITS. Therefore, we can conclude that our approaches will be appropriate when the execution speed is not critical.

5.7 Notes

The generalized vehicle routing problem (GVRP) is an extension of the vehicle routing problem, and consists of designing optimal delivery or collection routes from a given depot to a number of predefined, mutually exclusive and exhaustive clusters, subject to capacity restrictions. The GVRP was introduced by Ghiani and Improta [48] and reduces to the generalized traveling salesman problem if there is no restriction on the capacity of the vehicle.

Within this chapter, we described the following results concerning the GVRP:

- An efficient transformation of the GVRP into the classical capacitated arc routing problem.

- Four formulations of the problem based on mixed integer programming.
- Special cases of the proposed formulations: the generalized multiple traveling salesman problem, the generalized traveling salesman problem and the clustered generalized vehicle routing problem;
- two algorithmic approaches for solving the GVRP: a hybrid algorithm based on a genetic algorithm applied on the global graph and a novel memetic algorithm that combines a genetic algorithm (GA) with a powerful local search (LS) procedure.

The material presented in this chapter is based on the results published by Bektaş et al. [9]. Ghiani and Improta [48] and Pop et al. [163, 165, 168].

Chapter 6

The Generalized Fixed-Charge Network Design Problem (GFCNDP)

Network design problems that arise in telecommunication applications have originated new challenges in the field of combinatorial optimization. Hierarchical telecommunication networks are physical organizations of communications facilities, each higher level covering a wider or more general area of operation than the next lower level. Hierarchical networks consist of layers of networks and are well-suited for coping with changing and increasing demands. Hierarchical telecommunication networks exist for historical reasons and enable economy of scale in the central high speed networks, the backbone-networks.

A network consists of nodes and links interconnecting the nodes. A hierarchical network consists of disjoint sets of nodes called clusters. In addition, each cluster contains at least one hub node. The backbone consists of hub nodes interconnected by backbone links. Similarly, the nodes in each cluster are interconnected by cluster links.

When dealing with designing hierarchical network problems, the following interrelated subproblems have to be solved [199]:

- Hub location.
- Clustering of the nodes.
- Interconnection of the nodes in the backbone and cluster networks.
- Routing.

These four subproblems have often been considered, and were mostly solved independently. Only a few of them have been considered simultaneously. The problem of treating all mentioned subproblems in an integrated way is called the *hierarchical network design problem.* For more information on hierarchical network design we refer to [20, 98, 199].

An example of a hierarchical network is shown in Figure 6.1.

In the above figure, each cluster contains one hub, which is marked by a filled square. The backbone links are the links interconnecting the hubs and the cluster links are the links which interconnect nodes internally in a cluster. Note, that there are no links between non-hub-nodes in different clusters.

This chapter deals with the problem of designing a backbone mesh network interconnecting a given number of clusters. The problem is an important subproblem when constructing hierarchical telecommunication networks. It is called the generalized

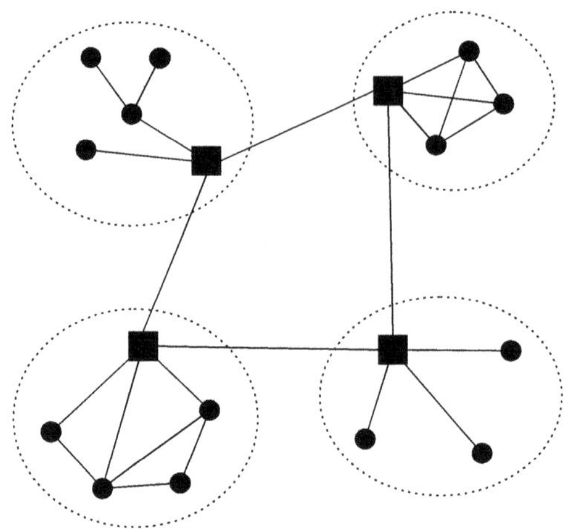

Figure 6.1. Example of a hierarchical network design.

fixed-charge network design (GFCND) problem and was introduced by Thomadsen and Stidsen [200].

The GFCND problem belongs to the class of generalized network design problems which are obtained from the classical network design problems in a natural way by considering a related problem relative to a given partition of the nodes of the graph into node sets.

6.1 Definition of the GFCNDP

Given an n-node undirected graph $G = (V, E)$ and $V_1, \dots, V_m$ a partition of V into m *clusters*, we assume that the nodes in each cluster are internally connected by a cluster-network and the clusters, on the other hand, are not connected but have to be connected by a backbone-network, in order to be able to communicate. We assume as well, that between any two selected nodes (hubs) from different clusters there exists an inter-cluster edge having a positive cost. In addition to the costs of edge establishment, each inter-cluster edge has a required capacity that depends on the communication demand and on the way the demands are routed.

The *GFCNDP* seeks to find the cheapest backbone network connecting exactly one hub from each of the given clusters. The problem involves two related decisions

- Selecting a hub from each cluster $|V_i|$, for all $i = 1, \dots, m$.
- Finding the cheapest backbone network connecting exactly the selected hubs.

In Figure 6.2 a) five node sets (clusters) are presented. The clusters may have a different number of nodes. We assume that all nodes from each cluster are connected by an intercluster network.

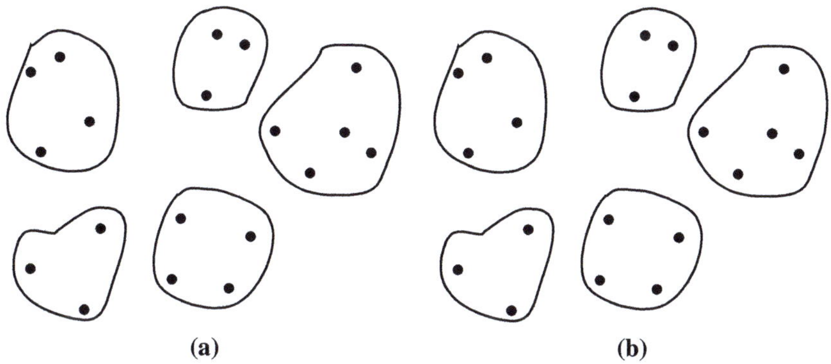

Figure 6.2. The generalized fixed-charge network design (a) node sets (clusters), (b) backbone network.

The clusters, on the other hand, are not connected but have to be connected by a backbone network, in order to be able to communicate. In Figure 6.2 b) such a possible backbone network is presented.

6.2 Integer programming formulations of the GFCNDP

Given an undirected graph $G = (V, E)$, the associated directed graph is obtained by replacing each edge $e = \{i, j\} \in E$ with the opposite arcs (i, j) and (j, i) with the same weight as the edge $\{i, j\} \in E$. In order to model the GFCND problem as a mixed integer programming problem, we define the following related sets, parameters and decision variables

Sets and notations:

V – the set of all the nodes, $i \in V$.

$V_1, \dots, V_m$ – a partition of V into m subsets called *clusters*.

E – the set of edges (unordered node-pairs), $e \in E$.

A – the set of arcs (ordered node-pairs), $a \in A$.

D – the set of commodities which represent the demands to be satisfied between any unordered cluster-pairs, $d \in D$.

s_a and t_a – denote the start and terminating nodes of arc a.

s_d and t_d – denote the start and terminating clusters of demand d.

Parameters:

c_e – the fixed cost of the edge e, $e \in E$.

g_a – the capacity cost of arc a, which is the cost per unit of demand using that arc.

b_d – the communication demand volume for demand d.

Decision Variables:

- The binary decision variables

$$z_i = \begin{cases} 1 & \text{if the node } i \text{ is selected as a hub-node} \\ 0 & \text{otherwise} \end{cases},$$

$$x_e = \begin{cases} 1 & \text{if the edge } e \in E \text{ is chosen between two hub-nodes} \\ 0 & \text{otherwise} \end{cases},$$

$$w_a = \begin{cases} 1 & \text{if the arc } a \in A \text{ is chosen between two hub-nodes} \\ 0 & \text{otherwise} \end{cases}.$$

- The fractional continuous flow variables: f_{ij}^d reflect transportation decisions for each arc $(i, j) \in A$ and commodity $d \in D$.

Our first model uses the binary decision variables defined above and the multicommodity flow variables f_a^d.

$$\min \sum_{e \in E} c_e x_e + \sum_{d \in D, a \in A} g_a b_d f_a^d$$

$$s.t. \quad z(V_k) = 1, \quad \forall\, k \in K = \{1, \ldots, m\}, \tag{6.1}$$

$$\sum_{a \in \delta^+(i)} f_a^d - \sum_{a \in \delta^-(i)} f_a^d = \begin{cases} z_i, & i \in s_d \\ -z_i, & i \in t_d \\ 0 & \text{otherwise} \end{cases}, \quad d \in D, i \in V, \tag{6.2}$$

$$x_e \le z_i, \quad \forall\, e \in E \text{ and } i \in e, \tag{6.3}$$

$$f_{ij}^d \le w_{ij}, \quad \forall\, a = (i, j) \in A,\ d \in D, \tag{6.4}$$

$$w_{ij} + w_{ji} = x_e, \quad \forall\, e = \{i, j\} \in E, \tag{6.5}$$

$$f_a^d \ge 0, \quad \forall\, a = (i, j) \in A,\ d \in D, \tag{6.6}$$

$$x, z, w \in \{0, 1\}. \tag{6.7}$$

In the above mixed integer programming formulation of the GFCNDP, the objective function represents the the combined edge establishment costs and edge capacity requirement costs, the constraints (6.1) guarantee that exactly one hub node is selected from each cluster, the flow conservation constraints (6.2) guarantee that demands are satisfied between each origin-destinations clusters. The right hand-side is z_i if the demand originates at the cluster s_d, $-z_i$ if the demand ends in the cluster t_d and 0 otherwise. The restrictions (6.3) guarantee that only edges between selected hubs will be established. The restrictions (6.4) not only ensure that arc capacities are respected, but they force the flow of any commodity to be 0 if the arc is not chosen in the design and the constraints (6.5) relate the connection between the variables attached to the edges and those attached to the arcs. Finally the constraints (6.7) and (6.8) define the domains of the variables.

We will call this model the *directed multicommodity flow model*. We let P_{mcflow} denote the projection of the feasible set of the linear programming relaxation of this model into the (x, z)-space.

We obtain a closely related formulation by eliminating the binary variables w_{ij}. The resulting formulation consists of the constraints (6.1), (6.2), (6.3), (6.6), (6.7) plus

$$f_{ij}^{d} + f_{ji}^{d'} \leq x_e, \quad \forall\, d \in D \quad \text{and} \quad \forall\, e \in E. \tag{6.8}$$

We refer to this model as the *bidirectional flow formulation* of the GFCNDP and let P_{bdflow} denote its set of feasible solutions in (x, z)-space. Observe that, since we have eliminated the variables w_a in constructing the bidirectional flow formulation, this model is defined on the undirected graph $G = (V, E)$, even though for each demand $d \in D$ we permit flows f_{ij}^d and f_{ji}^d in both directions on edge $e = \{i, j\}$. This model was introduced by Thomadsen and Stidsen [200] and, based on it, the authors proposed a branch-cut-and-price algorithm for solving the GFCNDP.

In the bidirectional flow formulation, the constraints (6.8) which we called the bidirectional flow inequalities, link the flow of different demands flowing in different directions on the edge $\{i, j\}$.

The following result holds:

Proposition 6.1. $P_{\text{mcflow}} = P_{\text{bdflow}}$.

Proof. If $(w, x, z, f) \in P_{\text{mcflow}}$, then, using the constraints (6.4) we have that

$$f_{ij}^{d} + f_{ji}^{d'} \leq w_{ij} + w_{ji} = x_e, \quad \forall\, e = \{i, j\} \in E \quad \text{and} \quad \forall\, d, d' \in D.$$

On the other hand, assume that $(x, z, f) \in P_{\text{bdflow}}$. From the constraints (6.8), we get

$$\max_{d} f_{ij}^{d} + \max_{d'} f_{ji}^{d'} \leq x_e, \quad \forall\, e = \{i, j\} \in E.$$

Hence, we can choose w such that $\max_d f_{ij}^d \leq w_{ij}$ and $x_e = w_{ij} + w_{ji}$ for all $e = (i, j) \in E$. For example, take

$$w_{ij} = \frac{1}{2}(x_e + \max_{d} f_{ij}^{d} - \max_{d'} f_{ji}^{d'}).$$

Clearly, $(w, x, z, f) \in P_{\text{mcflow}}$. □

Let G' be the graph obtained from G after replacing all nodes of a cluster V_i with a supernode representing V_i. For convenience, we identify V_i with the supernode that represents it. We will call this graph the global graph and we assume that G' with vertex set $\{V_1, \ldots, V_m\}$ is complete. We will denote the set of nodes of this global graph by V_g, $V_g = \{1, \ldots, m\}$, the set of edges by E_g, $E_g = \{e_g = (i, j) \mid i, j \in \{1, \ldots, m\}\}$ and the set of arcs by A_g.

Our last formulation aims to distinguish between *global*, i.e. inter-cluster connections, and *local* ones, i.e. connections between nodes from different clusters. The local-global approach introduced by Pop [142] was applied successfully to provide strong integer programming formulations, exact algorithms, solution procedures in the case of the generalized minimum spanning tree problem [160, 150] and the generalized traveling salesman problem [79].

We introduce the binary variables y_{ij} $(i, j \in \{1, \ldots, m\})$ to describe the global connections, so $y_{ij} = 1$ if cluster V_i is connected to cluster V_j and $y_{ij} = 0$ otherwise.

This polytope, denoted by P_{FCND}, can be represented by the following polynomial number of constraints:

$$\sum_{a_g \in \delta^+(i)} F^d_{a_g} - \sum_{a_g \in \delta^-(i)} F^d_{a_g} = \begin{cases} 1, & i \in s_{gd} \\ -1, & i \in t_{gd} \\ 0 & \text{otherwise} \end{cases}, \forall\, d \in D, i \in V_g, \tag{6.9}$$

$$F^d_{ij} + F^d_{ji} \leq y_{e_g}, \qquad \forall\, e_g = (i, j) \in E_g, d \in D, \tag{6.10}$$

$$y_{ij} \in \{0, 1\}, \qquad \forall\, i, j \in V_g, \tag{6.11}$$

$$F^d_{a_g} \geq 0, \qquad \forall\, a_g \in A_g, d \in D, \tag{6.12}$$

where the continuous decision variables $F^d_{a_g}$ correspond to the fraction of demand d routed on the arc a_g and the relation to the decision variables f^d_a is given by $F^d_{a_g} = \sum_{\{a=(i,j)\in A \mid i \in V_k, j \in V_l\}} f^d_a$, $a_g = (k, l) \in A_g$.

If the 0-1 vector y corresponds to a fixed-charge network design problem on the contracted graph, the corresponding local solution $x, z \in \{0, 1\}^{|E|}$ that minimizes the costs can be obtained by solving the following integer linear programming problem:

$$\begin{aligned} \min \sum_{e \in E} & c_e x_e \\ s.t. \quad z(V_k) &= 1, && \forall\, k \in K = \{1, \ldots, m\}, \\ x(V_l, V_r) &= y_{lr}, && \forall\, l, r \in K = \{1, \ldots, m\},\ l \neq r, \\ x(i, V_r) &\leq z_i, && \forall\, r \in K, \forall\, i \in V \setminus V_r, \\ x_e, z_i &\in \{0, 1\}, && \forall\, e = \{i, j\} \in E,\ \forall\, i \in V, \end{aligned}$$

where $x(V_l, V_r) = \sum_{i \in V_l, j \in V_r} x_{ij}$ and $x(i, V_r) = \sum_{j \in V_r} x_{ij}$.

The observations presented so far lead to our final formulation, called the *local-global formulation* of the GFCND problem as an 0-1 mixed integer programming problem:

$$\min \sum_{e \in E} c_e x_e + \sum_{d \in D, a \in A} g_a b_d f_a^d$$

$$s.t. \qquad z(V_k) = 1, \quad \forall\, k \in \{1,\ldots,m\},$$

$$\sum_{a_g \in \delta^+(i)} F_{a_g}^d - \sum_{a_g \in \delta^-(i)} F_{a_g}^d = \begin{cases} 1, & i \in s_{gd} \\ -1, & i \in t_{gd} \\ 0 & \text{otherwise} \end{cases}, \forall\, d \in D, i \in V_g$$

$$F_{ij}^d + F_{ji}^d \leq y_{e_g}, \quad \forall\, e_g = (i,j) \in E_g, d \in D$$

$$F_{a_g}^d = \sum_{\substack{\{a=(i,j)\in A \mid \\ i \in V_k, j \in V_l\}}} f_a^d, \quad \forall\, a_g = (k,l) \in A_g$$

$$x(V_l, V_r) = y_{lr}, \quad \forall\, l, r \in K = \{1,\ldots,m\},\ l \neq r \tag{6.13}$$

$$x(i, V_r) \leq z_i, \quad \forall\, r \in K, \forall\, i \in V \setminus V_r \tag{6.14}$$

$$f_a^d \geq 0, \quad \forall\, a = (i,j) \in A,\ d \in D$$

$$x, z, y \in \{0,1\}.$$

The integrality restriction regarding the decision variables z can be relaxed and replaced by the restrictions $z_i \geq 0,\ i \in V$. Obviously, if the decision variables $x, y \in \{0,1\}$ then restrictions (6.13) and (6.14) imply that $z \in \{0,1\}$.

We observe as well that the local-global model of the GFCND problem has a polynomial number of restrictions and variables, and the number of integer decision variables is less compared to the other MIP models of the problem.

6.3 Solving the GFCNDP

Despite the importance of the GFCND problem in network design not much research on the problem has been done. As far as we know, only two approaches have been developed in order to solve the GFCND problem, a branch-cut-and-price algorithm proposed by Thomadsen and Stidsen [200] and a solution procedure based on the local-global formulation of the problem developed by Pop and Pop Sitar [160].

Due to the substantial number of decision variables and constraints that occur in the bidirectional flow formulation of the GFCNDP, Thomadsen and Stidsen [200] described a branch-cut-and-price algorithm that takes advantage of the fact that only few non-zero variables and few binding constraints exist.

The most widely used method for solving MIPs, however, is the branch-and-cut algorithm, employed by most state-of-the-art MIP solvers. Although the user can extend the existing branch-and-cut solvers by adding problem specific plugins, these solvers feature very effective generic algorithms, which can be used to solve many general MIPs without further problem knowledge.

Nowadays, branch-and-price is a well-established method to solve huge and extremely difficult combinatorial optimization problems. Its success relies on exploiting structures contained in the problem with use of a decomposition and being able to solve the occurring pricing problems rather efficiently with problem specific algorithms. Several general acceleration strategies for the branch-and-price process were proposed [22, 116].

Branch-cut-and-price is a combination of branch-and-price and branch-and-cut. The variables are given implicitly and added to the problem only when needed. Additionally, cutting planes are added to the problem during the solving process, in order to strengthen the LP relaxation. Variables and cutting planes are created alternately.

The branch-cut-and-price algorithm proposed by Thomadsen and Stidsen [200] is specified next.

Algorithm 6.2 (The branch-cut-and-price algorithm for solving the GFCNDP (Thomadsen and Stidsen [200])).

```
VARS = initial set of variables
CONS = initial set of constraints
BRANCHES = set consisting of an initial branch
INCUMBENT = +∞
do
  Select and remove BRANCH ∈ BRANCHES
  do
    LP-lowerbound = SOLVE LP (CONS, VARS, BRANCH)
    Add variables with negative reduced cost to VARS
      Add violated constraints to CONS
   while variables or constraints added
   if LP solution is integer and LP-lowerbound < INCUMBENT:
      Update incumbent: INCUMBENT = LP-lowerbound
      Fathom
   else if LP-lowerbound ≥ INCUMBENT:
      Fathom
   else
      Select branching variable υ
      Add the two branchings υ = 1 and υ = 0 to BRANCHES
 while BRANCHES ≠ ∅
```

6.4 Computational results

In order to solve the GFCND problem with the branch-cut-and-price algorithm, Thomadsen and Stidsen [200] generated test instances with 50–300 nodes, in steps of 25 nodes and with 10–30 clusters, in steps of 10 clusters.

Each test instance consists of a number of nodes located randomly in the unit square $[0, 1] \times [0, 1]$. The clusters were created in the same way as in [43], using the procedure called CLUSTERING, and is described in the following. This procedure sets the number of clusters $m = [\frac{n}{5}]$, identifies the m farthest nodes from each other, and assigns each remaining node to its nearest center. Obviously, some real-world problems may have different cluster structures. However, the solution proposed in this paper is able to handle any cluster structure.

The costs of the edges c_e, $e = \{i, j\} \in E$ were set equal to the Euclidean distance between the nodes i and j and for each pair of clusters V_k and V_l, $k, l \in \{1, \dots, m\}$ a demand d is defined. The value of the communication demand, b_d, is set equal to a uniformly distributed random variable between 0 and 20 times the product of the number of nodes in each of the two clusters.

The tests were performed on a 1 000 MHz SUN Fire V440, running Solaris 9, using CPLEX as a LP-solver. The obtained results are given in Table 6.1.

Table 6.1. Branch-cut-and-price algorithm computational results [200].

Pb. size		Average cluster degree	Time (s)		
m	n		Total	Root-LP	Branch
10	50	2.4	42	40	2
	75	2.8	73	73	0
	100	3.4	234	234	0
	125	2.6	605	604	0
	150	3.8	654	653	0
	175	3.8	767	767	0
	200	4.2	2 785	2 784	0
	225	3.6	3 318	3 281	37
	250	3.2	5 499	5 497	0
	275	4.0	4 658	4 656	0
	300	3.4	7 510	7 508	0
20	50	2.9	1 499	1 182	317
	75	2.8	4 904	4 903	0
	100	3.2	11 978	11 976	0
	125	3.5	34 385	29 799	4 586
	150	3.5	34 361	34 358	0
30	50	2.6	23 511	16 108	7 404

Table 6.2. Computational results for solving the GFCND problem.

Problem size		*Local-global MIP*	*Branch-cut-and-price*
m	n	CPU	CPU
10	40	0.15 s	–
	60	0.47 s	–
	80	1.00 s	–
	100	1.90 s	234 s
	120	5.58 s	–
	140	2.93 s	–
	160	6.19 s	–
	180	6.13 s	–
	200	6.60 s	2 785 s
20	40	1.72 s	–
	60	8.82 s	–
	80	5.78 s	–
	100	5.96 s	11 978 s
	120	48.08 s	–
	140	12.92 s	–
30	50	9.75 s	23 511
	90	106.25 s	–

Analyzing the computational results we can observe that the branch-cut-and-price algorithm can find optimal solutions of the GFCND problem for test instances with up to 10 clusters and 300 nodes, 20 clusters and 150 nodes, and 30 clusters and 50 nodes. It is also clear that most of the computational time is spent on solving the root-LP problem.

Using the advantages of the local-global formulation of the GFCND problem, its compact form and the reduced number of variables and constraints, we solved the problem with CPLEX as a MIP solver. Due to the fact that the authors Thomadsen and Stidsen [200] did not make their instances available, we generated the test instances as in the case of the GMSTP with 40–200 nodes, in steps of 20 nodes and 10–30 clusters in steps of 10 clusters.

The tests were performed on an Intel Core 2 quad with 4 GB memory. The computational results obtained when solving the MIP local-global model of the GFCND problem with CPLEX are reported in Table 6.2 and are compared with the results obtained by Thomadsen and Stidsen [200].

The first two columns contain the number of clusters and nodes for the given test instances and the third and forth columns contain the total computation time needed to solve the problem to optimality, using the local-global MIP, respectively using the branch-cut-and-price algorithm proposed by [200]. In the last column '–' means that for those instances, [200] did not provide computational results.

Comparing the computational results for the common instances, we can see that the necessary times in order to provide the optimal solution in the case of the local-global mixed integer programming formulation of the problem, using CPLEX, were less than in the case of the branch-cut-and-price algorithm.

6.5 Notes

The GFCNDP consists of finding the cheapest backbone mesh network that interconnects one hub from a given number of clusters. This problem is an important subproblem when constructing hierarchical telecommunication networks and was introduced by Thomadsen and Stidsen [200].

In this chapter the following results have been presented:

- two novel formulations of the GFCNDP based on integer programming: a directed multicommodity flow model from which we obtained a closely related formulation called the bidirectional flow formulation described by Thomadsen and Stidsen [200] and the local-global model;
- two approaches for solving the GFCND problem: a branch-cut-and-price algorithm, proposed by Thomadsen and Stidsen [200], and a solution procedure, based on the local-global formulation of the problem developed by Pop and Pop Sitar [169].

The material presented in this chapter is based on the results published by Pop and Pop Sitar [169], and Thomadsen and Stidsen [200].

Further work may be directed towards the design of exact methods and metaheuristic algorithms for solving the GFCND problem.

Chapter 7

The Generalized Minimum Edge-Biconnected Network Problem (GMEBCNP)

As we have seen in Chapter 2, when we aim at designing communication networks, a minimum spanning tree is usually the cheapest network that will allow a given set of sites to communicate. However, such a network is not robust against failures, since it might not survive the break of even a single link or site. For many communication structures, an important issue, besides the minimization of connection costs, is reliability. The network should be robust against failures in connections or against switching nodes, in the sense that any two nodes do not loose connection in case of simultaneous failures up to a certain maximum number. To accomplish this, redundant communication routes must exist for any pair of nodes.

In order to deal with this issue, the terms *vertex-connectivity* and *edge-connectivity* have been introduced in graph theory, to describe this kind of robustness.

The classical edge-biconnectivity augmentation (E2AUG) problem consists of identifying the cheapest set of augmented edges, so that a single edge deletion does not disconnect the graph. The problem was introduced and investigated by Eswaran and Tarjan [35]. They also showed that the problem is $\mathcal{NP}$-hard by a reduction from the Hamiltonian cycle problem.

In this section, we are concerned with the generalized version of the edge-biconnectivity augmentation (E2AUG) problem called the generalized minimum edge-biconnected network problem (GMEBCNP). Given an undirected graph, whose nodes are partitioned into a number of clusters, the GMEBCNP is to find a minimum cost subgraph containing exactly one node from each cluster and being connected in an edge-biconnected way.

The GMEBCNP was introduced by Huygens [84] and arises in the design of backbones in large communication networks.

A similar problem is the generalized minimum vertex-biconnected network problem (GMVBCNP), where the goal is to find a minimum cost subgraph containing exactly one node from each cluster and satisfying the vertex-biconnectivi-ty constraint. For more information on this problem, we refer to [136].

7.1 Definition and complexity of the GMEBCNP

Given an n-node weighted undirected graph $G = (V, E)$ whose edges are associated with non-negative costs and $V_1, \dots, V_m$ a partition of V into m *clusters*, the *generalized minimum edge-biconnected network problem* (GMEBCNP) seeks to find a minimum-cost subgraph S, connecting a subset of nodes, which includes *exactly* one node from each cluster $V_i, i \in \{1, \dots, m\}$ and which contains no bridges. The bridges of a connected graph are the graph edges whose removal disconnects the graph. More generally, a bridge (also known as a cut-edge or cut arc or an isthmus) is an edge whose deletion increases the number of connected components, see [72] for further details. We will call such a subgraph a *generalized edge-biconnected subgraph.*

In the next figure we present an example of a feasible solution of the GMEBCNP.

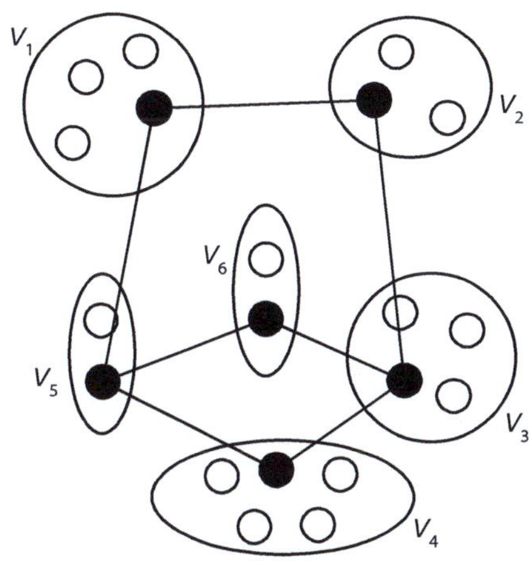

Figure 7.1. Example of a solution of the GMEBCNP.

The GMEBNP is an $\mathcal{NP}$-hard problem, since in the particular case where $|V_i| = 1$, $\forall\ i = 1, \dots, m$, the problem of finding a minimum-cost biconnected network, spanning all the nodes of a graph is an $\mathcal{NP}$-hard problem. See Garey and Johnson [45].

The generalized minimum edge-biconnected network problem is closely related to some other generalized network design problems such as the generalized traveling salesman problem (GTSP). Because every solution of the GTSP is obviously edge-biconnected and therefore is also a solution of the GMEBCNP, its solution value can be seen as an upper bound of the GMEBCNP.

7.2 A mixed integer programming model of the GMEBCNP

In this section, we present the mixed integer programming formulation described by Hu et al. [82]. The model is a multicommodity flow mixed integer programming formulation based on the local-global approach which we originally suggested for the GMSTP (see Chapter 2, Section 2.3.5). The following decision binary variables are considered

$$x_{ij} = \begin{cases} 1 & \text{if edge } e = (i,j) \in E \text{ is included in the selected subgraph} \\ 0 & \text{otherwise} \end{cases},$$

$$z_i = \begin{cases} 1 & \text{if the node } i \in V \text{ is included in the selected subgraph} \\ 0 & \text{otherwise} \end{cases},$$

$$y_{lr} = \begin{cases} 1 & \text{if cluster } V_l \text{ is connected to cluster } V_r \text{ in the global graph} \\ 0 & \text{otherwise} \end{cases},$$

$$f_{lr}^k = \begin{cases} 1 & \text{if a flow } f \text{ of commodity } k \text{ exists from cluster } V_l \text{ to cluster } V_r \\ 0 & \text{otherwise} \end{cases},$$

where $l, r \in \{1, \ldots, m\}$ and $k \in \{2, \ldots, m\}$.

The formulation consists of two parts. The first part, called the multi-commodity flow part, operates on the global graph and is based on sending from the root cluster V_1 two units of flow f to every other cluster, using edge-disjoint routes. The flows for different clusters are distinguished by their commodity k. The second part, called the local-global part, described in Section 2.5.3 relates the local variables x_{ij} and z_i to the global variables y_{lr}, describing the connections between clusters.

$$\min \sum_{e \in E} c_e x_e$$

$$\text{s.t.} \quad \sum_{l=1}^{m} f_{lr}^k - \sum_{r=1}^{m} f_{rp}^k = \begin{cases} -2 & \text{if } r = 1 \\ 2 & \text{if } r = k \,, \\ 0 & \text{else} \end{cases} \quad \forall\, r \in \{1, \ldots, m\}, \; \forall\, k \in \{2, \ldots, m\} \tag{7.1}$$

$$f_{lr}^k + f_{rl}^k \le 1 \qquad \forall\, l, r \in \{1, \ldots, m\}, \; \forall\, k \in \{2, \ldots, m\} \tag{7.2}$$

$$\sum_{i \in V_k} z_i = 1, \qquad \forall\, k \in K = \{1, \ldots, m\}, \tag{7.3}$$

$$x(V_l, V_r) = y_{lr}, \qquad \forall\, l, r \in K = \{1, \ldots, m\}, \; l \neq r \tag{7.4}$$

$$x_{ij} \le z_i, \qquad \forall\, i \in V_l, \; \forall\, j \in V \setminus V_l, \; \forall\, l \in \{1, \ldots, m\}, \tag{7.5}$$

$$y_{lr} \ge f_{lr}^k, \qquad \forall\, l, r \in \{1, \ldots, m\} \; l \neq r \;\; \forall\, k \in \{2, \ldots, m\}, \tag{7.6}$$

$$f_{lr}^k \ge 0, \qquad \forall\, l, r \in \{1, \ldots, m\}, \; \forall\, k \in \{2, \ldots, m\}, \tag{7.7}$$

$$y_{lr} \in \{0, 1\}, \qquad \forall\, l, r \in \{1, \ldots, m\}, \tag{7.8}$$

$$x_e, z_i \in \{0, 1\}, \qquad \forall\, e = (i, j) \in E, \; \forall\, i \in V, \tag{7.9}$$

where $x(V_l, V_r) = \sum_{i \in V_l, j \in V_r} x_{ij}$.

In this model, the constraints (7.1) ensure that two commodities k of flow f are produced in the root cluster V_1, preserved in any other cluster they are not assigned and consumed by cluster V_k. In order to achieve the edge-biconnectivity, the constraints (7.2) forbid the transportation of two commodities assigned for the same cluster over the same connection. Constraints (7.3) guarantee that from each cluster we select exactly one node, restrictions (7.4) force us to select only edges between nodes belonging to the clusters which are connected in the global structure, and the inequalities (7.5) ensure that only edges incident to selected nodes are chosen. The valid global structure is obtained by inequalities (7.6), which ensure global connections to be included in the solution if a flow is active on it. Finally, the restrictions (7.7) and (7.8) are the integrality constraints.

7.3 Solving the GMEBCNP

Despite the importance of the GMEBCNP in the real world, especially in applications to the design of backbones in large communication networks, not much research on the problem has been done. Huygens [84] provided integer programming formulations of the problem, along with facet defining inequalities and as far as we know, only one specific algorithm has been developed for solving the GMEBCNP, namely, the improved Variable Neighborhood Search (VNS) approach [82], in which several variants are proposed, which we will describe in this section.

The variable neighborhood search (VNS) is a metaheuristic, introduced by Mladenovic [124] for solving combinatorial and global optimization problems. Its basic idea is to systematically change the neighborhood within the local search. The VNS had a very rapid development, with a dozen of papers being published on this topic. For a survey on VNS, and its extensions and applications, we refer to [70, 71, 125].

Hu et al. [82] proposed four types of neighborhoods, focusing on different aspects of the GMEBCNP. In order to create an initial solution of good quality, they considered an adaptation of the Christofides heuristic. It starts with building a solution of the generalized minimum spanning tree problem (GMSTP), using an Kruskal based heuristic. Next, the generalized spanning tree is augmented until it becomes a feasible solution of the GMEBCNP.

7.3.1 Simple Node Optimization Neighborhood (SNON)

The simple node optimization neighborhood consists of all solutions that differ from a current solution by exactly one spanned node. A move within this neighborhood is realized by exchanging a node within a cluster with another one, also removing the connections of the initial node, and adding the corresponding edges to the new node. In the next picture, we present such a move, exchanging the spanned node v_2 from V_2 with another node v_2' belonging to the same cluster.

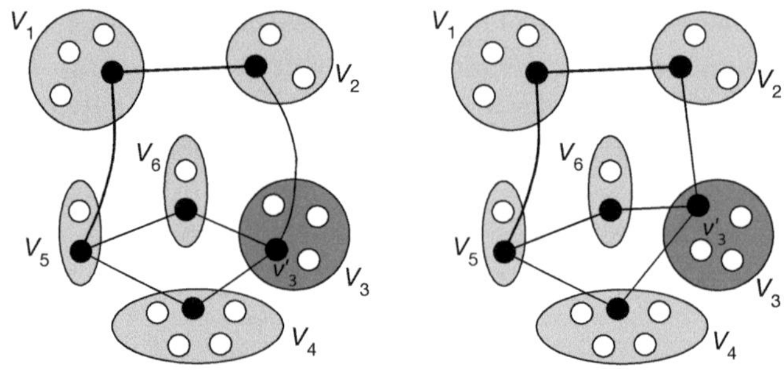

Figure 7.2. A SNON move, changing the spanned node $v_3 \in V_3$ with another node belonging to the same cluster.

7.3.2 Node Re-Arrangement Neighborhood (NRAN)

The node rearrangement neighborhood consists of all solutions that differ from a current solution by exchanging exactly two nodes. This situation is illustrated in the next figure.

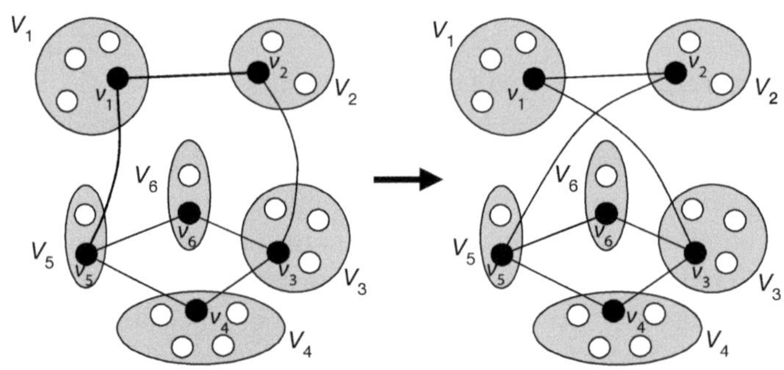

Figure 7.3. A NRAN move, swapping the nodes v_1 and v_2.

7.3.3 Edge Augmentation Neighborhood

The edge augmentation neighborhood mainly considers modifications on the edges and consists of all solutions that differ from a current solution by adding a new edge and removing other edges that now are redundant. Such a move within this neighborhood is presented in the next figure.

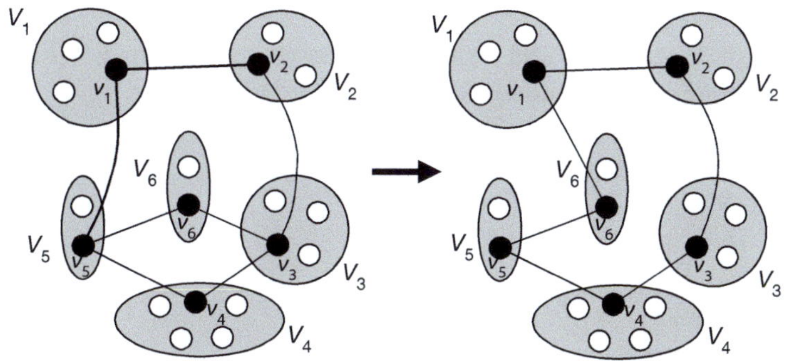

Figure 7.4. A EAN move, adding the edge (v_1, v_6) and removing the redundant edges (v_1, v_5) and (v_3, v_6).

7.3.4 Node Exchange Neighborhood

The node exchange neighborhood is a combination of the simple node optimization neighborhood and edge augmentation neighborhood, and addresses modifications of the spanned nodes as well as the edges connected them. The neighborhood consists of all solutions that differ from the current solution by exactly one spanned node and an arbitrary number of edges. This situation is illustrated in Figure 7.5.

7.3.5 Variable Neighborhood Descent

In their proposed VNS scheme, Hu et al. [82] used the variable neighborhood descent (VND) as local improvement. They provided three VND variants. The first one, denoted VND1, applies the previously described neighborhood structures. The second one, denoted VND2, applies the above mentioned neighborhoods and in addition two neighborhoods, based on the graph reduction technique [82]. Finally, the last variant called the self-adapted variable neighborhood descent (SAVND), applies the same neighborhood structures as VND2, but instead of a given static order, the neighborhoods are rearranged automatically during the search process.

The pseudocode of the general VNS is presented in Algorithm 7.1.

Algorithm 7.1 (VNS algorithm for solving the GMEBCNP).

```
Create initial solution S
repeat
    k = 0
    repeat
        S' = S
```

```
        if k > 0 then
            Shaking(S', k): add k randomly chosen edges from E \ T
            remove the redundant edges
        VND1 (S') or VND2 (S') or SAVND (S')
        if solution S' is better then S then
            S = S'
            k = 1
        else
            k=k+1
      until k = k_max
    until a termination criterion is met
return solution S
```

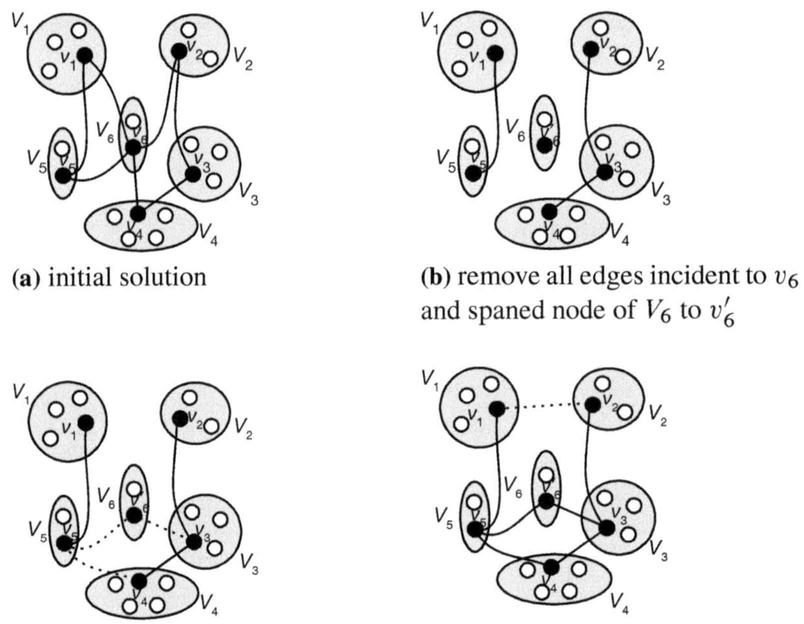

(**a**) initial solution

(**b**) remove all edges incident to v_6 and spaned node of V_6 to v_6'

(**c**) add cheapest edges between all pairs of graph components

(**d**) add cheapest edges between all pairs of graph components

Figure 7.5. A NEN move, changing the spanned node $v_6 \in V_6$, removing all adjacent edges and reaugmenting the subgraph in order to get biconnectivity.

7.4 Computational results

Hu et al. [82] have tested their VNS variants for solving the GMEBCNP on two types of instances, namely, Euclidean TSPlib instances with geographical center clustering according to [43] and random instances introduced by Ghosh [49], originally in the case of the GMSTP. For more information on these instances, we refer to Section 2.6 or [82].

The experiments were performed on a Pentium 4, 2.6 GHz PC with 1 GB RAM and, in order to compute the average values and standard deviations, Hu et al. performed 30 independent runs for each instance and algorithm variant.

After analyzing the computational results presented in Tables 7.1 and 7.2, we observe that, in the case of TSPlib instances, from among the described VNS variants SAVNS performs best, whereas for the random instances, the results are ambiguous. For the group Euclidean and non-Euclidean instances SAVNS performs best and in the case of random Euclidean the VNS2 variant performs best.

7.5 Notes

The generalized minimum edge-biconnected network problem (GMEBCNP) is an extension of the edge-biconnectivity augmentation problem and consists of finding the minimum cost subgraph containing exactly one node from each cluster and connecting in an edge-biconnected way. The GMEBCNP was introduced by Huygens [84] and arises in the design of backbones in large communication networks.

Within this chapter, we have presented an integer programming model of the problem and an improved Variable Neighborhood Search (VNS) approach, in which several variants were proposed.

The material presented in this chapter is based on the results published by Hu et al. [82].

Table 7.1. Results on TSPlib instances with geographical clustering.

		VNS1 (I)				VNS2 (II)				SAVNS (A)				Wilcoxon tests		
Instance	time	$C(T_{\text{best}})$	$\overline{C(T)}$	std dev	$\overline{t_{\text{best}}}$	$C(T_{\text{best}})$	$\overline{C(T)}$	std dev	$\overline{t_{\text{best}}}$	$C(T_{\text{best}})$	$\overline{C(T)}$	std dev	$\overline{t_{\text{best}}}$	$\gamma_{I,II}$	$\gamma_{II,A}$	$\gamma_{I,A}$
gr137	150 s	440.0	440.8	2.4	6.4 s	440.0	440.0	0.0	31.7 s	440.0	440.0	0.0	16.2 s	0.12	NA	0.12
kroa150	150 s	11 532.0	11 532.8	2.4	6.4 s	11 532.0	11 532.4	4.0	56.1 s	11 532.0	11 532.0	0.0	28.4 s	0.79	0.06	0.12
krob200	300 s	13 177.0	13 309.4	79.9	75.5 s	13 177.0	13 201.2	22.2	137.6 s	13 177.0	13 206.4	23.8	120.4 s	<0.01	0.14	<0.01
ts225	300 s	68 346.0	68 769.7	305.9	72.4 s	68 346.0	68 658.1	212.2	140.7 s	68 346.0	68 563.7	166.3	123.8 s	0.08	0.07	<0.01
gil262	300 s	1 078.0	1 157.1	54.9	146.5 s	1 059.0	1 076.3	15.9	192.7 s	1 059.0	1 070.8	10.1	168.3 s	<0.01	0.27	<0.01
pr264	300 s	29 948.0	31 639.6	1 449.3	138.0 s	29 810.0	30 434.8	725.5	238.3 s	29 810.0	29 879.3	56.0	226.4 s	<0.01	<0.01	<0.01
pr299	450 s	22 853.0	23 953.9	792.4	331.1 s	22 644.0	22 829.6	215.9	307.7 s	22 644.0	22 659.1	12.7	300.7 s	<0.01	<0.01	<0.01
lin318	450 s	21 506.0	23 101.1	840.3	409.4 s	21 028.0	21 750.4	444.4	338.5 s	20 795.0	21 321.0	228.7	299.4 s	<0.01	<0.01	<0.01
rd400	600 s	6 846.0	7 275.4	238.0	414.3 s	6 755.0	6 961.7	103.2	425.8 s	6 745.0	6 833.2	42.0	427.6 s	<0.01	<0.01	<0.01
fl417	600 s	10 234.0	10 636.4	286.8	198.7 s	10 085.0	10 547.9	234.2	329.9 s	9 708.0	9 881.9	160.8	431.2 s	0.08	<0.01	<0.01
gr431	600 s	1 337.0	1 408.2	50.8	434.3 s	1 303.0	1 346.2	27.0	346.0 s	1 284.0	1 312.3	18.0	467.3 s	<0.01	<0.01	<0.01
pr439	600 s	65 830.0	72 752.7	3 857.2	497.2 s	60 972.8	67 727.8	3 302.3	460.1 s	60 642.0	62 276.9	1 710.7	471.7 s	<0.01	<0.01	<0.01
pcb442	600 s	25 545.0	26 051.2	197.2	570.5 s	22 439.0	23 882.8	947.0	372.2 s	22 148.0	22 612.6	325.9	468.4 s	<0.01	<0.01	<0.01

Table 7.2. Results on the Ghosh's instances with 600 s CPU time.

		VNS1 (I)				VNS2 (II)				SAVNS (A)				Wilcoxon tests		
Instance	time	$C(T_{\text{best}})$	$\overline{C(T)}$	std dev	$\overline{t_{\text{best}}}$	$C(T_{\text{best}})$	$\overline{C(T)}$	std dev	$\overline{t_{\text{best}}}$	$C(T_{\text{best}})$	$\overline{C(T)}$	std dev	$\overline{t_{\text{best}}}$	$\gamma_{I,II}$	$\gamma_{II,A}$	$\gamma_{I,A}$
Group Eucl 125	600 s	159.5	159.5	0.0	29.8 s	159.5	159.5	0.0	93.1 s	159.5	159.5	0.0	38.2 s	NA	NA	NA
	600 s	163.5	163.5	0.0	25.1 s	163.5	163.5	0.0	35.9 s	163.5	163.5	0.0	14.4 s	NA	NA	NA
	600 s	166.1	166.1	0.0	9.8 s	166.1	166.1	0.0	27.0 s	166.1	166.1	0.0	6.5 s	NA	NA	NA
Group Eucl 500	600 s	684.4	717.6	25.8	432.8 s	640.4	678.5	17.8	379.9 s	629.8	644.8	9.1	460.2 s	<0.01	<0.01	<0.01
	600 s	765.1	799.0	35.4	505.4 s	657.4	698.9	24.2	438.2 s	645.3	663.8	11.9	527.0 s	<0.01	<0.01	<0.01
	600 s	747.3	761.7	37.2	332.8 s	651.7	693.7	25.4	486.3 s	643.0	655.6	8.9	505.8 s	<0.01	<0.01	<0.01
Group Eucl 600	600 s	105.1	105.1	0.0	44.9 s	105.1	105.5	1.9	315.3 s	105.1	105.1	0.0	40.7 s	0.50	0.50	NA
	600 s	105.2	105.2	0.0	10.7 s	105.2	105.9	2.7	398.4 s	105.2	105.2	0.0	8.3 s	0.01	0.01	NA
	600 s	107.5	107.5	0.0	3.3 s	107.5	107.5	0.0	48.0 s	107.5	107.5	0.0	1.5 s	NA	NA	NA
Random Eucl 250	600 s	3 571.1	3 746.6	123.2	156.2 s	2 995.1	3 226.6	116.9	503.4 s	3 235.2	3 449.3	145.7	546.2 s	<0.01	<0.01	<0.01
	600 s	3 309.5	3 661.8	166.9	156.1 s	2 662.1	2 968.7	189.4	491.2 s	2 750.2	3 299.3	161.4	550.2 s	<0.01	<0.01	<0.01
	600 s	3 051.9	3 403.2	270.1	126.9 s	2 698.6	2 888.1	98.4	500.1 s	2 793.2	3 111.9	255.0	560.1 s	<0.01	<0.01	<0.01
Random Eucl 400	600 s	943.8	1 027.2	71.5	99.9 s	841.0	926.9	61.2	401.9 s	943.8	1 028.4	48.7	59.9 s	<0.01	<0.01	0.28
	600 s	813.3	1 059.5	138.3	193.0 s	813.3	875.7	56.8	310.9 s	813.3	1 043.2	109.0	89.1 s	<0.01	<0.01	0.20
	600 s	814.9	858.7	40.5	53.7 s	794.4	836.1	39.8	385.2 s	814.9	860.7	27.8	100.0 s	<0.01	<0.01	0.15
Random Eucl 600	600 s	599.8	725.1	103.9	183.5 s	599.8	605.1	7.5	371.2 s	599.8	699.2	90.4	77.3 s	<0.01	<0.01	0.47
	600 s	785.4	823.7	55.3	226.7 s	643.1	762.0	50.6	428.7 s	643.1	813.4	53.9	151.4 s	<0.01	<0.01	0.40
	600 s	695.3	778.9	60.1	256.1 s	596.5	674.3	52.3	399.4 s	695.3	743.7	94.2	88.3 s	<0.01	<0.01	<0.01
Non Eucl 200	600 s	180.1	244.9	34.4	335.7 s	172.6	224.5	23.4	323.2 s	179.8	219.0	31.0	239.8 s	0.01	0.06	<0.01
	600 s	133.8	216.7	33.8	189.5 s	140.2	198.3	31.1	251.2 s	179.3	215.9	19.0	139.7 s	0.02	0.01	0.48
	600 s	156.0	179.8	28.4	124.8 s	135.5	190.6	29.0	272.0 s	154.4	175.6	20.1	98.3 s	0.04	<0.01	0.38
Non Eucl 500	600 s	902.6	1 121.3	123.3	206.1 s	830.5	996.7	88.5	243.0 s	771.0	960.2	99.3	387.8 s	<0.01	0.04	<0.01
	600 s	897.5	1 008.2	128.9	175.2 s	860.7	1 016.4	87.2	230.7 s	734.3	921.2	106.9	340.7 s	0.27	<0.01	0.01
	600 s	785.3	1 025.8	109.5	242.0 s	844.0	990.8	83.0	186.4 s	710.4	929.0	120.8	382.4 s	0.10	<0.01	<0.01
Non Eucl 600	600 s	102.9	138.6	25.1	330.8 s	93.0	129.9	14.2	315.2 s	91.6	125.0	20.7	301.0 s	0.09	0.02	<0.01
	600 s	107.3	132.0	19.0	277.3 s	102.3	130.2	13.6	332.8 s	94.6	118.5	19.9	275.7 s	0.45	<0.01	<0.01
	600 s	85.4	119.4	19.0	300.4 s	94.8	133.1	17.8	404.7 s	79.5	113.4	18.5	232.6 s	<0.01	<0.01	0.13

Bibliography

[1] R. Ahuja, T. Magnanti, J. Orlin, Network Flows, Prentice-Hall, Englewood Cliffs, 1993.

[2] J. R. Araque, G. Kudva, T. L. Morin and J. F. Pekny, A branch and cut algorithm for vehicle routing problem, Annals of Operations Research, Vol. 50(1), pp. 37–59, 1994.

[3] A. A. Assad and B. L. Golden, Arc Routing Methods and Applications, in M. G. Ball, T. L. Magnanti, C. L. Monma, G. L. Nemhauser, editors, Network Routing, Vol. 8 of Handbooks in Operations Research and Management Science, chapter 5, pp. 375–483, Elsevier, 1995.

[4] A. A. Assad, B. L. Golden, W. Pearn, The Capacitated Chinese Postman Problem: Lower Bounds and Solvable Cases, American Journal of Mathematical and Management Science, Vol. 7(1,2), pp. 63–88, 1987.

[5] R. Baldacci, E. Bartolini and G. Laporte, Some applications of the Generalized Vehicle Routing Problem, Journal of the Operational Research Society, Vol. 61(7), pp. 1072–1077, 2010.

[6] M. O. Ball, T. L. Magnanti, C. L. Monma, G. L. Nemhauser (Eds.), Network Routing, Handbooks in Operations Research and Management Science 8, Elsevier, Amsterdam, 1995.

[7] R. Bar-Yehuda and S. Even, A linear time approximation algorithm for the weighted vertex cover problem, Journal of Algorithms, Vol. 2(2), pp. 198–203, 1981.

[8] A. Behzad and M. Modarres, A new efficient transformation of the generalized traveling salesman problem into traveling salesman problem, in Proceedings of the 15th International Conference of Systems Engineering, 2002.

[9] T. Bektaş, G. Erdoğan and S. Røpke, Formulations and Branch-and-Cut Algorithms for the Generalized Vehicle Routing Problem, Transportation Science, Vol. 45 (3), pp. 299–316, 2011.

[10] D. Ben-Arieh, G. Gutin, M. Penn, A. Yeo and A. Zverovich, Transformations of generalized ATSP into ATSP, Operations Research Letters, Vol. 31(3), pp. 357–365, 2003.

[11] D. J. Bertsimas, The Probabilistic Minimum Spanning Tree Problem, Networks, Vol. 20(3), pp. 245–275, 1990.

[12] E. Bonabeau, M. Dorigo, G. Tehraulaz, Swarm intelligence from natural to artificial systems, Oxford, UK, Oxford University Press, 1999.

[13] B. Bontoux, C. Artigues and D. Feillet, A Memetic Algorithm with a Large Neighborhood Crossover Operator for the Generalized Traveling Salesman Problem, Computers & Operations Research, Vol. 37(11), pp. 1844–1852, 2010.

[14] V. Cacchiani, A. E. F. Muritiba, M. Negreiros and P. Toth, A Multistart Heuristic for the Equality Generalized Traveling Salesman Problem, Networks, Vol. 57(3), pp. 231–239, 2011.

[15] S. Camazine, J. L. Deneubourg, N. R. Franks, J. Sneyd, G. Theraulaz and E. Bonabeau, Self organization in biological systems, Princeton, Princeton University Press, 2001.

[16] V. Černý, A thermodynamical approach to the traveling salesman problem: an efficient simulated annealing algorithm, Journal of Optimization Theory and Applications, Vol. 45, pp. 41–55, 1985.

[17] L. Chapleau, J. A. Ferland, G. Lapalme, J.-M. Rousseau, A parallel insert method for the capacitated arc routing problem, Operations Research Letters, Vol. 3(2), pp. 95–99, 1984.

[18] J. A. Chisman, The clustered traveling salesman problem, Computers & Operations Research, Vol. 2, pp. 115–119, 1975.

[19] B. Crevier, J.-F. Cordeau and G. Laporte, The multi-depot vehicle routing problem with inter-depot routes, European Journal of Operational Research, Vol. 176(2), pp. 756–773, 2007.

[20] J. R. Current, C. S. ReVelle and J. L. Cohon, The hierarchical network design problem, European Journal of Operational Research, Vol. 27(1), pp. 57–66, 1986.

[21] M. Demange, J. Monnot, P. C. Pop and B. Ries, Selective graph coloring in some special classes of graphs, in Proceedings of ISCO 2012, Lecture Notes in Computer Science (to appear).

[22] G. Desaulniers, J. Desrosiers and M. M. Solomon, Acceleration strategies in column generation methods for vehicle routing and crew scheduling problems, in C. C. Ribeiro and P. Hansen, editors, Essays and Surveys in Metaheuristics, pages 309–324, Kluwer, Boston, 2001.

[23] M. Desrochers and G. Laporte, Improvements and extensions to the Miller-Tucker-Zemlin subtour elimination constraints, Operations Research Letters, Vol. 10(1), pp. 27–36, 1991.

[24] M. Desrochers, J. K. Lenstra, M. W. P. Savelsbergh and F. Soumis, Vehicle routing with time windows: optimization and approximation, in Vehicle Routing: Methods and Studies, B. L. Golden and A. A. Assad (editors), Elsevier, 1998.

[25] E. W. Dijkstra, A note on the two problems in connection with graphs, Numerische Mathematik, Vol. 1, pp. 269–271, 1959.

[26] V. Dimitrijević and Z. Šarić, An efficient transformation of the generalized traveling salesman problem into the traveling salesman problem on digraphs, Information Science, Vol. 102(1–4), pp. 105110, 1997.

[27] C. Ding, Y. Cheng and M. He, Two-Level genetic algorithm for clustered traveling salesman problem with application in large-scale TSPs, Tsinghua Science & Technology, Vol. 12(4), pp. 459–465, 2007.

[28] M. Dorigo and L. M. Gambardella, Ant Colony System: A cooperative learning approach to the Traveling Salesman Problem, IEEE Transactions on Evolutionary Computation, Vol. 1(1), pp. 53–66, 1997.

[29] M. Dror and M. Haouari, Generalized Steiner problems and other variants, Journal of Combinatorial Optimization, Vol 4(4), pp. 415–436, 2000.

[30] M. Dror, H. Haouari and J. Chaouachi, Generalized spanning trees, European Journal of Operational Research, Vol. 120(3), pp. 583–592, 2000.

[31] C. W. Duin, A. Volgenant and S. Voss, Solving group Steiner problems as Steiner problems, European Journal of Operational Research, Vol. 154(1), pp. 323–329, 2004.

[32] K. Easton, A. Martinoli and R. Goodman, Distributed Collective Building of Two-Dimensional Structures Using Autonomous Robots, Technical Report, Caltech Center for Neuromorphic Systems Engineering, 2001.

[33] H. A. Eiselt, M. Gendreau and G. Laporte, Arc routing problems, Part 1: The Chinese postman problem, Operations Research, Vol. 43(2), pp. 231–242, 1995.

[34] H. A. Eiselt, M. Gendreau and G. Laporte, Arc routing problems, Part 1: The Chinese postman problem, Operations Research, Vol. 43(3), pp. 399–414, 1995.

[35] K. P. Eswaran and R. E. Tarjan, Augmentation problems, SIAM Journal on Computing, Vol. 5(4), pp. 653–665, 1976.

[36] D. Feillet, P. Dejax and M. Gendreau, Traveling salesman problems with profits, Transportation Science, Vol. 39(2), pp. 188–205, 2005.

[37] C. Feremans, Generalized Spanning Trees and Extensions, PhD thesis, Universite Libre de Bruxelles, 2001.

[38] C. Feremans, M. Labbé and G. Laporte, A comparative analysis of several formulations for the generalized minimum spanning tree problem, Networks Vol. 39(1), pp. 29–34, 2002.

[39] C. Feremans, M. Labbé and G. Laporte, Generalized network design problems, European Journal of Operational Research Vol. 148, pp. 1–13, 2003.

[40] C. Feremans, M. Labbé and G. Laporte, The generalized minimum spanning tree problem: Polyhedral analysis and branch-and-cut algorithm, Networks, Vol. 43(2), pp. 71–86, 2004.

[41] C. Feremans, A. Grigoriev and R. Sitters, The geometric generalized minimum spanning tree problem with grid clustering, 4OR, Vol. 4(4), pp. 319–329, 2006.

[42] M. Fischetti, J. J. Salazar and P. Toth, The symmetric generalized traveling salesman polytope, Networks, Vol. 26, pp. 113–123, 1995.

[43] M. Fischetti, J. J. Salazar and P. Toth, A branch-and-cut algorithm for the symmetric generalized traveling salesman problem, Operations Research, Vol. 45(3), pp. 378–394, 1997.

[44] D. Gale, A Theorem on Flows in Networks, Pacific Journal of Mathematics Vol. 7, pp. 1073–1082, 1957.

[45] M. R. Garey and D. S. Johnson, Computers and Intractability: A Guide to the Theory of NP-Completeness, Freeman, San Francisco, California, 1979.

[46] N. Garg and G. Konjevod, A Polylogarithmic Approximation Algorithm for the Group Steiner Tree Problem, Journal of Algorithms, Vol. 37, pp. 66–84, 2000.

[47] M. Gerla and L. Frata, Tree structured fiber optics MAN's, IEEE Journal on Selected Areas in Communication SAC Vol. 6, pp. 934–943, 1988.

[48] G. Ghiani, G. Improta, An efficient transformation of the generalized vehicle routing problem, European Journal of Operational Research, Vol. 122(1), pp. 11–17, 2000.

[49] D. Ghosh, Solving medium to large sized Euclidean generalized minimum spanning tree problems, Technical Report NEP-CMP-2003–08–02, Indian Institute of Management, Research and Publication Department, Ahmedabad, India, 2003.

[50] D. Ghosh, A Probabilistic Tabu Search Algorithm for the Generalized Minimum Spanning Tree Problem, Technical Report NEP-CMP-2003–07–02, Indian Institute of Management, Research and Publication Department, Ahmedabad, India, 2003.

[51] F. Glover, Future paths for integer programming and links to artificial intelligence, Decision Sciences, Vol. 8, pp. 156–166, 1977.

[52] F. W. Glover and G. A. Kochenberger, Handbook of Metaheuristics, Kluwer Academic Publishers, 2002.

[53] B. Golden, S. Raghavan and D. Stanojevic, Heuristic search for the generalized minimum spanning tree problem, INFORMS Journal on Computing, Vol. 17(3), pp. 290–304, 2005.

[54] B. Golden, S. Raghavan and D. Stanojevic, The prize-collecting generalized minimum spanning tree problem, Journal of Heuristics, Vol. 14(1), pp. 69–93, 2008.

[55] B. L. Golden and R. T. Wong, Capacitated arc routing problems, Networks, Vol. 11, pp. 305–315, 1981.

[56] B. L. Golden and A. A. Assad, A decision-theoretic framework for comparing heuristics, European Journal of Operational Research, Vol. 18, pp. 167–171, 1984.

[57] M. X. Goemans, Course Notes: Approximation Algorithms, 1994.

[58] M. X. Goemans and D. J. Bertsimas, Survivable networks, linear programming and parsimonious property, Mathematical Programming, Vol. 60(1–3), pp. 145–166, 1993.

[59] M. Grotschel, L. Lovász and A. Scrijver, The ellipsoid method and its consequences in combinatorial optimization, Combinatorica, Vol. 1, pp. 169–179, 1981.

[60] M. Grötschel, L. Lovász and A. Schrijver, Geometric Algorithms and Combinatorial Optimization, Springer-Verlag, Berlin, 1988.

[61] M. Guntsch, J. Branke, M. Middendorf, H. Schmeck, ACO strategies for dynamic TSP, Abstract Proceedings of ANTS 2000, pp. 59–62, 2000.

[62] M. Guntsch and M. Middendorf, Pheromone modification strategies for ant algorithms applied to dynamic TSP, In Applications of Evolutionary Computing, Proceedings of EvoWorkshops, Springer Verlag 2001.

[63] M. Guntsch, M. Middendorf and H. Schmeck, An Ant Colony Optimization approach to dynamic TSP, in Proceedings of GECCO 2001, pp. 860–867, 2001.

[64] G. Gutin, A. P. Punnen (Eds.), Traveling Salesman Problem and Its Variations, Kluwer Academic Publishers, Dordrecht, 2002.

[65] G. Gutin and D. Karapetyan, A memetic algorithm for the generalized traveling salesman problem, Natural Computing, Vol. 9(1), pp. 47–60, 2010.

[66] G. Gutin, D. Karapetyan and N. Krasnogor, A Memetic Algorithm for the Generalized Traveling Salesman Problem, in Proceedings of NICSO 2007, pp. 199–210, 2007.

[67] G. Hadjicharalambous, P. C. Pop, E. Pyrga, G. Tsaggouris and C. D. Zaroliagis, The Railway Traveling Salesman Problem. Algorithmic Methods for Railway Optimization, Lecture Notes in Computer Science, Vol. 4359, pp. 264–275, 2007.

[68] M. Haouari, J. Chaouachi and M. Dror, Solving the generalized minimum spanning tree problem by a branch-and-bound algorithm, Journal of the Operational Research Society, Vol. 56(4), pp. 382–389, 2005.

[69] M. Haouari and J. S. Chaouachi, Upper and Lower Bounding Strategies for the Generalized Minimum Spanning Tree Problem, European Journal of Operational Research, Vol. 172,(2) pp. 632–647, 2006.

[70] P. Hansen and N. Mladenović, Variable neighborhood search: Principles and applications, European Journal of Operational Research, Vol. 130(3), pp. 449–467, 2001.

[71] P. Hansen and N.Mladenović, Variable neighborhood search, in F. W. Glover and G. A. Kochenberger, editors, Handbook of Metaheuristics, Dordrecht, 2003. Kluwer.

[72] F. Harary, Graph Theory, Reading, MA: Addison-Wesley, 1994.

[73] P. Henry-Labordere, The record balancing problem: A dynamic programming solution of a generalized traveling salesman problem, RAIRO Operations Research, B2, pp. 43–49, 1969.

[74] A. Hertz, G. Laporte and M. Mittaz, A tabu search heuristic for the capacitated arc routing problem, Operations Research, Vol. 48(1), pp. 129–135, 2000.

[75] D. Hochbaum, Approximation algorithms for set covering and vertex cover problem, SIAM Journal of Computing, Vol. 11, pp. 555–556, 1982.

[76] B. Hu, M. Leitner and G. R. Raidl, Computing Generalized Minimum Spanning Trees with Variable Neighborhood Search. In Proceedings of the 18th Mini Euro Conference on Variable Neighborhood Search, P. Hansen, N. Mladenovic, J. A. M. Perez, B. M. Batista, J. M. Moreno-Vega, eds. Tenerife, Spain, 2005.

[77] B. Hu, M. Leitner and G. R. Raidl, Combining variable neighborhood search with integer linear programming for the generalized minimum spanning tree problem, Journal of Heuristics, Vol. 14(5), pp. 473–499, 2008.

[78] B. Hu and G. Raidl, Solving the railway traveling salesman problem via a transformation into the classical traveling salesman problem, In Fatos Xhafa, Francisco Herrera, Ajith Abraham, Mario Koppen, Jose Manuel Benitez, editors, 8th International Conference on Hybrid Intelligent Systems - HIS 2008, pp. 73–77, Barcelona, Spain, 2008.

[79] B. Hu and G. Raidl, Effective neighborhood structures for the generalized traveling salesman problem, in Proc. of Evolutionary Computation in Combinatorial Optimisation - EvoCOP 2008, LNCS, Vol. 4972, pp. 36–47, 2008.

[80] B. Hu, Hybrid Metaheuristics for Generalized Network Design Problems, PhD thesis, Vienna University of Technology, 2008.

[81] B. Hu and G. R. Raidl, An evolutionary algorithm with solution archive for the generalized minimum spanning tree problem, in Alexis Quesada-Arencibia et al., editors, Proc. of EUROCAST 2011 - 13th International Conference on Computer Aided Systems Theory, Las Palmas de Gran Canaria, Spain, February 6–11, 2011, pp. 256–259, 2011.

[82] B. Hu, M. Leitner, G. R. Raidl, The Generalized Minimum Edge Biconnected Network Problem: Efficient Neighborhood Structures for Variable Neighborhood Search, Networks, Vol 55(3), pp. 256–275, 2010.

[83] F. Hutter, H. H., Hoos, K. Leyton-Brown and T. Stützle, ParamILS: An Automatic Algorithm Configuration Framework, Journal of Artificial Intelligence Research (JAIR), Vol. 36, pp. 267–306, 2009.

[84] D. Huygens, Version generalisee du probleme de conception de reseau 2-arete-connexe, Master's thesis, Universite Libre de Bruxelles, 2002.

[85] ILOG CPLEX 7.0, 2000.

[86] S. Iordache, Consultant-Guided Search - A New Metaheuristic for Combinatorial Optimization Problems, in Proceedings of the 12th Genetic and Evolutionary Computation Conference, ACM Press, 2010.

[87] S. Iordache, Consultant-Guided Search Algorithms with Local Search for the Traveling Salesman Problem, in PPSN XI - International Conference Parallel Problem Solving from Nature, LNCS, Vol. 6239, Krakow, Poland, Springer, 2010, pp. 81–90.

[88] S. Iordache, Consultant-Guided Search Algorithms for the Quadratic Assignment, in Hybrid Metaheuristics - 7th International Workshop, HM 2010, LNCS, Vol. 6373, Vienna, Austria, Springer, 2010, pp. 148–159.

[89] H. Ishii, S. Shiode, T. Nishida and Y. Namasuya, Stochastic spanning tree problem, Discrete Applied Mathematics, Vol. 3(4), pp. 263–273, 1981.

[90] D. Johnson, C. Aragon, L. McGeoch and C. Schevon, Optimization by Simulated Annealing: An Experimental Evaluation, Part I, Graph Partitioning, Operations Research, Vol. 37(6), pp. 865–892, 1989.

[91] D. S. Johnson and L. McGeoch, Experimental Analysis of Heuristics for STSP, in: Gutin, G., Punnen, A. (eds.), The Traveling Salesman Problem and its Variations, Kluwer, Dordrecht, 2002.

[92] I. Kara, T. Bektaş, Integer linear programming formulation of the generalized vehicle routing problem, 5-th EURO/INFORMS Joint International Meeting, July 6–10, 2003, Istanbul, Turkey.

[93] I. Kara, G. Laporte and T. Bektaş, A note on the lifted Miller-Tucker-Zemlin subtour elimination constraints for the capacitated vehicle routing problem, European Journal of Operational Research, Vol. 158(3), pp. 793–795, 2004.

[94] I. Kara and P. C. Pop, New Mathematical Models of the Generalized Vehicle Routing Problem and Extensions, in Proceedings of the Applied Mathematical Programming and Modelling Conference (APMOD 2008), Bratislava, Slovak Republic, May 28–31, 2008.

[95] D. Karapetyan and G. Gutin, Efficient local search algorithms for known and new neighborhoods for the generalized traveling salesman problem, European Journal of Operational Research, Vol. 219(2), pp. 234–251, 2012.

[96] N. Karmarkar, A new polynomial-time algorithm in linear programming, Combinatorica, Vol. 4, pp. 373–395, 1984.

[97] S. Kirkpatrick, C. D. Gelatt and M. D. Vecchi, Optimization by Simulated Annealing, Science, Vol. 220, No. 4598, pp. 671–680, 1983.

[98] J. G. Klincewicz, Hub location in backbone/tributary network design: a review, Location Science, Vol. 6(1–4), pp. 307–335, 1998.

[99] A. M. C. A. Koster, Frequency assignment models and algorithms, Ph.D. dissertation, Proefschrift Universiteit Maastricht, 1999.

[100] J. B. Kruskal, On the shortest spanning subtree of a graph and the traveling salesman problem, Proceedings of the American Mathematical Society, Vol. 7, pp. 48–50, 1956.

[101] G. Laporte, Location-routing problems, In B. L. Golden and A. A. Assad (eds.), Vehicle Routing: Methods and Studies, North-Holland, Amsterdam, pp. 163–197, 1988.

[102] G. Laporte, S. Chapleau, P. E. Landry and H. Mercure, An algorithm for the design of mail box collection routes in urban areas, Transportation Research B, Vol. 23(4), pp. 271–280, 1989.

[103] G. Laporte and Y. Nobert, Generalized Traveling Salesman through n sets of nodes: an integer programming approach, INFOR, Vol. 21(1), pp. 61–75, 1983.

[104] G. Laporte and I. H. Osman, Routing Problems: A bibliography, Annals of Operations Research, Vol. 61(1), pp. 227–262, 1995.

[105] G. Laporte, A. Asef-Vaziri and C. Sriskandarajah, Some applications of the generalized traveling salesman problem, Journal of the Operational Research Society, Vol. 47(12), pp. 1461–1467, 1996.

[106] G. Laporte and U. Palekar, Some applications of the clustered traveling salesman problem, Journal of the Operational Research Society, Vol. 53(9), pp. 972–976, 2002.

[107] G. Laporte, What you should know about the Vehicle Routing Problem, Naval Research Logistics, Vol. 54(8), pp. 811–819, 2007.

[108] A. Larsen, O. B. G. Madsen and M. Solomon, Partially Dynamic Vehicle Routing - Models and Algorithms, Journal of the Operational Research Society, Vol. 53(6), pp. 637–646, 2002.

[109] E. L. Lawler, J. K. Lenstra, A. H. G. Rinooy Kan, D. B. Shmoys (Eds.), Travelling Salesman Problem: A Guided Tour of Combinatorial Optimization, Wiley, Chichester, 1985.

[110] C. Lemaréchal, Lagrangian relaxation, Lecture Notes in Computer Science, Vol. 2241, pp. 112–156, 2001.

[111] W.-J. Li, J. Tsao and O. Ulular, The shortest path with at most l nodes in each of the series/parallel clusters, Networks, Vol. 26, pp. 263–271, 1995.

[112] Y. Lien, E. Ma and B. W. Wah, Transformation of the Generalized Traveling Salesman Problem into the Standard Traveling Salesman Problem, Information Sciences, Vol. 74(1–2), pp. 177–189, 1993.

[113] S. Lin and B. W. Kernighan, An effective heuristic algorithm for the traveling-salesman problem, Operations Research, Vol. 21, pp. 498–516, 1973.

[114] L. Lovász, On some connectivity properties of Eulerian graphs, Acta Mathematica Academiae Scientiarum Hungaricae, Vol. 28, pp. 129–138, 1976.

[115] L. Lovász and A. Schrijver, Cones of matrices and set-functions and 0–1 optimization, SIAM Journal on Optimization, 1(2), pp. 166–190, 1991.

[116] M. E. Lübbecke and J. Desrosiers, Selected topics in column generation, Operations Research, Vol. 53(6), pp. 1007–1023, 2005.

[117] N. Maculan, The Steiner tree problem in graphs, Surveys in Combinatorial Optimization, S. Martello, G. Laporte, M. Minoux, C. C. Ribeiro (Ed.), Annals of Discrete Mathematics, Vol. 31, pp. 185–212, 1987.

[118] T. L. Magnanti and L. A. Wolsey, Optimal trees, Handbooks in Operations Research & Management Science, M. O. Ball, T. L. Magnanti C. L. Monma, G. L. Nemhauser (Editors) Elsevier Science, Amsterdam, Vol. 7, chap. 9, pp. 503–615, 1995.

[119] J. E. Mendoza, B. Castanier, C. Gueret, A. L. Medaglia and N. Velasco, A memetic algorithm for the multi-compartment vehicle routing problem with stochastic demand, Computers & Operations Research, Vol. 37(11), pp. 1886–1898, 2010.

[120] K. Menger, „Das botenproblem", in Ergebnisse eines Mathematischen Kolloquiums 2 (K. Menger, editor), Teubner, Leipzig, pp. 11–12, 1932.

[121] N. Metropolis, A. W. Rosenbluth, M. N. Rosenbluth, A. H. Teller and E. Teller, Equation of State Calculations by Fast Computing Machines, Journal of Chemical Physics, Vol. 21(6), pp. 1087–1092, 1953.

[122] C. E. Miller, A. W. Tucker and R. A. Zemlin, Integer programming formulation of traveling salesman problems, Journal of the Association for Computing Machinery, Vol. 7(4), pp. 326–329, 1960.

[123] J. E. Mitchell, Branch-and-cut algorithms for combinatorial optimization problems, Handbook of Applied Optimization, pp. 65–77, Oxford University Press, ISBN: 0–19–512594–0, January 2002.

[124] N. Mladenović, A Variable neighborhood algorithm - a new metaheuristic for combinatorial optimization, Abstracts of papers presented at Optimization Days, Montreal, p. 112, 1995.

[125] N. Mladenović and P. Hansen, Variable neighborhood search, Computers & Operations Research, Vol. 24, pp. 1097–1100, 1997.

[126] L. Moccia, J-F. Cordeau and G. Laporte, An incremental neighborhood tabu search heuristic for the generalized vehicle routing problem with time windows, Journal of the Operational Research Society, Vol. 63, pp. 232–244, 2012.

[127] P. Moscato, On Evolution, Search, Optimization, Genetic Algorithms and Martial Arts: Towards Memetic Algorithms, Caltech Concurrent Computation Program, Report 826, 1989.

[128] Y. S. Myung, C. H. Lee and D. W. Tcha, On the generalized minimum spanning tree problem, Networks, Vol. 26(4), pp. 231–241, 1995.

[129] G. Nagy and S. Salhi, Location-routing issues: Models and methods, European Journal of Operational Research, Vol. 177(2), pp. 649–672, 2007.

[130] S. C. Narula and C. A. Ho, Degree-constrained Minimum Spanning Tree, Computer & Operations Research, Vol. 7, pp. 239–249, 1980.

[131] S. U. Ngueveu, C. Prins and R. W. Calvo, An effective memetic algorithm for the cumulative capacitated vehicle routing problem, Computers & Operations Research, Vol. 37(11), pp. 1877–1885, 2010.

[132] C. E. Noon, The Generalized Traveling Salesman Problem, PhD thesis, University of Michigan, 1988.

[133] C. E. Noon and J. C. Bean, A Lagrangian based approach for the asymmetric generalized traveling salesman problem, Operations Research, Vol. 39, pp. 623–632, 1991.

[134] C. E. Noon and J. C. Bean, An efficient transformation of the generalized traveling salesman problem, Technical Report 91–26, 1991.

[135] T. Öncan, J.-F. Cordeau and G. Laporte, A Tabu Search Heuristic for the Generalized Minimum Spanning Tree Problem, European Journal of Operational Research 191(2), pp. 306–319, 2008.

[136] A. Pagacz, B. Hu, G. R. Raidl, A memetic algorithm with population management for the generalized minimum vertex-biconnected network problem, in F. Xhafa et al., editors, 2nd International Conference on Intelligent Networking and Collaborative Systems, Workshop on Information Network Design, pp. 356–361, Conference Publishing Services, 2010.

[137] C. Pintea, P. C. Pop and C. Chira, Reinforcing Ant Colony System for the Generalized Traveling Salesman Problem, in Proceedings of International Conference Bio-Inspired Computing-Theory and Applications (BIC-TA), Wuhan, China, Vol. Evolutionary Computing Section, pp. 245–252, 2006.

[138] C. Pintea, P. C. Pop and D. Dumitrescu, An Ant-based Technique for the Dynamic Generalized Traveling Salesman Problem, in Proceedings of the 7th International Conference on Systems Theory and Scientific Computation, Athens, Greece, August 24–26, pp. 255–259, 2007.

[139] C. M. Pintea, P. C. Pop, C. Chira, The generalized traveling salesman problem solved with ant algorithms, Journal of Universal Computer Science, Vol. 13, pp. 1065–1075, 2007.

[140] C. Pintea and D. Dumitrescu, Solving a combinatorial problem using a local optimization in ant based system, Technical Report, 1274, Optimization-online, 2006.

[141] C.-M. Pintea, C. Chira, D. Dumitrescu and P. C. Pop, A Sensitive Metaheuristic for Solving a Large Optimization Problem, in Proceedings of the SOFSEM Conference, Novy Smokovec, High Tatras, Slovakia, Lecture Notes in Computer Science, Vol. 4910, pp. 551–559, 2008.

[142] P. C. Pop, The generalized minimum spanning tree problem, Twente University Press, the Netherlands, 2002.

[143] P. C. Pop, New Models of the Generalized Minimum Spanning Tree Problem, Journal of Mathematical Modelling and Algorithms, Vol. 3(2), pp. 153–166, 2004.

[144] P. C. Pop and C. Pop Sitar, A note on the complexity of the Generalized Minimum Spanning Tree Problem, Studia Universitatis Babes-Bolyai, Series Informatics, Vol. XLIX, Number 2, pp. 75–80, 2004.

[145] P. C. Pop and I. Zelina, Heuristic Methods for the Generalized Minimum Spanning Tree Problem, Acta Universitatis Apulensis, Vol. 8, pp. 385–396, 2004.

[146] P. C. Pop, Cercetări Operaţionale, Editura Risoprint, Cluj-Napoca, 2005.

[147] P. C. Pop, Approximation Theory in Combinatorial Optimization, Application to the Generalized Minimum Spanning Tree Problem, Revue d'Analyse Numerique et Approximation, Tome 34(1), pp. 93–102, 2005.

[148] P. C. Pop, G. J. Still, An easy way to obtain strong duality results in linear, linear semidefinite and linear semi-infinite programming, Mathematica, Vol. 47 (70), No. 1, pp. 105–112, 2005.

[149] P. C. Pop, A. Horvat Marc and C. Pop Sitar, The At Least version of the Generalized Minimum Spanning Tree Problem, Carpathian Journal of Mathematics, Vol. 22(1–2), pp. 129–135, 2006.

[150] P. C. Pop, W. Kern and G. J. Still, A new relaxation method for the generalized minimum spanning tree problem. European Journal of Operational Research, Vol. 170(3), pp. 900–908, 2006.

[151] P. C. Pop, On the Prize-Collecting Generalized Minimum Spanning Tree Problem, Annals of Operations Research, Vol. 150(1), 193–204, 2007.

[152] P. C. Pop, New integer programming formulations of the generalized traveling salesman problem, American Journal of Applied Sciences, Vol. 4(11), pp. 932–937, 2007.

[153] P. C. Pop, C. Pop Sitar, I. Zelina and I. Tascu, Exact algorithms for generalized combinatorial optimization problems, Lecture Notes in Computer Science, vol. 4616, pp. 154–162, 2007.

[154] P. C. Pop, C. Sabo, C. Pop Sitar and M. Crăciun, A Simulated Annealing Based Approach for Solving the Generalized Minimum Spanning Tree Problem, Creative Mathematics and Informatics, Vol. 16, pp. 42–53, 2007.

[155] P. C. Pop, C. Pintea, C. Pop Sitar and D. Dumitrescu, A Bio-Inspired Approach for a Dynamic Railway Problem, IEEE Proceedings of the 9th International Symposium on Symbolic and Numeric Algorithms for Scientific Computing, pp. 449–453, IEEE Computer Society Press, Timisoara, Romania, September 26–29, 2007.

[156] P. C. Pop, C. Pintea and C. Pop Sitar, An Ant colony based approach to the Railway Travelling Salesman Problem, in Proceedings of the Evo Conference, Valencia, Spain, Lecture Notes in Computer Science, Vol. 4448, pp. 702–711, 2007.

[157] P. C. Pop, C. Zaroliagis and G. Hadjicharalambous, A Cutting Plane Approach to Solve the Railway Traveling Salesman Problem. Studia Univ. "Babes-Balyai", Mathematica, Vol. LIII, No.1, pp. 63–73, 2008.

[158] P. C. Pop, C. M. Pintea, I. Zelina and D. Dumitrescu, Solving the Generalized Vehicle Routing Problem with an ACS-based Algorithm, American Institute of Physics, Vol. 1117, pp. 157–162, 2009.

[159] P. C. Pop, Efficient Transformations of the Generalized Combinatorial Optimization Problems into Classical Variants, in Proceedings of the 9-th Balkan Conference on Operational Research, Constanta, Romania, 2–6 September 2009.

[160] P. C. Pop, A survey of different integer programming formulations of the generalized minimum spanning tree problem, Carpathian Journal of Mathematics, Vol. 25(1), pp. 104–118, 2009.

[161] P. C. Pop, O. Matei and C. Sabo, A New Approach for Solving the Generalized Traveling Salesman Problem, in Proceedings of HM 2010, Editors M. J. Blesa et al., Lecture Notes in Computer Science, Springer, Vol. 6373, pp. 62–72, 2010.

[162] P. C. Pop, O. Matei, C. Pop Sitar and C. Chira, A genetic algorithm for solving the generalized vehicle routing problem, in Proceedings of HAIS 2010, Part II, Editors E. S. Corchado Rodriguez et al., Lecture Notes in Artificial Intelligence, Springer, Vol. 6077, pp. 119–126, 2010.

[163] P. C. Pop, I. Kara and A. Horvat Marc, New Mathematical Models of the Generalized Vehicle Routing Problem and Extensions, Applied Mathematical Modelling, Vol. 36(1), pp, 97–107, 2012.

[164] P. C. Pop, O. Matei and C. Sabo, A Memetic Algorithm for Solving the Generalized Minimum Spanning Tree Problem, Advances in Intelligent and Soft Computing, Springer, Vol. 96, pp. 187–194, 2011.

[165] P. C. Pop, C. Pop Sitar, I. Zelina, V. Lupşe and C. Chira, Heuristic algorithms for solving the generalized vehicle routing problem, International Journal of Computers, Communications & Control, Vol. 6(1), pp. 158–166, 2011.

[166] P. C. Pop and S. Iordache, A Hybrid Heuristic Approach for Solving the Generalized Traveling Salesman Problem, in Proceedings of GECCO 2011, Association for Computing Machinery, pp. 481–488, 2011, ISBN: 978–1–4503–0557–0.

[167] P. C. Pop, O. Matei and H. Valean, An Efficient Soft Computing Approach to the Generalized Vehicle Routing Problem, Advances in Intelligent and Soft Computing, Springer, Vol. 87, pp. 281–289, 2011.

[168] P. C. Pop, O. Matei and C. Pop Sitar, An Improved Genetic Algorithm for Solving the Generalized Vehicle Routing Problem, to appear in Neurocomputing, 2012.

[169] P. C. Pop and C. Pop Sitar, New models of the generalized fixed-charge network design problem, Carpathian Journal of Mathematics, Vol. 28 (1), pp. 143–150, 2012.

[170] P. C. Pop and C. Pop Sitar, A new efficient transformation of the generalized vehicle routing problem into the classical vehicle routing problem, Yugoslav Journal of Operations Research, Vol. 21, No. 2, 2011.

[171] R. C. Prim, Shortest connection networks and some generalizations, Bell Systems Technical Journal, Vol. 36, pp. 1389–1401, 1957.

[172] C. Prins and S. Bouchenoua, A Memetic Algorithm Solving the VRP, the CARP and General Routing Problems with Nodes, Edges and Arcs, Studies in Fuzziness and Soft Computing, Vol. 166, pp. 65–85, 2005.

[173] J. J. Prisco, Fiber optic regional area networks in New York and Dallas, IEEE Journal on Selected Areas in Communication SAC-4, pp. 750–757, 1986.

[174] G. R. Raidl, J. Puchinger, C. Blum, Metaheuristic Hybrids, in M. Gendreau and J. Y. Potvin: Handbook of Metaheuristics, 2nd edition, International Series in Operations Research & Management Science, Vol. 146, Springer, pp. 469–496, 2010.

[175] S. Raghavan, On modeling the generalized minimum spanning tree, Technical report, University of Maryland, 2002.

[176] G. Reich and P. Widmayer (Eds.), Beyond Steiner's problem: A VLSI oriented generalization, Graph Theoretic Concepts in Computer Science WG89, Lecture Notes in Computer Science, Vol. 411, Springer, pp. 196–210, 1990.

[177] G. Reinelt, TSPLIB- A Traveling Salesman Problem Library, ORSA Journal on Computing, Vol. 3(4), pp. 376–384, 1991.

[178] J. Renaud and F. F. Boctor, An efficient composite heuristic for the Symmetric Generalized Traveling Salesman Problem, European Journal of Operational Research, Vol. 108(3), pp. 571–584, 1998.

[179] J. Renaud, F. F. Boctor and G. Laporte, A fast composite heuristic for the symmetric traveling salesman problem, INFORMS Journal on Computing, Vol. 8(2), pp. 134–143, 1996.

[180] P. P. Repoussis, C. D. Tarantilis and G. Ioannou, A Hybrid Metaheuristic for a Real Life Vehicle Routing Problem, in Numerical Methods and Applications, Lecture Notes in Computer Science, Vol. 4310, pp. 247–254, 2007.

[181] J. P. Saskena, Mathematical model of scheduling clients through welfare agencies, Journal of the Canadian Operational Research Society, Vol. 8, pp. 185–200, 1970.

[182] F. Schulz, D. Wagner and K. Weihe, Dijkstras algorithm on-line: An empirical case study from public railroad transport, ACM Journal of Experimental Algorithmics, Vol. 5(12), pp. 571–584, 2000.

[183] P. Schuurman and G. J. Woeginger, Approximation schemes - A tutorial, Report, TU Graz, Austria, 2001.

[184] J. Silberholz and B. Golden, The generalized traveling salesman problem: A new genetic algorithm approach. Extending the Horizons, Advances in Computing, Optimization and Decision Technologies, Vol. 37(4), pp. 165–181, 2007.

[185] X. H. Shi, Y. C. Liang, H. P. Lee, C. Lu, Q. X. Wang, Particle swarm optimization-based algorithms for TSP and generalized TSP, Information Processing Letters, Vol. 103, pp. 169–176, 2007.

[186] S. J. Shyu, P. Y. Yin, B. M. T. Lin and M. Haouari, Ant-tree: an ant colony optimization approach to the generalized minimum spanning tree problem, Journal of Experimental & Theoretical Artificial Intelligence 15(1), pp. 103–112, 2003.

[187] P. Slavik, On the Approximation of the Generalized Traveling Salesman Problem, Working Paper, University of Buffalo, 1997.

[188] L. V. Snyder and M. S. Daskin, A random-key genetic algorithm for the generalized traveling salesman problem, European Journal of Operational Research, Vol. 174(1), pp. 38–53, 2006.

[189] G. Sollin, Probleme de l'arbre minimum (unpublished manuscript prepared for the C. Berge's Paris seminar, February 8, 1961).

[190] M. M. Solomon, Algorithms for the Vehicle Routing Problem with Time Windows, Transportation Science, Vol. 29(2), pp. 156–166, 1995.

[191] Srivastava, S. S. S. Kumar, R. C. Garg, P. Sen, Generalized traveling salesman problem through n sets of nodes, CORS Journal, Vol. 7, pp. 97–101, 1969.

[192] D. Stanojevic, Optimization of contemporary telecommunications networks: generalized spanning trees and WDM optical networks, PhD thesis, University of Maryland, 2005.

[193] T. Stützle and H. H. Hoos, $\mathcal{M}ax - \mathcal{M}in$ Ant System, Future Generation Computer Systems, Elsevier Science Publishers, Vol. 16(9), pp. 889–914, 2000.

[194] T. Stützle and H. H. Hoos, The Max-Min Ant System and local search for the traveling salesman problem, in T. Bäck, Z. Michalewicz and X. Yao, editors, Proceedings of the 1997 IEEE International Conference on Evolutionary Computation (ICEC'97), IEEE Press, Piscataway, NJ, pp. 309–314, 1997.

[195] L. A. Szekely and H. Wang, On subtrees of trees, Research report, University of Carolina, 2004.

[196] M. F. Tasgetiren, P. N. Suganthan and Q.-K. Pan, A discrete particle swarm optimization algorithm for the generalized traveling salesman problem, in Proceedings of the 9th annual conference on Genetic and evolutionary computation, pp. 158–167, 2007.

[197] M. F. Tasgetiren, P. N. Suganthan and Q.-K. Pan, An ensemble of discrete differential evolution algorithms for solving the generalized traveling salesman problem, Applied Mathematics and Computation, Vol. 215(9), pp. 3356–3368, 2010.

[198] S. Thienel, ABACUS, A branch-and-cut system, Ph.D. dissertation, Universität zu Köln, 1995.

[199] T. Thomadsen, Hierarchical Network Design, PhD thesis, Technical University of Denmark, 2005.

[200] T. Thomadsen and T. Stidsen, The generalized fixed-charge network design problem, Computers & Operations Research, Vol. 34(4), pp. 997–1007, 2007.

[201] P. Toth and D. Vigo, The vehicle routing problem, SIAM, Philadelphia, 2002.

[202] A. van Breedam, An analysis of the behavior of the heuristics of the vehicle routing problem for a selection of problems, with vehicle-related, customer-related and time-related constraints, Ph.D. Dissertation, University of Antwerp, Belgium, 1994.

[203] V. V. Vazirani, Approximation Algorithms, Springer Verlag, Berlin, 2001.

[204] W. Xu, Quadratic minimum spanning tree problems and related topics, Ph.D. dissertation, University of Maryland, 1984.

[205] M. Yannakakis, Expressing combinatorial optimization problems by linear programs, Journal of Computer and System Sciences, Vol. 43(3), pp. 441–466, 1991.

[206] Z. Wang, C. H. Che and A. Lim, Tabu search for generalized minimum spanning tree problem, Q. Yang and G. Webb (Eds.), PRICAI 2006, LNAI, Springer-Verlag Berlin Heidelberg, Vol. 4099, pp. 918–922, 2006.

[207] H. P. Williams, Model solving in mathematical programming, Wiley, Chichester, 1992.

[208] http://www.iwr.uni-heidelberg.de/groups/comopt/software/ TSPLIB95/vrp/, (2008).

[209] http://nhse.cs.rice.edu/softlib/catalog/tsplib.html

Index

2-opt local search 86
3-opt local search 86

algorithm 3
  ant colony 119
  branch-cut-and-price 173
  genetic 147
  hybrid 147
  Lin–Kernighan 60
  memetic 147
  non-deterministic 4
  polynomial time 4
algorithms
  approximation 2, 29
  average-case polynomial-time 29
  exact 2
  genetic 6
  heuristic 2
  memetic 6
  metaheuristic 6
  superpolynomial-time 29
ant colony optimization 6

bidirectional flow formulation 22

children 12
clustered generalized vehicle routing problem 144
combinatorial optimization 1
conflict graph 43
consultant-guided search 78
cutting plane method 111

directed generalized cutset formulation 17
directed multicommodity flow model 20
directed subpacking formulation 18
dynamic programming 12
dynamic railway traveling salesman problem 121

generalized
  combinatorial optimization problems 7
  cutset formulation 15
  edge-biconnected subgraph 179
  fixed-charge network design problem 8
  minimum edge-biconnected network problem 8, 178
  minimum spanning arborescence problem 17
  minimum spanning tree problem 8, 9
  minimum vertex-biconnected network problem 178
  multicut formulation 15
  multiple traveling salesman problem 141
  network design problems 7
  spanning tree 10
  subtour elimination formulation 14, 66
  traveling salesman problem 8
  vehicle routing problem 8, 128
generalized fixed-charge network design 168
generalized Hamiltonian tour 61
genetic programming 6
global
  connections 12
global
  tree 12

hierarchical telecommunication network 167
hybrid heuristic 78

integer programming 2
iterated local search 6

Lagrangian relaxation 6
layered network 64

leaves 13
linear programming 1
relaxation 3
local connections 12
local search 147
local-global approach 12
local-global formulation 26, 70
lower bound 31

minimum spanning tree problem 9

parsimonious property 37
performance
average-case 29
experimental 29
worst-case 29
personal preference 78
polynomial
reduction 5
transformation 4
primal-dual methods 32
prize collecting generalized minimum spanning tree problem 13
problem
capacitated arc routing 140
decision 4
drilling 92
edge-biconnectivity augmentation 178
quadratic assignment 78
set cover 10

railway traveling salesman problem 8, 100
reinforcing ant colony system 72
relaxation 31
rooting procedure 46
rounding 32
running time 3

sabbatical leave 78
semidefinite programming 7
sensitive robot metaheuristic 92
set cover 10
set cover problem 10
simulated annealing 6, 48
size reduction method 103
spanning tree polytope 24
static railway traveling salesman problem 122
support graph 44

tabu search 6, 43
time-expanded digraph 100
traveling salesman problem 60

undirected multicommodity flow model 22

variable neighborhood descent 183
variable neighborhood search 6, 181
vehicle routing problem 128

Bei Fragen zur Produktsicherheit wenden Sie sich bitte an:
If you have any questions regarding product safety,
please contact:

Walter de Gruyter GmbH
Genthiner Straße 13
10785 Berlin
productsafety@degruyterbrill.com